RAPID REVIEW

BIOCHEMISTRY

Rapid Review Series

SERIES EDITOR
Edward F. Goljan, MD

BEHAVIORAL SCIENCE, SECOND EDITION
Vivian M. Stevens, PhD; Susan K. Redwood, PhD; Jackie L. Neel, DO;
Richard H. Bost, PhD; Nancy W. Van Winkle, PhD; Michael H. Pollak, PhD

BIOCHEMISTRY, THIRD EDITION
John W. Pelley, PhD; Edward F. Goljan, MD

GROSS AND DEVELOPMENTAL ANATOMY, THIRD EDITION
N. Anthony Moore, PhD; William A. Roy, PhD, PT

HISTOLOGY AND CELL BIOLOGY, SECOND EDITION
E. Robert Burns, PhD; M. Donald Cave, PhD

MICROBIOLOGY AND IMMUNOLOGY, THIRD EDITION
Ken S. Rosenthal, PhD; Michael J. Tan, MD

NEUROSCIENCE
James A. Weyhenmeyer, PhD; Eve A. Gallman, PhD

PATHOLOGY, THIRD EDITION
Edward F. Goljan, MD

PHARMACOLOGY, THIRD EDITION
Thomas L. Pazdernik, PhD; Laszlo Kerecsen, MD

PHYSIOLOGY
Thomas A. Brown, MD

LABORATORY TESTING IN CLINICAL MEDICINE
Edward F. Goljan, MD; Karlis Sloka, DO

USMLE STEP 2
Michael W. Lawlor, MD, PhD

USMLE STEP 3
David Rolston, MD; Craig Nielsen, MD

BIOCHEMISTRY THIRD EDITION

John W. Pelley, PhD
Associate Professor
Department of Cell Biology and Biochemistry
Texas Tech University Health Sciences Center
School of Medicine
Lubbock, Texas

Edward F. Goljan, MD
Professor of Pathology
Department of Pathology
Oklahoma State University Center for Health Sciences
College of Osteopathic Medicine
Tulsa, Oklahoma

MOSBY
ELSEVIER

1600 John F. Kennedy Blvd.
Ste 1800
Philadelphia, PA 19103-2899

RAPID REVIEW BIOCHEMISTRY, Third Edition   ISBN: 978-0-323-06887-1

Notice

Knowledge and best practice in this field are constantly changing. As new research and experience broaden our knowledge, changes in practice, treatment and drug therapy may become necessary or appropriate. Readers are advised to check the most current information provided (i) on procedures featured or (ii) by the manufacturer of each product to be administered, to verify the recommended dose or formula, the method and duration of administration, and contraindications. It is the responsibility of the practitioner, relying on their own experience and knowledge of the patient, to make diagnoses, to determine dosages and the best treatment for each individual patient, and to take all appropriate safety precautions. To the fullest extent of the law, neither the Publisher nor the Authors assumes any liability for any injury and/or damage to persons or property arising out of or related to any use of the material contained in this book.

The Publisher

Library of Congress Cataloging-in-Publication Data

Pelley, John W.
 Rapid review biochemistry / John W. Pelley, Edward F. Goljan. – 3rd ed.
   p. ; cm. – (Rapid review series)
 Rev. ed. of: Biochemistry. 2nd ed. c2007.
 ISBN 978-0-323-06887-1
 1. Biochemistry–Outlines, syllabi, etc. 2. Biochemistry–Examinations, questions, etc. I. Goljan, Edward F. II. Pelley, John W. Biochemistry. III. Title. IV. Series: Rapid review series.
  [DNLM: 1. Metabolism–Examination Questions. 2. Biochemical Phenomena–Examination Questions. 3. Nutritional Physiological Phenomena–Examination Questions. QU 18.2 P389r 2011]
  QP518.3.P45 2011
  612'.015–dc22

                                                                     2009045666

Acquisitions Editor: James Merritt
Developmental Editor: Christine Abshire
Publishing Services Manager: Hemamalini Rajendrababu
Project Manager: K Anand Kumar
Design Direction: Steve Stave

Printed in the United States of America

Last digit is the print number: 9 8 7 6 5 4 3 2 1

SERIES PREFACE

The first and second editions of the *Rapid Review Series* have received high critical acclaim from students studying for the United States Medical Licensing Examination (USMLE) Step 1 and consistently high ratings in *First Aid for the USMLE Step 1*. The new editions will continue to be invaluable resources for time-pressed students. As a result of reader feedback, we have improved on an already successful formula. We have created a learning system, including a print and electronic package, that is easier to use and more concise than other review products on the market.

SPECIAL FEATURES

Book

- **Outline format:** Concise, high-yield subject matter is presented in a study-friendly format.
- **High-yield margin notes:** Key content that is most likely to appear on the examination is reinforced in the margin notes.
- **Visual elements:** Full-color photographs are used to enhance students' study and recognition of key pathology images. Abundant two-color schematics and summary tables enhance the study experience.
- **Two-color design:** Colored text and headings make studying more efficient and pleasing.

New Online Study and Testing Tool

- **More than 350 USMLE step 1–type multiple-choice questions:** Clinically oriented, multiple-choice questions mimic the current USMLE format, including high-yield images and complete rationales for all answer options.
- **Online benefits:** New review and testing tool delivered by the USMLE Consult platform, the most realistic USMLE review product on the market. Online feedback includes results analyzed to the subtopic level (discipline and organ system).
- **Test mode:** A test can be created from a random mix of questions or generated by subject or keyword using the timed test mode. USMLE Consult simulates the actual test-taking experience using NBME's FRED interface, including style and level of difficulty of the questions and timing information. Detailed feedback and analysis highlights strengths and weaknesses and enables more focused study.
- **Practice mode:** A test can be created from randomized question sets or fashioned by subject or keyword for a dynamic study session. The practice mode features unlimited attempts at each question, instant feedback, complete rationales for all answer options, and a detailed progress report.
- **Online access:** Online access allows students to study from an Internet-enabled computer wherever and whenever it is convenient. This access is activated through registration on www.studentconsult.com with the pin code printed inside the front cover.

Student Consult

- **Full online access:** The complete text and illustrations of this book can be obtained at www.studentconsult.com.
- **Save content to a PDA:** Through our unique Pocket Consult platform, students can clip selected text and illustrations and save them to a PDA for study on the fly!
- **Free content:** An interactive community center with a wealth of additional valuable resources is available.

ACKNOWLEDGMENT OF REVIEWERS

The publisher expresses sincere thanks to the medical students who provided many useful comments and suggestions for improving the text and the questions. Our publishing program will continue to benefit from the combined insight and experience provided by your reviews. For always encouraging us to focus on our target, the USMLE Step 1, we thank the following:

Thomas A. Brown, West Virginia University School of Medicine

Patricia C. Daniel, PhD, Kansas University Medical Center

John A. Davis, PhD, Yale University School of Medicine

Daniel Egan, Mount Sinai School of Medicine

Steven J. Engman, Loyola University Chicago Stritch School of Medicine

Michael W. Lawlor, Loyola University Chicago Stritch School of Medicine

Craig Wlodarek, Rush Medical College

ACKNOWLEDGMENTS

In a way, an author begins to work on a book long before he sits down at a word processor. Lessons learned in the past from my own teachers and mentors, discussions with colleagues and students, and daily encouragement from family and friends have contributed greatly to the writing of this book.

My wife, MJ, has been a constant source of love and support. Her sensitivity made me aware that I was ready to write this book, and she allowed me to take the time I needed to complete it.

The many caring, intelligent students whom I have taught at Texas Tech over the years have inspired me to hone my thinking, teaching, and writing skills, all of which affected the information that went into the book and the manner in which it was presented.

John A. Davis, MD, PhD, Çağatay H. Ersahin, MD, PhD, Anna M. Szpaderska, DDS, PhD are thanked for their input in previous editions, which continues to add value to the book.

The editorial team at Elsevier was superb. Ruth Steyn and Sally Anderson improved the original manuscript to make my words sound better than I could alone. My highest praise and gratitude are reserved for Susan Kelly, who provided her editorial expertise and professionalism for the first edition. She has become a valued colleague and trusted friend. Likewise, my efforts to update and refine the content of this third edition have been greatly enhanced by my interactions with Dr. Goljan, the Series Editor, and Christine Abshire, the Developmental Editor.

My compliments to Jim Merritt, who undertook a difficult coordination effort to get all of the authors on the "same page" for the very innovative re-launch of the Rapid Review Series second edition and for continuing to see the maturation of this series in the third edition. He and Nicole DiCicco are to be commended for being so helpful and professional.

John W. Pelley, PhD

I would like to acknowledge the loving support of my wife, Joyce, and my tribe of grandchildren for the inspiration to keep on teaching and writing.

Edward F. Goljan, MD
"Poppie"

CONTENTS

Chapter **1** CARBOHYDRATES, LIPIDS, AND AMINO ACIDS: METABOLIC FUELS AND BIOSYNTHETIC PRECURSORS 1

Chapter **2** PROTEINS AND ENZYMES 10

Chapter **3** MEMBRANE BIOCHEMISTRY AND SIGNAL TRANSDUCTION 24

Chapter **4** NUTRITION 35

Chapter **5** GENERATION OF ENERGY FROM DIETARY FUELS 54

Chapter **6** CARBOHYDRATE METABOLISM 63

Chapter **7** LIPID METABOLISM 81

Chapter **8** NITROGEN METABOLISM 98

Chapter **9** INTEGRATION OF METABOLISM 113

Chapter **10** NUCLEOTIDE SYNTHESIS AND METABOLISM 124

Chapter **11** ORGANIZATION, SYNTHESIS, AND REPAIR OF DNA 129

Chapter **12** GENE EXPRESSION 138

Chapter **13** DNA TECHNOLOGY 151

COMMON LABORATORY VALUES 161

INDEX 165

CARBOHYDRATES, LIPIDS, AND AMINO ACIDS: METABOLIC FUELS AND BIOSYNTHETIC PRECURSORS

I. Carbohydrates

A. Overview
1. Glucose provides a significant portion of the energy needed by cells in the fed state.
2. Glucose is maintained in the blood as the sole energy source for the brain in the nonstarving state and as an available energy source for all other tissues.

B. Monosaccharides
1. They are aldehydes (aldoses) or ketones (ketoses) with the general molecular formula $(CH_2O)_x$, where $x = 3$ or more.
2. They are classified by the number of carbon atoms and the nature of the most oxidized group (Table 1-1).
   a. Most sugars can exist as optical isomers (D or L forms), and enzymes are specific for each isomer.
   b. In human metabolism, most sugars occur as D forms.
3. Pyranose sugars (e.g., glucose, galactose) contain a six-membered ring, whereas furanose sugars (e.g., fructose, ribose, deoxyribose) contain a five-membered ring.
4. Reducing sugars are open-chain forms of five and six carbon sugars that expose the carbonyl group to react with reducing agents.

C. Monosaccharide derivatives
1. Monosaccharide derivatives are important metabolic products, although excesses or deficiencies of some contribute to pathogenic conditions.
2. Sugar acids
   a. Ascorbic acid (vitamin C) is required in the synthesis of collagen.
      (1) Prolonged deficiency of vitamin C causes scurvy (i.e., perifollicular petechiae, corkscrew hairs, bruising, gingival inflammation, and bleeding).
   b. Glucuronic acid reacts with bilirubin in the liver, forming conjugated (direct) bilirubin, which is water soluble.
   c. Glucuronic acid is a component of glycosaminoglycans (GAGs), which are major constituents of the extracellular matrix.
3. Deoxy sugars
   a. 2-Deoxyribose is an essential component of the deoxyribonucleotide structure.
4. Sugar alcohols (polyols)
   a. Glycerol derived from hydrolysis of triacylglycerol is phosphorylated in the liver to form glycerol phosphate, which enters the gluconeogenic pathway.
      (1) Liver is the only tissue with glycerol kinase to phosphorylate glycerol.
   b. Sorbitol derived from glucose is osmotically active and is responsible for damage to the lens (cataract formation), Schwann cells (peripheral neuropathy), and pericytes (retinopathy), all associated with diabetes mellitus.
   c. Galactitol derived from galactose contributes to cataract formation in galactosemia.

Blood sugar is analogous to the battery in a car; it powers the electrical system (neurons) and is maintained at a proper "charge" of 70 to 100 mg/dL by the liver.

Scurvy: vitamin C deficiency produces abnormal collagen.

Glucuronic acid: reacts with bilirubin to produce conjugated bilirubin

2-Deoxyribose: component of deoxyribonucleotide structure

Glycerol 3-phosphate: substrate for gluconeogenesis and for synthesizing triacylglycerol

Sorbitol: cataracts, neuropathy, and retinopathy in diabetes mellitus

TABLE 1-1. Monosaccharides Common in Metabolic Processes

CLASS/SUGAR*	CARBONYL GROUP	MAJOR METABOLIC ROLE
Triose (3 Carbons)		
Glyceraldehyde	Aldose	Intermediate in glycolytic and pentose phosphate pathways
Dihydroxyacetone	Ketose	Reduced to glycerol (used in fat metabolism); present in glycolytic pathway
Tetrose (4 Carbons)		
Erythrose	Aldose	Intermediate in pentose phosphate pathway
Pentose (5 Carbons)		
Ribose	Aldose	Component of RNA; precursor of DNA
Ribulose	Ketose	Intermediate in pentose phosphate pathway
Hexose (6 Carbons)		
Glucose	Aldose	Absorbed from intestine with Na^+ and enters cells; starting point of glycolytic pathway; polymerized to form glycogen in liver and muscle
Fructose	Ketose	Absorbed from intestine by facilitated diffusion and enters cells; converted to intermediates in glycolytic pathway; derived from sucrose
Galactose	Aldose	Absorbed from intestine with Na^+ and enters cells; converted to glucose; derived from lactose
Heptose (7 Carbons)		
Sedoheptulose	Ketose	Intermediate in pentose phosphate pathway

*Within cells, sugars usually are phosphorylated, which prevents them from diffusing out of the cell.

5. Amino sugars
   a. Replacement of the hydroxyl group with an amino group yields glucosamine and galactosamine.
   b. *N*-acetylated forms of these compounds are present in GAGs.
6. Sugar esters
   a. Sugar forms glycosidic bonds with phosphate or sulfate.
   b. Phosphorylation of glucose after it enters cells effectively traps it as glucose-6-phosphate, which is further metabolized.
7. Glycosylation
   a. Refers to the reaction of sugar aldehyde with protein amino groups to form a nonreversible covalent bond.
   b. Excessive glycosylation in diabetes leads to endothelial membrane alteration, producing microvascular disease.
   c. In arterioles, glycosylation of the basement membrane renders them permeable to protein, producing hyaline arteriolosclerosis.

D. **Common disaccharides**
   1. Disaccharides are hydrolyzed by digestive enzymes, and the resulting monosaccharides are absorbed into the body.
   2. Maltose = glucose + glucose
      a. Starch breakdown product
   3. Lactose = glucose + galactose
      a. Milk sugar
   4. Sucrose = glucose + fructose
      a. Table sugar
      b. Sucrose, unlike glucose, fructose, and galactose, is a nonreducing sugar.

E. **Polysaccharides**
   1. Polysaccharides function to store glucose or to form structural elements.
   2. Sugar polymers are commonly classified based on the number of sugar units (i.e., monomers) that they contain (Table 1-2).

Phosphorylation of glucose: traps it in cells for further metabolism

Glycosylation of basement membranes of small vessels renders them permeable to proteins.

Hemoglobin A_{1c}: formed by glucose reaction with terminal amino groups and used clinically as a measure of long-term blood glucose concentration

Disaccharides are not absorbed directly but hydrolyzed to monosaccharides first.

The glycosidic bond linking two sugars is designated α or β.

Maltose = glucose + glucose

Lactose = glucose + galactose

Sucrose = glucose + fructose

Reducing sugars: open-chain forms undergo a color reaction with Fehling's reagent indicating that the sugar does not have a glycosidic bond.

TABLE 1-2. Types of Carbohydrates

TYPE	NUMBER OF MONOMERS	EXAMPLES
Monosaccharides	1	Glucose, fructose, ribose
Disaccharides	2	Lactose, sucrose, maltose
Oligosaccharides	3-10	Blood group antigens, membrane glycoproteins
Polysaccharides	>10	Starch, glycogen, glycosaminoglycans

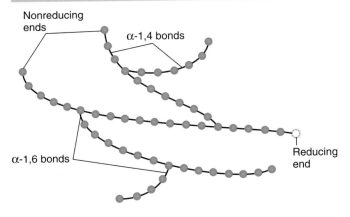

Nonreducing ends

α-1,4 bonds

α-1,6 bonds

Reducing end

1-1: Schematic depiction of glycogen's structure. Each glycogen molecule has one reducing end *(open circle)* and many nonreducing ends. Because of the many branches, which are cleaved by glycogen phosphorylase one glucose unit *(closed circles)* at a time, glycogen can be rapidly degraded to supply glucose in response to low blood glucose levels.

3. Starch, the primary glucose storage form in plants, has two major components, both of which can be degraded by human enzymes (e.g., amylase).
   a. Amylose has a linear structure with α-1,4 linkages.
   b. Amylopectin has a branched structure with α-1,4 linkages and α-1,6 linkages.
4. Glycogen, the primary glucose storage form in animals, has α-glycosidic linkages, similar to amylopectin, but it is more highly branched (Fig. 1-1).
   a. Glycogen phosphorylase cleaves the α-1,4 linkages in glycogen, releasing glucose units from the nonreducing ends of the many branches when the blood glucose level is low.
   b. Liver and muscle produce glycogen from excess glucose during the well-fed state.
5. Cellulose
   a. Structural polysaccharide in plants
   b. Glucose polymer containing β-1,4 linkages
   c. Although an important component of fiber in the diet, cellulose supplies no energy because human digestive enzymes cannot hydrolyze β-1,4 linkages (i.e., insoluble fiber).
6. Hyaluronic acid and other GAGs
   a. Negatively charged polysaccharides contain various sugar acids, amino sugars, and their sulfated derivatives.
   b. These structural polysaccharides form a major part of the extracellular matrix in humans.

II. Lipids
A. Overview
1. Fatty acids, the simplest lipids, can be oxidized to generate much of the energy needed by cells in the fasting state (excluding brain cells and erythrocytes).
2. Fatty acids are precursors in the synthesis of more complex cellular lipids (e.g., triacylglycerol).
3. Only two fatty acids are essential and must be supplied in the diet: linoleic acid and linolenic acid.

B. Fatty acids
1. Fatty acids (FAs) are composed of an unbranched hydrocarbon chain with a terminal carboxyl group.
2. In humans, most fatty acids have an even number of carbon atoms, with a chain length of 16 to 20 carbon atoms (Table 1-3).

Glycogen: storage form of glucose

Glycogen phosphorylase: important enzyme for glycogenolysis and release of glucose

Cellulose: important form of fiber in diet; cannot be digested in humans

Hyaluronic acid and GAGs: important components of the extracellular matrix

Digestive enzymes: cleave α-glycosidic bonds in starch but not β-glycosidic bonds in cellulose (insoluble fiber)

Fatty acids: greatest source of energy for cells (excluding brain cells and erythrocytes)

Essential fatty acids: linoleic acid and linolenic acid

TABLE 1-3. **Common Fatty Acids in Humans**

COMMON NAME	CARBON CHAIN LENGTH: NUMBER OF ATOMS
Palmitic	16
Stearic	18
Palmitoleic	16
Oleic	18
Linoleic (essential)	18
Linolenic (essential)	18
Arachidonic	20

a. Short-chain (2 to 4 carbons) and medium-chain (6 to 12 carbons) fatty acids occur primarily as metabolic intermediates in the body.
   (1) Dietary short- and medium-chain fatty acids (sources: coconut oil, palm kernel oil) are directly absorbed in the small intestine and transported to the liver through the portal vein.
   (2) They also diffuse freely without carnitine esterification into the mitochondrial matrix to be oxidized.
b. Long-chain fatty acids (14 or more carbons) are found in triacylglycerols (fat) and structural lipids.
   (1) They require the carnitine shuttle to move from the cytosol into the mitochondria.
3. Unsaturated fatty acids contain one or more double bonds.
   a. Double bonds in most naturally occurring fatty acids have the *cis* (not *trans*) configuration.
   b. *Trans* fatty acids are formed in the production of margarine and other hydrogenated vegetable oils and are a risk factor for atherosclerosis.
   c. The distance of the unsaturated bond from the terminal carbon is indicated by the nomenclature n-3 (ω-3) for 3 carbons and n-6 (ω-6) for 6 carbons.
   d. Oxidation of unsaturated fatty acids in membrane lipids yields breakdown products that cause membrane damage, which can lead to hemolytic anemia (e.g., vitamin E deficiency).

C. Triacylglycerols
   1. Highly concentrated energy reserve
   2. Formed by esterification of fatty acids with glycerol
   3. Excess fatty acids in the diet and fatty acids synthesized from excess dietary carbohydrate and protein are converted to triacylglycerols and stored in adipose cells.

D. Phospholipids
   1. Phospholipids are derivatives of phosphatidic acid (diacylglycerol with a phosphate group on the third glycerol carbon)
      a. Major component of cellular membranes.
      b. Named for the functional group esterified to the phosphate (Table 1-4).
   2. Fluidity of cellular membranes correlates inversely with the melting point of the fatty acids in membrane phospholipids.
   3. Phospholipases cleave specific bonds in phospholipids.
      a. Phospholipases A_1 and A_2 remove fatty acyl groups from the first and second carbon atoms (C1 and C2) during remodeling and degradation of phospholipids.
         (1) Corticosteroids decrease phospholipase A_2 activity by inducing phospholipase A_2 inhibitory proteins, thereby decreasing the release of arachidonic acid.
      b. Phospholipase C liberates diacylglycerol and inositol triphosphate, two potent intracellular signals.
      c. Phospholipase D generates phosphatidic acid from various phospholipids.
   4. Lung surfactant
      a. Decreases surface tension in the alveoli; prevents small airways from collapsing
      b. Contains abundant phospholipids, especially phosphatidylcholine
      c. Respiratory distress syndrome (RDS), hyaline membrane disease
         (1) Associated with insufficient lung surfactant production leading to partial lung collapse and impaired gas exchange
         (2) Most frequent in premature infants and in infants of diabetic mothers

E. Sphingolipids
   1. Sphingolipids are derivatives of ceramide, which is formed by esterification of a fatty acid with the amino group of sphingosine.
   2. Sphingolipids are localized mainly in the white matter of the central nervous system.

Short- or medium-chain fatty acids: directly reabsorbed

Long-chain fatty acids: require carnitine shuttle

Carnitine deficiency reduces energy available from fat to support glucose synthesis, resulting in nonketotic hypoglycemia.

n-3 (ω-3) unsaturated fatty acids: 3 carbons from terminal

n-6 (ω-6) unsaturated fatty acids: 6 carbons from terminal

Trans fatty acids: margarine, risk factor for atherosclerosis

Triacylglycerol: formed by esterification of fatty acids, as in glycerol

Phospholipids: major component of cellular membranes

Corticosteroids reduce arachidonic acid release from membranes by inactivating phospholipase A_2

Diacylglycerol and inositol triphosphate: potent intracellular signals

Lung surfactant: decreases surface tension and prevents collapse of alveoli; deficient in respiratory distress syndrome

TABLE 1-4. **Phospholipids**

FUNCTIONAL GROUP	PHOSPHOLIPID TYPE
Choline	Phosphatidylcholine (lecithin)
Ethanolamine	Phosphatidylethanolamine (cephalin)
Serine	Phosphatidylserine
Inositol	Phosphatidylinositol
Glycerol linked to a second phosphatidic acid	Cardiolipin

TABLE 1-5. Sphingolipids

FUNCTIONAL GROUP	SPHINGOLIPID TYPE
Phosphatidylcholine	Sphingomyelin
Galactose or glucose	Cerebroside
Sialic acid-containing oligosaccharide	Ganglioside

3. Different sphingolipids are distinguished by the functional group attached to the terminal hydroxyl group of ceramide (Table 1-5).
4. Hereditary defects in the lysosomal enzymes that degrade sphingolipids cause sphingolipidoses (i.e., lysosomal storage diseases), such as Tay-Sachs disease and Gaucher's disease.
5. Sphingomyelins
   a. Phosphorylcholine attached to ceramide
   b. Found in cell membranes (e.g., nerve tissue, blood cells)
   c. Signal transduction
6. Cerebrosides
   a. One galactose or glucose unit joined in β-glycosidic linkage to ceramide
   b. Found largely in myelin sheath
7. Gangliosides
   a. Oligosaccharide containing at least one sialic acid (*N*-acetyl neuraminic acid) residue linked to ceramide
   b. Found in myelin sheath

F. Steroids
1. Steroids are lipids containing a characteristic fused ring system with a hydroxyl or keto group on carbon 3.
2. Cholesterol
   a. Most abundant steroid in mammalian tissue.
   b. Important component of cellular membranes; modulates membrane fluidity
   c. Precursor for synthesis of steroid hormones, skin-derived vitamin D, and bile acids
3. The major steroid classes differ in total number of carbons and other minor variations (Fig. 1-2).
   a. Cholesterol: 27 carbons
   b. Bile acids: 24 carbons (derived from cholesterol)
   c. Progesterone and adrenocortical steroids: 21 carbons
   d. Androgens: 19 carbons
   e. Estrogens: 18 carbons (derived from aromatization of androgens)

G. Eicosanoids
1. Eicosanoids function as short-range, short-term signaling molecules.
   a. Two pathways generate three groups of eicosanoids from arachidonic acid, a 20-carbon polyunsaturated n-6 (ω-6) fatty acid.
   b. Arachidonic acid is released from membrane phospholipids by phospholipase A_2 (Fig. 1-3).
2. Prostaglandins (PGs)
   a. Formed by the action of cyclooxygenase on arachidonic acid
   b. Prostaglandin H_2 (PGH_2), the first stable prostaglandin produced, is the precursor for other prostaglandins and for thromboxanes.
   c. Biologic effects of prostaglandins are numerous and often related to their tissue-specific synthesis.
      (1) Promote acute inflammation
      (2) Stimulate or inhibit smooth muscle contraction, depending on type and tissue
      (3) Promote vasodilation (e.g., afferent arterioles) or vasoconstriction (e.g., cerebral vessels), depending on type and tissue
      (4) Pain (along with bradykinin) in acute inflammation
      (5) Production of fever
3. Thromboxane A_2 (TXA_2)
   a. Produced in platelets by the action of thromboxane synthase on PGH_2
   b. TXA_2 strongly promotes arteriole contraction and platelet aggregation.
   c. Aspirin and other nonsteroidal anti-inflammatory drugs (NSAIDs) acetylate and inhibit cyclooxygenase, leading to reduced synthesis of prostaglandins

Sphingolipids: defects in lysosomal enzymes produce lysosomal storage disease.

Sphingomyelins: found in nerve tissue and blood

Cerebrosides: found in the myelin sheath

Gangliosides: found in the myelin sheath

Sphingolipidoses (e.g., Tay-Sachs disease): defective in lysosomal enzymes; cause accumulation of sphingolipids; lysosomal storage disease

Cholesterol: most abundant steroid in mammalian tissue

Cholesterol: precursor for steroid hormones, vitamin D, and bile acids

Eicosanoids: short-term signaling molecules

Prostaglandins: formed by action of cyclooxygenase on arachidonic acid

PGH_2: precursor prostaglandin

Prostaglandin action is specific to the tissue, such as vasodilation in afferent arterioles and vasoconstriction in cerebral vessels.

TXA_2: platelet aggregation; vasoconstriction; bronchoconstriction

1-2: Steroid structures. A characteristic four-membered fused ring with a hydroxyl or keto group on C_3 is a common structural feature of steroids. The five major groups of steroids differ in the total number of carbon atoms. Cholesterol *(upper left)*, obtained from the diet and synthesized in the body, is the precursor for all other steroids.

Prostaglandins: effects include acute inflammation and smooth muscle contraction and relaxation (vasoconstriction and vasodilation); inhibited by aspirin and NSAIDs

LTB_4: neutrophil chemotaxis and adhesion

LTC_4, $LTCD_4$, $LTCE_4$: found in nerve tissue and blood

Zileuton: inhibits lipoxygenase

Montelukast, zafirlukast: leukotriene receptor antagonists

Essential amino acids cannot be synthesized by the body and must be consumed in the diet.

(anti-inflammatory effect) and of TXA_2 (antithrombotic effect due to reduced platelet aggregation).

4. Leukotrienes (LTs)
   a. Noncyclic compounds whose synthesis begins with the hydroxylation of arachidonic acid by lipoxygenase
   b. Leukotriene B_4 (LTB$_4$) is a strong chemotactic agent for neutrophils and activates neutrophil adhesion molecules for adhesion to endothelial cells.
   c. Slow-reacting substance of anaphylaxis (SRS-A), which contains LTC_4, LTD_4, and LTE_4, is involved in allergic reactions (e.g., bronchoconstriction).
   d. Antileukotriene drugs include zileuton, which inhibits lipoxygenase, and zafirlukast and montelukast, which block leukotriene receptors on target cells.
      (1) These drugs are used in the treatment of asthma, because LTC_4, LTD_4, and LTE_4 are potent bronchoconstrictors.

III. **Amino Acids**
   A. **Overview**
      1. Amino acids constitute the building blocks of proteins and are precursors in the biosynthesis of numerous nonprotein, nitrogen-containing compounds, including heme, purines, pyrimidines, and neurotransmitters (e.g., glycine, glutamate).
      2. Ten of the 20 common amino acids are synthesized in the body; the others are essential and must be supplied in the diet.
   B. **Structure of amino acids**
      1. All amino acids possess an α-amino group (or imino group), α-carboxyl group, a hydrogen atom, and a unique side chain linked to the α-carbon.

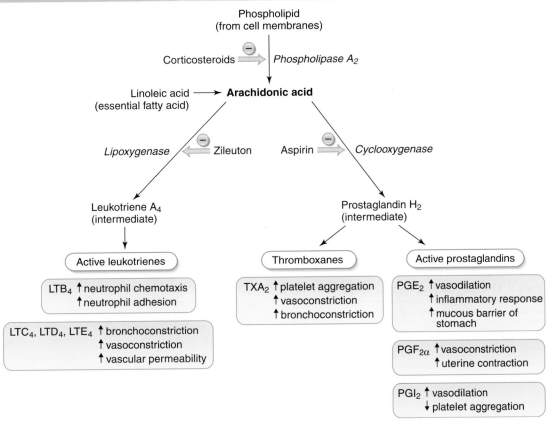

1-3: Overview of eicosanoid biosynthesis and major effects of selected leukotrienes, thromboxanes, and prostaglandins. The active components of the slow-reacting substance of anaphylaxis (SRS-A) are the leukotrienes LTC_4, LTD_4, and LTE_4. PGI_2, also known as prostacyclin, is synthesized in endothelial cells. The therapeutic effects of aspirin and zileuton result from their inhibition of the eicosanoid synthetic pathways. By inhibiting phospholipase A_2, corticosteroids inhibit the production of all of the eicosanoids. $PGF_{2\alpha}$, prostaglandin $F_{2\alpha}$; PGH_2, prostaglandin H_2; TXA_2, thromboxane A_2.

   a. Unique side chain (R group) distinguishes one amino acid from another.
   b. The 20 common amino acids found in proteins are classified into three major groups based on the properties of their side chains.
      (1) Side chains are hydrophobic (nonpolar), uncharged hydrophilic (polar), or charged hydrophilic (polar).
      (2) Hydrophobic amino acids are most often located in the interior lipid-soluble portion of the cell membrane; hydrophilic amino acids are located on the outer and inner surfaces of the cell membrane.
   c. Asymmetry of the α-carbon gives rise to two optically active isomers.
      (1) The L form is unique to proteins.
      (2) The D form occurs in bacterial cell walls and some antibiotics.
2. Hydrophobic (nonpolar) amino acids
   a. Side chains are insoluble in water (Table 1-6).
   b. Essential amino acids in this group are isoleucine, leucine, methionine, phenylalanine, tryptophan, and valine.
   c. Levels of isoleucine, leucine, and valine are increased in maple syrup urine disease.
   d. Phenylalanine accumulates in phenylketonuria (PKU).
3. Uncharged hydrophilic (polar) amino acids
   a. Side chains form hydrogen bonds (Table 1-7).
   b. Threonine is the only essential amino acid in this group.
   c. Tyrosine must be supplied to patients with PKU due to dietary limitation of phenylalanine.
4. Charged hydrophilic (polar) amino acids
   a. Side chains carry a net charge at or near neutral pH (Table 1-8).
   b. Essential amino acids in this group are arginine, histidine, and lysine.
   c. Arginine is a precursor for the formation of nitric oxide, a short-acting cell signal that underlies action as a vasodilator.

Side chain (R group) distinguishes one amino acid from another.

Isoleucine, leucine, valine: branched-chain amino acids; increased levels in maple syrup urine disease

PKU: phenylalanine metabolites accumulate and become neurotoxic; tyrosine must be added to diet.

Arginine and histidine stimulate growth hormone and insulin and are important for growth in children.

TABLE 1-6. **Hydrophobic (Nonpolar) Amino Acids**

AMINO ACID	DISTINGUISHING FEATURES
Glycine (Gly)	Smallest amino acid; inhibitory neurotransmitter of spinal cord; synthesis of heme; abundant in collagen
Alanine (Ala)	Alanine cycle during fasting; major substrate for gluconeogenesis
Valine (Val)*	Branched-chain amino acid; not degraded in liver; used by muscle; increased in maple syrup urine disease
Leucine (Leu)*	Branched-chain amino acid; not degraded in liver; ketogenic; used by muscle; increased in maple syrup urine disease
Isoleucine (Ile)*	Branched-chain amino acid; not degraded in liver; used by muscle; increased in maple syrup urine disease
Methionine (Met)*	Polypeptide chain initiation; methyl donor (as S-adenosylmethionine)
Proline (Pro)	Helix breaker; only amino acid with the side chain cyclized to an α-amino group; hydroxylation in collagen aided by ascorbic acid; binding site for cross-bridges in collagen
Phenylalanine (Phe)*	Increased in phenylketonuria (PKU); aromatic side chains (increased in hepatic coma)
Tryptophan (Trp)*	Precursor of serotonin, niacin, and melatonin; aromatic side chains (increased in hepatic coma)

*Essential amino acids.

TABLE 1-7. **Uncharged Hydrophilic (Polar) Amino Acids**

AMINO ACID	DISTINGUISHING FEATURES
Cysteine (Cys)	Forms disulfide bonds; sensitive to oxidation; component of glutathione, an important antioxidant in red blood cells; deficient in glucose-6-phosphate dehydrogenase (G6PD) deficiency
Serine (Ser)	Single-carbon donor; phosphorylated by kinases
Threonine (Thr)*	Phosphorylated by kinases
Tyrosine (Tyr)	Precursor of catecholamines, melanin, and thyroid hormones; phosphorylated by kinases; aromatic side chains (increased in hepatic coma); must be supplied in phenylketonuria (PKU); signal transduction (tyrosine kinase)
Asparagine (Asn)	Insufficiently synthesized by neoplastic cells; asparaginase used for treatment of leukemia
Glutamine (Gln)	Most abundant amino acid; major carrier of nitrogen; nitrogen donor in synthesis of purines and pyrimidines; NH_3 detoxification in brain and liver; amino group carrier from skeletal muscle to other tissues in fasting state; fuel for kidney, intestine, and cells in immune system in fasting state

*Essential amino acid.

TABLE 1-8. **Charged Hydrophilic (Polar) Amino Acids**

AMINO ACID	DISTINGUISHING FEATURES
Lysine (Lys)*	Basic; positive charge at pH 7; ketogenic; abundant in histones; hydroxylation in collagen aided by ascorbic acid; binding site for cross-bridges between tropocollagen molecules in collagen
Arginine (Arg)*	Basic; positive charge at pH 7; essential for growth in children; abundant in histones
Histidine (His)*	Basic; positive charge at pH 7; effective physiologic buffer; residue in hemoglobin coordinated to heme Fe^{2+}; essential for growth in children; zero charge at pH 7.40
Aspartate (Asp)	Acidic; strong negative charge at pH 7; forms oxaloacetate by transamination; important for binding properties of albumin
Glutamate (Glu)	Acidic; strong negative charge at pH 7; forms α-ketoglutarate by transamination; important for binding properties of albumin

*Essential amino acids.

C. Acid-base properties of amino acids
1. Overview
   a. Acidic groups (e.g., -COOH, $-NH_4^+$) are proton donors.
   b. Basic groups (e.g., $-COO^-$, $-NH_3$) are proton acceptors.
   c. Each acidic or basic group within an amino acid has its own independent pK_a.
   d. Whether a functional group is protonated or dissociated, and to what extent, depends on its pK_a and the pH according to the Henderson-Hasselbalch equation:

$$pH = pK_a + \log[A^-]/[HA]$$

2. Overall charge on proteins depends primarily on the ionizable side chains of the following amino acids:
   a. Arginine and lysine (basic): positive charge at pH 7

Henderson-Hasselbalch equation: used to calculate pH when $[A^-]$ and [HA] are given and to calculate $[A^-]$ and [HA] when pH is given

Amino acids and other weak acids establish an equilibrium between the undissociated acid form (HA) and the dissociated conjugate base (A^-):

$$HA \rightleftharpoons H^+ + A^-$$

A mixture of a weak acid and its conjugate base acts as a buffer by replenishing or absorbing protons and shifting the ratio of the concentrations of $[A^-]$ and $[HA]$.

The buffering ability of an acid-base pair is maximal when $pH = pK$, and buffering is most effective within ± 1 pH unit of the pK. The pH of the blood (normally 7.35 to 7.45) is maintained mainly by the CO_2/HCO_3^- buffer system; CO_2 is primarily controlled by the lungs and HCO_3^- is controlled by the kidneys.

- Hypoventilation causes an increase in arterial $[CO_2]$, leading to respiratory acidosis (decreased pH).
- Hyperventilation reduces arterial $[CO_2]$, leading to respiratory alkalosis (increased pH).
- Metabolic acidosis results from conditions that decrease blood HCO_3^-, such as an accumulation of lactic acid resulting from tissue hypoxia (shift to anaerobic metabolism) or of ketoacids in uncontrolled diabetes mellitus or a loss of HCO_3^- due to fluid loss in diarrhea or to impaired kidney function (e.g., renal tubular acidosis).
- Metabolic alkalosis results from conditions that cause an increase in blood HCO_3^-, including persistent vomiting, use of thiazide diuretics with attendant loss of H^+, mineralocorticoid excess (e.g., primary aldosteronism), and ingestion of bicarbonate in antacid preparations.

    b. Histidine (basic): positive charge at pH 7
       (1) In the physiologic pH range (7.34 to 7.45), the imidazole side group ($pK_a = 6.0$) is an effective buffer (Box 1-1).
       (2) Histidine has a zero charge at pH 7.40.
    c. Aspartate and glutamate (acidic): negative charge at pH 7

> Albumin: strong negative charge helps bind calcium in blood

       (1) Albumin has many of these acidic amino acids, which explains why it is a strong binding protein for calcium and other positively charged elements.
    d. Cysteine: negative charge at pH > 8
  3. Isoelectric point (pI)
    a. Refers to the pH value at which an amino acid (or protein) molecule has a net zero charge

> Physiologic pH: lysine, arginine, histidine carry (+) charge; aspartate and glutamate carry (−) charge.

    b. When pH > pI, the net charge on molecule is negative.
    c. When pH < pI, the net charge on molecule is positive.

D. Modification of amino acid residues in proteins
  1. Some R groups can be modified after amino acids are incorporated into proteins.
  2. Oxidation of the sulfhydryl group (-SH) in cysteine forms a disulfide bond (-S-S-) with a second cysteine residue.
    a. This type of bond helps to stabilize the structure of secreted proteins.
  3. Hydroxylation of proline and lysine yields hydroxyproline and hydroxylysine, which are important binding sites for cross-links in collagen.
    a. Hydroxylation requires ascorbic acid.

> Reduced cross-links in collagen in ascorbate deficiency produce more fragile connective tissue that is more susceptible to bleeding (e.g., bleeding gums in scurvy).

  4. Addition of sugar residues (i.e., glycosylation) to side chains of serine, threonine, and asparagine occurs during synthesis of many secreted and membrane proteins.
    a. Glycosylation of proteins by glucose occurs in patients with poorly controlled diabetes mellitus (e.g., glycosylated hemoglobin [HbA_{1c}], vessel basement membranes).
  5. Phosphorylation of serine, threonine, or tyrosine residues modifies the activity of many enzymes (e.g., inhibits glycogen synthase).

CHAPTER 2

PROTEINS AND ENZYMES

I. **Major Functions of Proteins**
   A. **Catalysis of biochemical reactions**
      1. Enzymes
   B. **Binding of molecules**
      1. Antibodies
      2. Hemoglobin (Hb)
   C. **Structural support**
      1. Elastin
      2. Keratin
      3. Collagen
   D. **Transport of molecules across cellular membranes**
      1. Glucose transporters
      2. Na^+/K^+-ATPase
   E. **Signal transduction**
      1. Receptor proteins
      2. Intracellular proteins (e.g., RAS)
   F. **Coordinated movement of cells and cellular structures**
      1. Myosin
      2. Dynein
      3. Tubulin
      4. Actin
II. **Hierarchical Structure of Proteins**
   A. **Overview**
      1. Primary structure is linear sequence.
      2. Secondary structure is α-helix and β-pleated sheets.
      3. Tertiary structure is a final, stable, folded structure, including supersecondary motifs.
      4. Quaternary structure is functional association of two or more subunits.
   B. **Primary structure**
      1. The primary structure is the linear sequence of amino acids composing a polypeptide.
      2. Peptide bond is the covalent amide linkage that joins amino acids in a protein.
      3. The primary structure of a protein determines its secondary (e.g., α-helices and β-sheets) and tertiary structures (overall three-dimensional structure).
      4. Mutations that alter the primary structure of a protein often change its function and may change its charge, as in the following example.
         a. The sickle cell mutation alters the primary structure and the charge by changing glutamate to valine.
         b. This alters the migration of sickle cell hemoglobin on electrophoresis.
   C. **Secondary structure**
      1. Secondary structure is the regular arrangement of portions of a polypeptide chain stabilized by hydrogen bonds.
      2. The α-helix is a spiral conformation of the polypeptide backbone with the side chains directed outward.
         a. Proline disrupts the α-helix because its α-imino group has no free hydrogen to contribute to the stabilizing hydrogen bonds.
      3. The β-sheet consists of laterally packed β-strands, which are extended regions of the polypeptide chain.

Specific folding of primary structure determines the final native conformation.

Proline: helix breaker

The β-sheets are resistant to proteolytic digestion.

Leucine zippers and zinc fingers: supersecondary structures commonly found in DNA-binding proteins

4. Motifs are combinations of secondary structures occurring in different proteins that have a characteristic three-dimensional shape.
   a. Supersecondary structures often function in the binding of small ligands and ions or in protein-DNA interactions.
   b. The zinc finger is a supersecondary structure in which Zn^{2+} is bound to 2 cysteine and 2 histidine residues.
      (1) Zinc fingers are commonly found in receptors that have a DNA-binding domain that interacts with lipid-soluble hormones (e.g., cortisol).
   c. The leucine zipper is a supersecondary structure in which the leucine residues of one α-helix interdigitate with those of another α-helix to hold the proteins together in a dimer.
      (1) Leucine zippers are commonly found in DNA-binding proteins (e.g., transcription factors).
5. Prions are infectious proteins formed from otherwise normal neural proteins through an induced change in their secondary structure.
   a. Responsible for encephalopathies such as kuru and Creutzfeldt-Jacob disease in humans
   b. Induce secondary structure change in the normal form on contact
   c. Structural change from predominantly α-helix in normal proteins to predominantly β-structure in prions
   d. Forms filamentous aggregates that are resistant to degradation by digestion or heat

D. Tertiary structure
1. Tertiary structure is the three-dimensional folded structure of a polypeptide, also called the native conformation.
   a. Composed of distinct structural and functional regions, or domains, stabilized by side chain interactions
   b. Supersecondary motifs associate during folding to form tertiary structure.
   c. Secreted proteins stabilized by disulfide (covalent) bonds.

E. Quaternary structure
1. Quaternary structure is the association of multiple subunits (i.e., polypeptide chains) into a functional multimeric protein.
2. Dimers containing two subunits (e.g., DNA-binding proteins) and tetramers (e.g., Hb) containing four subunits are most common.
3. Subunits may be held together by noncovalent interactions or by interchain disulfide bonds.

F. Denaturation
1. Denaturation is the loss of native conformation, producing loss of biologic activity.
2. Secondary, tertiary, and quaternary structures are disrupted by denaturing agents, but the primary structure is not destroyed; denaturing agents include the following.
   a. Extreme changes in pH or ionic strength
      (1) In tissue hypoxia, lactic acid accumulation in cells from anaerobic glycolysis causes denaturation of enzymes and proteins, leading to coagulation necrosis.
   b. Detergents
   c. High temperature
   d. Heavy metals (e.g., arsenic, mercury, lead)
      (1) With heavy metal poisonings and nephrotoxic drugs (e.g., aminoglycosides), denaturation of proteins in the proximal tubules leads to coagulation necrosis (i.e., ischemic acute tubular necrosis [ATN]).
3. Denatured polypeptide chains aggregate and become insoluble due to interactions of exposed hydrophobic side chains.
   a. In glucose 6-phosphate dehydrogenase (G6PD) deficiency, increased peroxide in red blood cells (RBCs) leads to denaturation of Hb (i.e., oxidative damage) and formation of Heinz bodies.

III. Enzymes: Protein Catalysts
A. Overview
1. Enzymes increase reaction rate by lowering activation energy but cannot alter the equilibrium of a reaction.
2. Coenzymes and prosthetic groups may participate in the catalytic mechanism.
3. The active site is determined by the folding of the polypeptide and may be composed of amino acids that are far apart.
4. Binding of substrate induces a change in shape of the enzyme and is sensitive to pH, temperature, and ionic strength.

Prions: infectious proteins formed by change in secondary structure instead of genetic mutation; responsible for kuru and Creutzfeldt-Jacob disease

Tertiary structure side-chain interactions: hydrophobic to center; hydrophilic to outside

Fibrous tertiary structure: structural function (e.g., keratins in skin, hair, and nails; collagen; elastin)

Globular tertiary structure: enzymes, transport proteins, nuclear proteins; most are water soluble

Quaternary structure: separate polypeptides functional as multimers of two or more subunits

Heavy metals, low intracellular pH, detergents, heat: disrupt stabilizing bonds in proteins, causing loss of function

G6PD deficiency: increased peroxide in RBCs leads to Hb denaturation, formation of Heinz bodies

K_m: measure of affinity for substrate

V_{max}: saturation of enzyme with substrate

5. Michaelis-Menton kinetics is hyperbolic, whereas cooperativity kinetics is sigmoidal; K_m is a measure of affinity for substrate, and V_{max} represents saturation of enzyme with substrate.
6. Inhibition can be reversible or irreversible.
   a. Inhibition is *not* regulation because the enzyme is inactivated when an inhibitor is bound.
7. Allosterism produces a change in the K_m due to binding of a ligand that alters cooperativity properties.
   a. The sigmoidal curve is displaced to the left for positive effectors and to the right for negative effectors.
8. Enzymes are regulated by compartmentation, allosterism, covalent modification, and gene regulation.

B. General properties of enzymes
1. Acceleration of reactions results from their decreasing the activation energy of reactions (Fig. 2-1).
2. High specificity of enzymes for substrates (i.e., reacting compounds) ensures that desired reactions occur in the absence of unwanted side reactions.
3. Enzymes *do not* change the concentrations of substrates and products at equilibrium, but they do allow equilibrium to be reached more rapidly.
4. No permanent change in enzymes occurs during the reactions they catalyze, although some undergo temporary changes.

Enzymes decrease activation energy but do not change equilibrium (spontaneity).

Enzymes are not changed permanently by the reaction they catalyze but can undergo a transition state.

Many coenzymes are vitamin derivatives.

C. Coenzymes and prosthetic groups
1. The activity of some enzymes depends on nonprotein organic molecules (e.g., coenzymes) or metal ions (e.g., cofactors) associated with the protein.
2. Coenzymes are organic nonprotein compounds that bind reversibly to certain enzymes during a reaction and function as a co-substrate.
   a. Many coenzymes are vitamin derivatives (see Chapter 4).
   b. Nicotine adenine dinucleotide (NAD^+), a derivative of niacin, participates in many oxidation-reduction reactions (e.g., glycolytic pathway).
   c. Pyridoxal phosphate, derived from pyridoxine, functions in transamination reactions (e.g., alanine converted to pyruvic acid) and some amino acid decarboxylation reactions (e.g., histidine converted to histamine).
   d. Thiamine pyrophosphate is a coenzyme for enzymes catalyzing oxidative decarboxylation of α-keto acids (e.g., degradation of branched-chain amino acids) and for transketolase (e.g., two-carbon transfer reactions) in the pentose phosphate pathway.
   e. Tetrahydrofolate (THF), derived from folic acid, functions in one-carbon transfer reactions (e.g., conversion of serine to glycine).
3. Prosthetic groups maintain stable bonding to the enzyme during the reaction.
   a. Biotin is covalently attached to enzymes that catalyze carboxylation reactions (e.g., pyruvate carboxylase).

Niacin: redox

Pyridoxine: transamination

Thiamine: decarboxylation

Biotin: carboxylation

Folate: single-carbon transfer

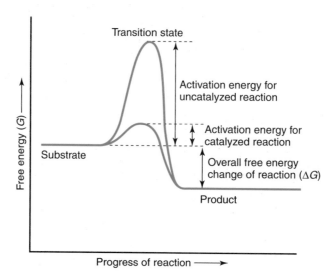

2-1: Energy profiles for catalyzed and uncatalyzed reactions. Catalyzed reactions require less activation energy and are therefore accelerated. The equilibrium of a reaction is proportional to the overall change in free energy (ΔG) between substrate and product, which must be negative for a reaction to proceed.

   b. Metal ion cofactors (metalloenzymes) associate noncovalently with enzymes and
      may help orient substrates or function as electron carriers.
      (1) Magnesium (Mg): kinases
      (2) Zinc (Zn): carbonic anhydrase, collagenase, alcohol dehydrogenase, superoxide
          dismutase (neutralizes O_2 free radicals)
      (3) Copper (Cu): oxidases (e.g., lysyl oxidase for cross-bridging in collagen
          synthesis), ferroxidase (converts Fe^{3+} to Fe^{2+} to bind to transferrin), cytochrome
          oxidase (transfers electrons to oxygen to form water)
      (4) Iron (Fe): cytochromes
      (5) Selenium (Se): glutathione peroxidase

> Metal ion cofactors: Mg, Zn, Cu, Fe, Se

D. Active site
   1. In the native conformation of an enzyme, amino acid residues that are widely separated
      in the primary structure are brought into proximity to form the three-dimensional active
      site, which binds and activates substrates.
   2. Substrate binding often causes a conformational change in the enzyme (induced fit)
      that strengthens binding.
   3. Transition state represents an activated form of the substrate that immediately precedes
      formation of product (see Fig. 2-1).
   4. Precise orientation of amino acid side chains in the active site of an enzyme depends on
      the amino acid sequence, pH, temperature, and ionic strength.
      a. Mutations or nonphysiologic conditions that alter the active site cause a change in
         enzyme activity.

> Active site: affected by amino acid sequence, pH, temperature, and ionic strength

E. Enzyme kinetics
   1. The reaction velocity (v), measured as the rate of product formation, always refers to the
      initial velocity after substrate is added to the enzyme.
   2. The Michaelis-Menten model involves a single substrate (S).
      a. Binding of substrate to enzyme (E) forms an enzyme-substrate complex (ES), which
         may react to form product (P) or dissociate without reacting:

$$E + S \rightleftharpoons ES \rightarrow E + P$$

   3. A plot of initial velocity at different substrate concentrations, [S] (constant enzyme
      concentration), produces a rectangular hyperbola for reactions that fit the Michaelis-
      Menten model (Fig. 2-2A).
      a. Maximal velocity, V_{max}, is reached when the enzyme is fully saturated with substrate
         (i.e., all of the enzyme exists as ES).
         (1) In a zero-order reaction, velocity is independent of [S].
         (2) In a first-order reaction, velocity is proportional to [S].
      b. K_m, the substrate concentration at which the reaction velocity equals one half of
         V_{max}, reflects the affinity of enzyme for substrate.
         (1) Low K_m enzymes have a high affinity for S (e.g., hexokinase).
         (2) High K_m enzymes have a low affinity for S (e.g., glucokinase).
   4. The Lineweaver-Burk plot, a double reciprocal plot of $1/v$ versus $1/[S]$ produces a
      straight line (see Fig. 2-2B).
      a. The y intercept equals $1/V_{max}$.
      b. The x intercept equals $1/K_m$.

> Michaelis-Menten model: hyperbolic curve, saturation at V_{max}, and K_m is substrate concentration for 50% V_{max}.
>
> Zero-order reaction: enzyme is saturated with substrate, and for first-order reaction, substrate concentration is below K_m.
>
> Low K_m: high affinity of enzyme for substrate (e.g., hexokinase); high K_m: low affinity of enzyme for substrate (e.g., glucokinase)

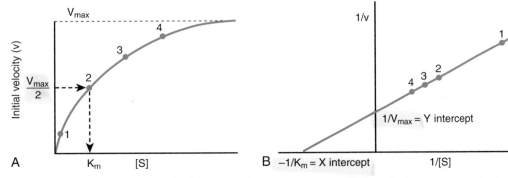

2-2: Enzyme kinetic curves. **A,** Initial velocity (v) versus substrate concentration [S] at constant enzyme concentration for an enzymatic reaction with Michaelis-Menten kinetics. **B,** Lineweaver-Burk double reciprocal plot obtained from the data points (1, 2, 3, 4) in graph **A.** K_m and V_{max} are determined accurately from the intersection of the resulting straight line with the horizontal and vertical axes, respectively.

5. Temperature and pH affect the velocity of enzyme-catalyzed reactions.
   a. Velocity increases as the temperature increases until denaturation causes loss of enzymatic activity.
   b. Changes in pH affect velocity by altering the ionization of residues at the active site and in the substrate.
      (1) Extremes of high or low pH cause denaturation.
   c. Velocity also increases with an increase in enzyme and substrate concentrations.
F. **Enzyme inhibition**
   1. Some drugs and toxins can reduce the catalytic activity of enzymes.
      a. Such inhibition is *not* considered to be physiologic regulation of enzyme activity.
   2. Competitive inhibitors are substrate analogues that compete with normal substrate for binding to the active site.
      a. Enzyme-inhibitor (EI) complex is unreactive (Fig. 2-3A).
      b. K_m is increased (*x* intercept in Lineweaver-Burk plot has smaller absolute value).
      c. V_{max} is unchanged (*y* intercept in Lineweaver-Burk plot is unaffected).
      d. Examples of competitive inhibitors
         (1) Methanol and ethylene glycol (antifreeze) compete with ethanol for binding sites to alcohol dehydrogenase. Infusing ethanol with methanol and ethylene glycol for the active site and reduces toxicity.
         (2) Methotrexate, a folic acid analogue, competitively inhibits dihydrofolate reductase; it prevents regeneration of tetrahydrofolate from dihydrofolate, leading to reduced DNA synthesis.
      e. High substrate concentration reverses competitive inhibition by saturating enzyme with substrate.
   3. Noncompetitive inhibitors bind reversibly away from the active site, forming unreactive enzyme-inhibitor and enzyme-substrate-inhibitor complexes (see Fig. 2-3B).
      a. K_m is unchanged (*x* intercept in Lineweaver-Burk plot is not affected).
      b. V_{max} is decreased (*y* intercept in Lineweaver-Burk plot is larger).
      c. Examples of noncompetitive inhibitors
         (1) Physostigmine, a cholinesterase inhibitor used in the treatment of glaucoma
         (2) Captopril, an angiotensin-converting enzyme (ACE) inhibitor used in the treatment of hypertension
         (3) Allopurinol, a noncompetitive inhibitor of xanthine oxidase, reduces formation of uric acid and is used in the treatment of gout.
      d. High substrate concentration does not reverse noncompetitive inhibition, because inhibitor binding reduces the effective concentration of active enzyme.
   4. Irreversible inhibitors permanently inactivate enzymes.
      a. Heavy metals (often complexed to organic compounds) inhibit by binding tightly to sulfhydryl groups in enzymes and other proteins, causing widespread detrimental effects in the body.
      b. Aspirin acetylates the active site of cyclooxygenase, irreversibly inhibiting the enzyme and reducing the synthesis of prostaglandins and thromboxanes (see Fig. 1-3 in Chapter 1).
      c. Fluorouracil binds to thymidylate synthase like a normal substrate but forms an intermediate that permanently blocks the enzyme's catalytic activity.
      d. Organophosphates in pesticides irreversibly inhibit cholinesterase.
   5. Overcoming enzyme inhibition
      a. Effects of competitive and noncompetitive inhibitors dissipate as the inhibitor is inactivated in the liver or eliminated by the kidneys.

Competitive inhibition: increased K_m and V_{max} unchanged; increased substrate reverses inhibition

Competitive inhibitors: methanol, ethylene glycol, methotrexate

Noncompetitive inhibition: decreased V_{max} and K_m unchanged; increased substrate does not reverse inhibition

Irreversible enzyme inhibitors: heavy metals, aspirin, fluorouracil, and organophosphates

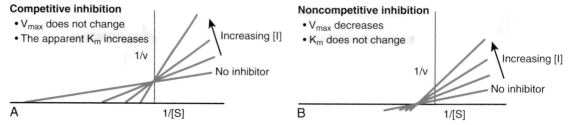

2-3: Effect of competitive and noncompetitive inhibitors (I) on Lineweaver-Burk plots. Notice that competitive inhibitor plots (**A**) intersect on the vertical axis (V_{max} is the same), whereas noncompetitive inhibitor plots (**B**) intersect on the horizontal axis (K_m is the same).

b. Effect of irreversible inhibitors, which cause permanent enzyme inactivation, are overcome only by synthesis of a new enzyme.

G. Cooperativity and allosterism

1. Cooperativity
   a. A change in the shape of one subunit due to binding of substrate induces increased activity by changing the shape of an adjacent subunit.
   b. Enzymes shift from the less active T form (tense form) to the more active R form (relaxed form) as additional substrate molecules are bound.
   c. Sigmoidal shape of the plot of velocity versus [S] characterizes cooperativity.
2. Allosterism occurs when binding of ligand by an enzyme at the allosteric site increases or decreases its activity.
   a. Allosteric effectors of enzymes are nonsubstrate molecules that bind to sites other than the active site.
   b. Positive effectors stabilize the more active R form (relaxed form), so that the K_m decreases (higher affinity for substrate).
      (1) The curve of velocity versus [S] is displaced to the left.
   c. Negative effectors stabilize the less active form (tense form), so that the K_m increases (lower affinity for substrate).
      (1) The curve of velocity versus [S] is displaced to the right.
3. Examples of allosteric enzymes in the glycolytic pathway are hexokinase, phosphofructokinase, and pyruvate kinase.
4. Regulated enzymes generally catalyze rate-limiting steps at the beginning of metabolic pathways (e.g., aminolevulinic acid [ALA], synthase at the beginning of heme synthesis).
   a. The end product of a regulated pathway is often an allosteric inhibitor of an enzyme near the beginning of the pathway. For example, carbamoyl phosphate synthetase II is inhibited by uridine triphosphate end product, and ALA synthase is inhibited by heme, the end product of porphyrin metabolism.

H. Cellular strategies for regulating metabolic pathways

1. Compartmentation of enzymes within specific organelles can physically separate competing metabolic pathways and control access of enzymes to substrates.
   a. Example: enzymes that synthesize fatty acids are located in the cytosol, whereas those that oxidize fatty acids are located in the mitochondrial matrix.
   b. Other examples: alkaline phosphatase (cell membranes), aspartate aminotransferase (mitochondria), γ-glutamyltransferase (smooth endoplasmic reticulum), and myeloperoxidase (lysosomes)
2. Change in gene expression leading to increased or decreased enzyme synthesis (i.e., induction or repression) can provide long-term regulation but has relatively slow response time (hours to days).
   a. Example: synthesis of fat oxidation enzymes in skeletal muscle is induced in response to aerobic exercise conditioning.
3. Allosteric regulation can rapidly (seconds to minutes) increase or decrease flow through a metabolic pathway.
   a. Example: cytidine triphosphate, the end product of the pyrimidine biosynthetic pathway, inhibits aspartate transcarbamoylase, the first enzyme in this pathway (i.e., feedback inhibition).
4. Reversible phosphorylation and dephosphorylation is a common mechanism by which hormones regulate enzyme activity.
   a. Kinases phosphorylate serine, threonine, or tyrosine residues in regulated enzymes; phosphatases remove the phosphate groups (i.e., dephosphorylation).
   b. Reversible phosphorylation and dephosphorylation, often under hormonal control (e.g., glucagon), increases or decreases the activity of key enzymes.
      (1) Example: glycogen phosphorylase is activated by phosphorylation (protein kinase A), whereas glycogen synthase is inhibited.
5. Enzyme cascades, in which a series of enzymes sequentially activate each other, can amplify a small initial signal, leading to a large response, as in the following example.
   a. Binding of glucagon to its cell-surface receptor on liver cells triggers a cascade that ultimately activates many glycogen phosphorylase molecules, each of which catalyzes production of numerous glucose molecules.
   b. This leads to a rapid increase in blood glucose.
6. Proenzymes (zymogens) are inactive storage forms that are activated as needed by proteolytic removal of an inhibitory fragment.

Homotropic effect: binding of substrate to one subunit increases binding of the substrate to other subunits.

Heterotropic effect: binding of different ligand alters binding of substrate to active site adjacent subunits.

Allosterism is a specific adaptation of the enzyme, in contrast with inhibition, which is nonspecific.

Feedback inhibition (allosteric regulation): end product of a pathway inhibits starting enzyme

a. Digestive proteases such as pepsin and trypsin are initially synthesized as proenzymes (e.g., pepsinogen, chymotrypsinogen) that are activated after their release into the stomach or small intestine.

b. In acute pancreatitis, activation of zymogens (e.g., alcohol, hypercalcemia) leads to autodigestion of the pancreas.

I. Isozymes (isoenzymes) and isoforms

1. Some multimeric enzymes have alternative forms, called isozymes, that differ in their subunit composition (derived from different alleles of the same gene) and can be separated by electrophoresis.
2. Different isozymes may be produced in different tissues.
   a. Creatine kinases
      (1) CK-MM predominates in skeletal muscle.
      (2) CK-MB predominates in cardiac muscle.
      (3) CK-BB predominates in brain, smooth muscle, and the lungs.
   b. Of the five isozymes of lactate dehydrogenase, LDH_1 predominates in cardiac muscle and RBCs, and LDH_5 predominates in skeletal muscle and the liver.
3. Different isozymes may be localized to different cellular compartments.
   a. Example: cytosolic and mitochondrial forms of isocitrate dehydrogenase
4. Isoforms are the various forms of the same protein, including isozyme forms (e.g., CK-MM isozymes are isoforms).
   a. Isoforms can be produced by post-translational modification (glycosylation), by alternative splicing, and from single nucleotide polymorphisms within the same gene.

J. Diagnostic enzymology

1. Plasma in normal patients contains few active enzymes (e.g., clotting factors).
2. Because tissue necrosis causes the release of enzymes into serum, the appearance of tissue-specific enzymes or isoenzymes in the serum is useful in diagnosing some disorders and estimating the extent of damage (Table 2-1).

IV. Hemoglobin and Myoglobin: O_2-Binding Proteins

A. Overview
1. Both hemoglobin and myoglobin bind oxygen, but cooperation between subunits allows hemoglobin to release most of its oxygen in the tissues.
2. Allosteric effectors that facilitate unloading of oxygen in the tissues include 2,3-bisphosphoglycerate (2,3-BPG) and pH (i.e., Bohr effect).
3. HbA_{1c} is a glycosylated form of hemoglobin that reflects the average blood glucose concentration.
4. Fetal hemoglobin (HbF) has higher affinity for O_2 than adult hemoglobin to facilitate transfer of oxygen from mother to fetus in the placenta.
5. Hemoglobinopathies involve physical changes (sickle cell Hb), functional changes (methemoglobin), and changes in amount synthesized (thalassemia).

B. Structure of Hb and myoglobin
1. Adult hemoglobin (HbA) is a tetrameric protein composed of two α-globin subunits and two β-globin subunits.
   a. A different globin gene encodes each type of subunit.
   b. All globins have a largely α-helical secondary structure and are folded into a compact, spherical tertiary structure.

TABLE 2-1. **Serum Enzyme Markers Useful in Diagnosis**

SERUM ENZYME	MAJOR DIAGNOSTIC USE
Alanine aminotransferase (ALT)	Viral hepatitis (ALT > AST)
Aspartate aminotransferase (AST)	Alcoholic hepatitis (AST > ALT)
	Myocardial infarction (AST only)
Alkaline phosphatase	Osteoblastic bone disease (e.g., fracture repair, Paget's disease, metastatic prostate cancer), obstructive liver disease
Amylase	Acute pancreatitis, mumps (parotitis)
Creatine kinase (CK)	Myocardial infarction (CK-MB)
	Duchenne muscular dystrophy (CK-MM)
γ-Glutamyltransferase (GGT)	Obstructive liver disease, increased in alcoholics
Lactate dehydrogenase (LDH, type I)	Myocardial infarction
Lipase	Acute pancreatitis (more specific than amylase)

2-4: Structure of heme, showing its relation to two histidines *(shaded areas)* in the globin chain. Heme is located within a crevice in the globin chains. Reduced ferrous iron (Fe^{2+}) forms four coordination bonds to the pyrrole rings of heme and one to the proximal histidine of globin. The sixth coordination bond position is used to bind O_2 or is unoccupied. The side chains attached to the porphyrin ring are omitted.

   2. One heme prosthetic group is located within a hydrophobic pocket in each subunit of Hb (total of four heme groups).
      a. The heme molecule is an iron-containing porphyrin ring (Fig. 2-4).
         (1) Defects in heme synthesis cause porphyria and sideroblastic anemias (e.g., lead poisoning).
      b. Iron normally is in reduced form (Fe^{2+}), which binds O_2.
      c. In methemoglobin, iron is in the oxidized form (Fe^{3+}), which *cannot* bind O_2.
         (1) This lowers the O_2 saturation, or the percentage of heme groups that are occupied by O_2.
         (2) An increase in methemoglobin causes cyanosis because heme groups *cannot* bind to O_2, which decreases the O_2 saturation without affecting the arterial P_{O_2} (amount of O_2 dissolved in plasma).
   3. Myoglobin is a monomeric heme-containing protein whose tertiary structure is very similar to that of α-globin or β-globin.
      a. The myoglobin monomer binds oxygen more tightly to serve as an oxygen reserve.

C. **Functional differences between Hb and myoglobin**
   1. Differences in the functional properties of hemoglobin (four heme groups) and myoglobin (one heme group) reflect the presence or absence of the quaternary structure in these proteins (Table 2-2).
   2. A sigmoidal O_2-binding curve for Hb indicates that binding (and dissociation) is cooperative (Fig. 2-5).
      a. Binding of O_2 to the first subunit of deoxyhemoglobin increases the affinity for O_2 of other subunits.
      b. During successive oxygenation of subunits, their conformation changes from the deoxygenated T form (low O_2 affinity) to the oxygenated R form (high O_2 affinity).
      c. Hb has high O_2 affinity at high P_{O_2} (in lungs) and low O_2 affinity at low P_{O_2} (in tissues), helping it to unload oxygen in the tissues.
   3. A hyperbolic O_2-binding curve for myoglobin indicates that it lacks cooperativity (as expected for a monomeric protein).
      a. Myoglobin is saturated at normal P_{O_2} in skeletal muscle and releases O_2 only when tissue becomes hypoxic, making it a good O_2-storage protein.

Hb has four heme groups to bind O_2; myoglobin has one heme group.

Hb: exhibits cooperativity

T form: low O_2 affinity

R form: high O_2 affinity

Myoglobin: lacks cooperativity

TABLE 2-2. **Comparison of Hemoglobin and Myoglobin**

CHARACTERISTIC	HEMOGLOBIN	MYOGLOBIN
Function	O_2 transport	O_2 storage
Location	In red blood cells	In skeletal muscle
Amount of O_2 bound at P_{O_2} in lungs	High	High
Amount of O_2 bound at P_{O_2} in tissues	Low	High
Quaternary structure	Yes (tetramer)	No (monomer)
Binding curve (% saturation vs P_{O_2})	Sigmoidal (cooperative binding of multiple ligand molecules)	Hyperbolic (binding of one ligand molecule in reversible equilibrium)
Number of heme groups	Four	One

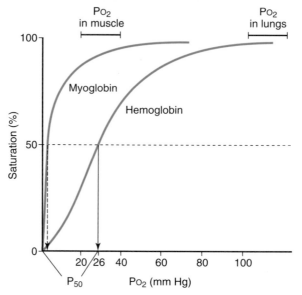

2-5: The O_2-binding curve for hemoglobin and myoglobin. P_{50}, the P_{O_2} corresponding to 50% saturation, is equivalent to K_m for an enzymatic reaction. The lower the value of P_{50}, the greater the affinity for O_2. The very low P_{50} for myoglobin ensures that O_2 remains bound, except under hypoxic conditions. Notice the sigmoidal shape of the hemoglobin curve, which is indicative of multiple subunits and cooperative binding. The myoglobin curve is hyperbolic, indicating noncooperative binding of O_2.

4. Carbon monoxide (CO)
   a. Hb and myoglobin have a 200-fold greater affinity for CO than for O_2.
   b. CO binds at the same sites as O_2, so that relatively small amounts rapidly cause hypoxia due to a decrease in O_2 binding to Hb (fewer heme groups occupied by O_2).
      (1) This lowers the O_2 saturation without affecting the arterial P_{O_2}.
   c. CO poisoning produces cherry red discoloration of the skin and organs.
      (1) It is treated with 100% O_2 or hyperbaric O_2.

D. Factors affecting O_2 binding by Hb
   1. Shift in the O_2-binding curve indicates a change in Hb affinity for O_2.
      a. Left shift indicates increased affinity, which promotes O_2 loading.
      b. Right shift indicates decreased affinity, which promotes O_2 unloading.
   2. Binding of 2,3-BPG, H^+ ions, or CO_2 to Hb stabilizes the T form and reduces affinity for O_2.
      a. The 2,3-BPG, a normal product of glycolysis in erythrocytes, is critical to the release of O_2 from Hb at P_{O_2} values found in tissues (Fig. 2-6).
         (1) The 1,3-BPG in glycolysis is converted into 2,3-BPG by a mutase.
      b. Elevated levels of H^+ and CO_2 (acidotic conditions) within erythrocytes in tissues also promote unloading of O_2.
         (1) The acidotic environment in tissue causes a right shift of the O_2-binding curve, ensuring release of O_2 to tissue.
         (2) Bohr effect is a decrease in the affinity of Hb for O_2 as the pH drops (i.e., increased acidity).
      c. Chronic hypoxia at high altitude increases synthesis of 2,3-BPG, causing a right shift of the O_2-binding curve.

CO and methemoglobin (Fe^{3+}) decrease O_2 saturation of blood and have a normal arterial P_{O_2}.

2,3-BPG stabilizes Hb in the T form to help unload O_2 to tissue.

Decreased O_2 binding: increased 2,3-BPG, H^+ ions, CO_2 (respiratory acidosis), and temperature

Bohr effect: decreased affinity of Hb for O_2 as pH drops

2-6: Effect of 2,3-bisphosphoglycerate (2,3-BPG) on O_2 binding by hemoglobin (Hb). In HbA stripped of 2,3-BPG, the O_2 affinity is so high that Hb remains nearly saturated at P_{O_2} values typical of tissues.

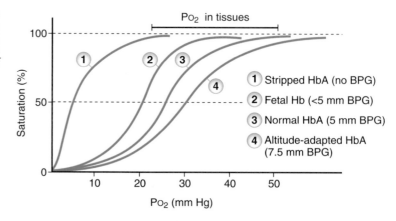

   d. Increase in temperature decreases O_2 affinity and promotes O_2 unloading from Hb during the accelerated metabolism that accompanies a fever.
      (1) Reduction of fever with antipyretics may be counterproductive, because neutrophils require molecular O_2 in the O_2-dependent myeloperoxidase system to kill bacteria.
   3. The following factors all promote increased O_2 affinity of Hb and cause a left shift of the O_2-binding curve:
      a. Decreased 2,3-BPG
      b. Hypothermia
      c. Alkalosis
      d. γ-Globin subunits (fetal hemoglobin, HbF)

E. Role of Hb and bicarbonate in CO_2 transport
   1. CO_2 produced in tissues diffuses into RBCs and combines with Hb or is converted to bicarbonate (HCO_3^-).
   2. About 20% of the CO_2 in blood is transported as carbamino Hb.
      a. CO_2 reacts with the N-terminal amino group of globin chains, forming carbamate derivative.
   3. About 70% of the CO_2 in blood is in the form of HCO_3^- (Fig. 2-7).
      a. Carbonic anhydrase within RBCs rapidly converts CO_2 from tissues to HCO_3^-, which exits the cell in exchange for Cl^- (chloride shift).
      b. In the lungs, the process reverses.
   4. About 10% of the CO_2 in blood is dissolved in plasma.

F. Other normal hemoglobins
   1. Several normal types of Hb are produced in humans at different developmental stages (Fig. 2-8).
   2. HbA_{1c}, a type of glycosylated Hb, is formed by a spontaneous binding (nonenzymatic glycosylation) of blood glucose to the terminal amino group of the β-subunits in HbA.
      a. In normal adults, HbA_{1c} constitutes about 5% of total Hb (HbA accounts for more than 95%).
      b. Uncontrolled diabetes mellitus (persistent elevated blood glucose) is associated with elevated HbA_{1c} concentration.
         (1) HbA_{1c} concentration indicates the levels of blood glucose over the previous 4 to 8 weeks, roughly the life span of an RBC and serves as a marker for long-term glycemic control.
   3. Fetal hemoglobin (HbF) has higher affinity for O_2 than HbA, permitting O_2 to flow from maternal circulation to fetal circulation in the placenta.
      a. Greater O_2 affinity of HbF results from its weaker binding of the negative allosteric effector 2,3-BPG compared with HbA (see Fig. 2-6).

> Right shift of O_2-binding curve: increased 2,3-BPG, acidotic state, high altitude, and fever promote O_2 unloading from Hb to tissues
>
> Left shift of O_2-binding curve: decreased 2,3-BPG, hypothermia, alkalosis, and HbF promote increased O_2 affinity of Hb
>
> HCO_3^-: major vehicle for carrying CO_2 in blood
>
> HbA_{1c}: marker for long-term glycemic control

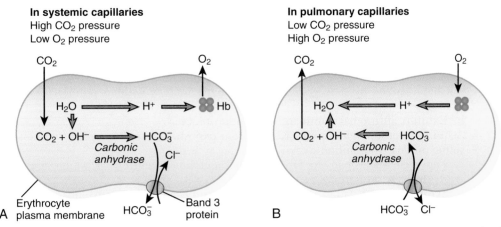

In systemic capillaries
High CO_2 pressure
Low O_2 pressure

In pulmonary capillaries
Low CO_2 pressure
High O_2 pressure

2-7: Relationship between CO_2 and O_2 transport in the blood. **A,** Most of the CO_2 that enters erythrocytes in peripheral capillaries is converted to HCO_3^- and H^+. The resulting decrease in intracellular pH leads to protonation of histidine residue in hemoglobin (Hb), reducing its O_2 affinity and promoting O_2 release. HCO_3^- exits the cell in exchange for Cl^- (i.e., chloride shift) by means of an ion-exchange protein (i.e., band 3 protein), shifting the equilibrium so that more CO_2 can enter. **B,** Within the lungs, reversal of these reactions leads to release of CO_2 and uptake of O_2.

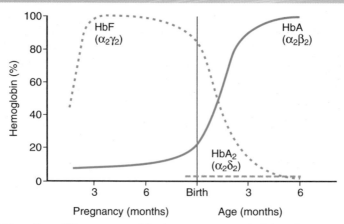

2-8: The hemoglobin (Hb) profile at different stages of development. In normal adults, HbA (consisting of two α-chains and two β-chains) constitutes more than 95% of total Hb. HbA$_2$ (two α-chains and two δ-chains) and HbF (two α-chains and two γ-chains) each contribute about 1% to 2% of total Hb. The β-chain production does not occur until after birth. HbA$_{1c}$, a glycosylated form of HbA, constitutes about 5% of the total Hb in normal adults, but the level is elevated in diabetics. HbA$_{1c}$ is an excellent marker for long-term glycemic control.

G. Hemoglobinopathies due to structural alterations in globin chains

1. Sickle cell hemoglobin (HbS) results from a mutation that replaces glutamic acid with valine at residue 6 in β-globin (β6 Glu → Val) and primarily affects individuals of African American descent.

   a. Deoxygenated HbS forms large linear polymers, causing normally flexible erythrocytes to become stiff and sickle shaped. Sickled cells plug venules, preventing capillaries from draining.

   b. Sickle cell anemia (homozygous condition)

      (1) Sickle cell anemia is an autosomal recessive (AR) disorder.

      (2) Hb profile: 85% to 95% HbS; small amounts of HbF and HbA$_2$ (no HbA).

      (3) Marked by severe hemolytic anemia, multiorgan pain due to microvascular occlusion by sickle cells, autosplenectomy, periodic attacks of acute symptoms (i.e., sickle cell crises), and osteomyelitis (*Salmonella*) and *Streptococcus pneumoniae* sepsis.

      (4) HbF inhibits sickling, and increased levels of HbF reduce the number of crises.

      (5) Hydroxyurea increases synthesis of HbF and reduces the number of sickle cell crises (i.e., occlusion of small vessels by sickle cells).

   c. Sickle cell trait is a heterozygous condition.

      (1) Hb profile: 55% to 60% HbA; 40% to 45% HbS; small amounts of HbA$_{1c}$, HbF, and HbA$_2$

      (2) Usually asymptomatic, except in the renal medulla, where O$_2$ tensions are low enough to induce sickling and renal damage (e.g., renal papillary necrosis)

2. Hemoglobin C (HbC) results from substitution of lysine for glutamate at position 6 in β-globin (β6 Glu → Lys).

   a. Although HbC and HbS are mutated at the same site, HbC is associated only with a mild chronic anemia in homozygotes.

3. Hereditary methemoglobinemia results from any one of several single amino acid substitutions that stabilize heme iron in the oxidized form (HbM).

   a. Characterized by slate gray cyanosis in early infancy without pulmonary or cardiac disease

   b. Exhibits autosomal dominant (AD) inheritance

4. Acquired methemoglobinemia results from exposure to nitrate and nitrite compounds, sulfonamides, and aniline dyes.

   a. These chemicals convert Hb to methemoglobin (heme Fe^{3+}), which does not bind O$_2$ (low O$_2$ saturation).

   b. Symptoms such as cyanosis (no response to O$_2$ administration), headache, and dizziness occur.

   c. Intravenous methylene blue (primary treatment) and ascorbic acid (ancillary treatment) help reduce Fe^{3+} to the Fe^{2+} state.

H. Hemoglobinopathies due to altered rates of globin synthesis (thalassemia)

1. Thalassemias are AR microcytic anemias caused by mutations that lead to the absence or reduced production of α-globin or β-globin chains.

Sickle cell anemia: severe hemolytic anemia; vaso-occlusive crises

Deoxygenated sickled RBCs: block circulation; HbF inhibits sickling

Sickle cell trait: usually asymptomatic except in the kidneys

HbA: 2α 2β chains

HbA$_2$: 2α 2δ chains

HbF: 2α 2γ chains

2. AR α-thalassemia
  a. It results from deletion of one or more of the four α-globin 1 genes on chromosome 16.
  b. It is most prevalent in Asian and African American populations.
  c. There are four types of α-thalassemia that range from mild to severe in their effect on the body.
  d. There is a silent carrier state.
    (1) Minimal deficiency of α-globin chains
    (2) No health problems experienced
  e. α-Thalassemia trait or mild α-thalassemia
    (1) Mild deficiency of α-globin chains
    (2) Patients have microcytic anemia, although many do not experience symptoms.
    (3) Often mistaken for iron deficiency anemia; patients incorrectly placed on iron medication
    (4) Hb electrophoresis is normal, because all normal Hb types require α-chains and all are equally decreased.

> Mild α-thalassemia (microcytic anemia): normal Hb electrophoresis

  f. Hemoglobin H disease
    (1) In this variant, three of four α-globin chain genes are deficient.
    (2) Deficiency is severe enough to cause severe anemia and serious health problems, such as an enlarged spleen, bone deformities, and fatigue.
    (3) Named for the abnormal hemoglobin H (β4 tetramers) that destroys red blood cells.

> Hemoglobin H (β4 tetramers) destroys red blood cells.

  g. Hydrops fetalis or α-thalassemia major (Hb Bart's disease)
    (1) In this variant, there is a total absence of α-globin chain genes.
    (2) Patients die before or shortly after birth.
    (3) HbF is replaced with γ4-tetramers (i.e., hemoglobin Barts)
3. β-Thalassemia
  a. It results from mutations affecting the rate of synthesis of β-globin alleles on chromosome 11
  b. Three types of β-thalassemia occur, and they range from mild to severe.
  c. It is most prevalent in Mediterranean and African American populations.
  d. Thalassemia minor or thalassemia trait
    (1) Mild deficiency of β-globin chains due to splicing defects
    (2) There are no significant health problems.
    (3) There is a mild microcytic anemia.
    (4) Hb electrophoresis shows a decrease in HbA, because β-globin chains are decreased, and a corresponding increase in HbA$_2$ and HbF, which do not have β-globin chains.

> Mild β-thalassemia (i.e., microcytic anemia): slightly decreased HbA; increased HbA$_2$ and HbF

  e. Thalassemia intermedia
    (1) There is a moderate deficiency of β-globin chains.
    (2) Patients present with moderately severe anemia, bone deformities, and enlargement of the spleen.
    (3) There is a wide range in the clinical severity of this condition.
    (4) Patients may need blood transfusions to improve quality of life but *not* to survive.
  f. Thalassemia major or Cooley's anemia
    (1) There is a complete deficiency of β-globin synthesis due to a nonsense mutation and production of a stop codon.
    (2) HbF and HbA$_2$ are produced, but there is no HbA production.
    (3) As HbF production decreases following birth, progressively severe anemia develops with bone distortions, splenomegaly, and hemosiderosis (iron overload from blood transfusions).
    (4) Extensive, lifelong blood transfusions lead to iron overload, which must be treated with chelation therapy.
    (5) Bone marrow transplantation can produce a cure.

> Cooley's anemia: no HbA produced; regular blood transfusions required

V. Collagen: Prototypical Fibrous Protein
A. Overview
1. Collagen is the most abundant protein in the body, and it is the major fibrous component of connective tissue (e.g., bone, cartilage).
2. Fibrous proteins (e.g., collagen, keratin, elastin) provide structural support for cells and tissues.
B. Collagen assembly
1. α-Chains, the individual polypeptides composing tropocollagen (Fig. 2-9A), consist largely of -Gly-X-Y- repeats.

TROPOCOLLAGEN TRIPLE HELIX

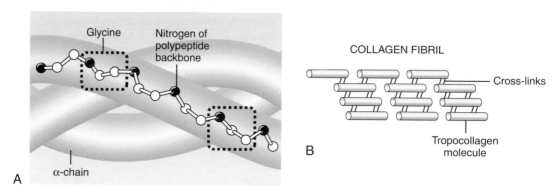

2-9: Collagen structure. **A,** Triple-stranded helix of tropocollagen is the structural unit of collagen. In all α-chains, much of the sequence contains glycine at every third position *(boxes)*. Proline and hydroxyproline (or hydroxylysine) commonly occupy the other two positions in the -Gly-X-Y- repeats. **B,** Notice the typical staggered array of linked tropocollagen molecules in the fibrils of fibrous collagen. The cross-links increase the tensile strength of collagen.

Tropocollagen, the basic structural unit of collagen, is a right-handed triple helix of α-chains.

Ascorbic acid: hydroxylation of proline and lysine in collagen synthesis; promotes cross-bridging

Collagen fibrils form spontaneously from tropocollagen and are stabilized by covalent cross-links between lysine and hydroxylysine residues on adjacent chains.

Deficient cross-linking reduces the tensile strength of collagen fibers.

Goodpasture's syndrome: hemoptysis, cough, fatigue from lung damage and nephritis symptoms and hematuria from kidney damage

Ehlers-Danlos syndrome: loose joints, hyperelastic skin, aortic dissection, colon rupture, collagen defects

Osteogenesis imperfecta: decreased synthesis of type I collagen; pathogenic fractures, blue sclera

Alport's syndrome: defective type IV collagen; nephritis, hearing loss, ocular defects

      a. Proline and hydroxyproline (or hydroxylysine) are often present in the X and Y positions, respectively.

      b. Hydroxylation of proline and lysine in α-chains occurs in the rough endoplasmic reticulum (RER) in a reaction that requires ascorbic acid (vitamin C).

  2. Procollagen triple helix assembles spontaneously from hydroxylated and glycosylated helical α-chains in the RER.

  3. Extracellular peptidases remove terminal propeptides from procollagen helix after it is secreted, yielding tropocollagen.

  4. Tropocollagen assembles to form collagen fibrils.

  5. Lysyl oxidase, an extracellular Cu^{2+}-containing enzyme, oxidizes the lysine side chain to reactive aldehydes that spontaneously form cross-links (see Fig. 2-9B).

      a. Cross-links increase tensile strength of collagen.

      b. Cross-link formation continues throughout life, causing collagen to stiffen with age.

      c. Increased cross-linking associated with aging decreases the elasticity of skin and joints.

C. Collagen types

  1. Fibrous collagens, which constitute about 70% of the total, have fibrillar structure.

      a. Type I: skin, bone, tendons, cornea

      b. Type II: cartilage, intervertebral disks

      c. Type III: blood vessels, lymph nodes, dermis, early phases of wound repair

      d. Type X: epiphyseal plates

  2. Type IV collagen forms flexible, sheetlike networks and is present within all basement membranes.

      a. In Goodpasture's syndrome, antibodies are directed against the basement membrane of pulmonary and glomerular capillaries.

D. Collagen disorders

  1. Ehlers-Danlos syndrome (multiple types of mendelian defects)

      a. Ehlers-Danlos syndrome is caused by mutations in α-chains, resulting in abnormalities in collagen structure, synthesis, secretion, or degradation.

        (1) Collagen types I and III are most often affected.

      b. It is associated with hyperextensive joints, hyperelasticity of skin, aortic dissection, rupture of the colon, and vessel instability resulting in skin hemorrhages.

  2. Osteogenesis imperfecta (i.e., brittle bone disease)

      a. Osteogenesis imperfecta is predominantly an AD disorder.

      b. It results from a deficiency in the synthesis of type I collagen.

      c. It is marked by multiple fractures, retarded wound healing, hearing loss, and blue sclera.

        (1) The blue color of sclera results from thinning of the sclera from loss of collagen, allowing visualization of the underlying choroidal veins.

  3. Alport's syndrome

      a. It is a mendelian disorder caused by defective type IV collagen.

      b. It is characterized by glomerulonephritis, sensorineural hearing loss, and ocular defects.

4. Scurvy
   a. It is caused by prolonged deficiency of vitamin C, which is needed for hydroxylation of proline and lysine residues in collagen.
   b. The tensile strength of collagen is decreased due to lack of cross-bridging of tropocollagen molecules.
      (1) Cross-bridges normally anchor at the sites of hydroxylation.
   c. Hemorrhages in the skin, bleeding gums leading to loosened teeth, bone pain, hemarthroses (i.e., vessel instability), perifollicular hemorrhage, and a painful tongue (i.e., glossitis) eventually develop.

Scurvy: tensile strength of collagen weakened due to lack of cross-bridges

CHAPTER 3

MEMBRANE BIOCHEMISTRY AND SIGNAL TRANSDUCTION

I. Basic Properties of Membranes

A. Overview

Membranes: phospholipids, sphingomyelin, cholesterol

1. Membranes are lipid bilayers containing phospholipids, sphingomyelin, and cholesterol.
2. Proteins can be integral, spanning both layers, or peripheral, loosely associated with either surface.
3. Membranes have fluid characteristics that are influenced by chain length and saturation, cholesterol content, and temperature.

B. Membrane components

1. Cell membrane lipids are arranged in two monolayers, or leaflets, to form the lipid bilayer, the basic structural unit of cellular membranes.
   a. Lipid composition differs within membranes of the same cell type, but phospholipids are the major lipid component of most membranes.
      (1) In membranes, the hydrophilic portion of the phospholipids is oriented facing outward toward the surrounding aqueous environment, and the hydrophobic portion is oriented facing inward toward the center of the bilayer.
   b. Cholesterol is present in the inner and outer leaflets.
   c. Phosphatidylcholine and sphingomyelin are found predominantly in the outer leaflet of the erythrocyte plasma membrane.
   d. Phosphatidylserine and phosphatidylethanolamine are found predominantly in the inner leaflet of the erythrocyte plasma membrane.
2. Proteins constitute 40% to 50% by weight of most cellular membranes.
   a. The particular proteins associated with each type of cellular membrane are largely responsible for its unique functional properties.
3. Carbohydrates in membranes are present only as extracellular moieties covalently linked to some membrane lipids (glycolipids) and proteins (glycoproteins).

Different composition of each leaflet of membrane bilayer; interface with aqueous phase on both sides

Cholesterol: present in inner and outer leaflets of cell membrane

C. Membrane proteins

1. Integral (intrinsic) proteins that span the entire bilayer, called transmembrane proteins, interface with the cytosol and the external environment.
   a. Examples: Transport proteins (e.g., glucose transporters), receptors for water-soluble extracellular signaling molecules (e.g., peptide hormones), and energy-transducing proteins (e.g., adenosine triphosphate [ATP] synthase)
2. Peripheral (extrinsic) proteins are loosely associated with the surface of either side of the membrane.
   a. Examples: Protein kinase C on the cytosolic side and certain extracellular matrix proteins on the external side
   b. Peripheral proteins are loosely bound and can be removed with salt and pH changes.
3. Lipid-anchored proteins are tethered to the inner or outer membrane leaflet by a covalently attached lipid group (e.g., isoprenyl group to RAS molecule).
   a. Alkaline phosphatase is anchored to the outer leaflet.
   b. RAS and other G proteins (i.e., key signal-transducing proteins) are anchored to the inner leaflet.

Integral proteins span the entire membrane.

D. Fluid properties of membranes

1. Membrane fluidity is controlled by several factors:
   a. Long-chain saturated fatty acids interact strongly with each other and decrease fluidity.
   b. *Cis* unsaturated fatty acids disrupt the interaction of fatty acyl chains and increase fluidity.

Anchoring of peripheral proteins with isoprenyl groups

RAS and other G proteins (i.e., key signal-transducing proteins): anchored to inner leaflet of cell membrane

c. Cholesterol prevents the movement of fatty acyl chains and decreases fluidity.

d. Higher temperatures favor a disordered state of fatty acids and increase fluidity.

2. Lateral movement is restricted by the presence of cell-cell junctions in the membrane or by interactions between membrane proteins and the extracellular matrix.

II. Movement of Molecules and Ions Across Membranes (Fig. 3-1 and Table 3-1)

A. Overview

1. Simple diffusion occurs down a concentration gradient without the aid of transport proteins, involving mainly gases and small, uncharged molecules such as water.

2. Facilitated diffusion occurs down a concentration gradient with the aid of transport proteins and involves ions (ion channels) and monosaccharides.

3. Primary active transport occurs against a concentration gradient using ATP energy.

4. Secondary active transport occurs against a concentration gradient by transporting a second molecule using ATP energy.

5. Several genetic defects, including cystic fibrosis, are due to abnormal transport proteins (defective cystic fibrosis transmembrane regulator).

B. Simple diffusion

1. Movement of molecules or ions down a concentration gradient requires no additional energy and occurs without aid of a membrane protein.

2. Limited substances cross membranes by simple diffusion.

Membrane components diffuse laterally.

Fluidity is increased by cis unsaturated fatty acids and high temperatures.

Simple diffusion: movement down a concentration gradient

Facilitated diffusion: movement down a concentration gradient with aid of transport proteins

Primary active transport: movement against a concentration gradient using ATP energy

Secondary active transport: movement against a concentration gradient using second molecule and ATP

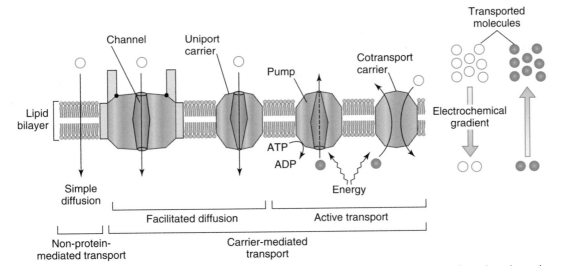

3-1: Overview of membrane-transport mechanisms. *Open circles* represent molecules that are moving down their electrochemical gradient by simple or facilitated diffusion. *Closed circles* represent molecules that are moving against their electrochemical gradient, which requires an input of cellular energy by active transport. Primary active transport is unidirectional and uses pumps, whereas secondary active transport requires cotransport carrier proteins.

TABLE 3-1. Mechanisms for Transporting Small Molecules and Ions Across Biomembranes

PROPERTY	PASSIVE DIFFUSION	FACILITATED DIFFUSION	PRIMARY ACTIVE TRANSPORT	SECONDARY ACTIVE TRANSPORT
Requires transport protein	−	+	+	+
Works against gradient	−	−	+	+
Coupled to ATP hydrolysis	−	−	+ (directly)	− (indirectly)
Powered by movement of cotransported ion	−	−	−	+
Examples of transported molecules	O_2, CO_2, many drugs, steroid hormones	Glucose and amino acids (most cells); Cl^- and HCO_3^- exchange (red blood cells)	Na^+/K^+, Ca^{2+}	Glucose and amino acids (intestine and kidney tubule); Ca^{2+} (cardiac muscle)

a. Gases (O_2, CO_2, nitric oxide)

b. Small uncharged polar molecules (water, ethanol, short-chain neutral fatty acids)

c. Lipophilic molecules (steroids)

3. Transport in either direction occurs, with net transport depending only on the direction of the gradient.

4. Rate of diffusion depends on the size of the transported molecule and gradient steepness.

a. Smaller molecules diffuse faster than larger molecules.

b. A steep concentration gradient produces faster diffusion than a shallow gradient.

C. **Facilitated diffusion**

1. Requires the aid of specialized membrane proteins that move molecules across the membrane down the concentration gradient without input of cellular energy.

2. Ion channels are protein-lined passageways through which ions flow at a high rate when the channel is open.

a. Many channels, which are usually closed, open in response to specific signals.

b. Nicotinic acetylcholine (ACh) receptor in the plasma membrane of skeletal muscle is a Na^+K^+ channel that opens on binding of an ACh.

3. Uniport carrier proteins facilitate diffusion of a single substance (e.g., glucose, particular amino acid).

a. Na^+-independent glucose transporters (GLUTs) are uniporters that passively transport glucose, galactose, or fructose from the blood into most cell types down a steep concentration gradient (Table 3-2).

b. Cycling of the uniporter between alternative conformations allows binding and release of the transported molecules (Fig. 3-2).

c. Direction of transport by the uniporter depends on the direction of the concentration gradient for the transported molecule.

4. Cotransport carrier proteins mediate movement of two different substances at the same time by facilitated diffusion or secondary active transport.

a. The direction of transport depends on the direction of the gradients for the transported molecules (similar to uniporters).

b. Symporters move both transported substances in the same direction.

c. Antiporters move the transported substances in opposite directions.

(1) Example: CL^-/HCO_3^- exchange protein (band 3 protein) in erythrocyte membrane, an antiporter that facilitates diffusion of Cl^- and HCO_3^-, functions in the transport of CO_2 from tissues to the lungs (see Fig. 2-7 in Chapter 2).

TABLE 3-2. Hexose Transport Proteins

TRANSPORTER	PRIMARY TISSUE LOCATION	SPECIFICITY AND PHYSIOLOGIC FUNCTIONS
GLUT1	Most cell types (e.g., brain, erythrocytes, endothelial cells, fetal tissues) but not kidney and small intestinal epithelial cells	Transports glucose (high affinity) and galactose but not fructose; mediates basal glucose uptake
GLUT2	Hepatocytes, pancreatic β cells, epithelial cells of small intestine and kidney tubules (basolateral surface)	Transports glucose (low affinity), galactose, and fructose; mediates high-capacity glucose uptake by liver at high blood glucose levels; serves as glucose sensor for β cells (insulin independent); exports glucose into blood after its uptake from lumen of intestine and kidney tubules
GLUT3	Neurons, placenta, testes	Transports glucose (high affinity) and galactose but not fructose; mediates basal glucose uptake
GLUT4	Skeletal and cardiac muscle, adipocytes	Mediates uptake of glucose (high affinity) in response to insulin stimulation, which induces translocation of GLUT4 transporters from the Golgi apparatus to the cell surface
GLUT5	Small intestine, sperm, kidney, brain, muscle, adipocytes	Transports fructose (high affinity) but not glucose or galactose
GLUT7	Membrane of endoplasmic reticulum (ER) in hepatocytes	Transports free glucose produced in ER by glucose-6-phosphatase to cytosol for release into blood by GLUT2
SGLUT1 (Na^+/K^+ symporter)	Epithelial cells of small intestine and kidney tubules (apical surface)	Cotransports glucose or galactose (but not fructose) and Na^+ in same direction; mediates uptake of sugar from lumen against its concentration gradient powered by coupled transport of Na^+ down its gradient

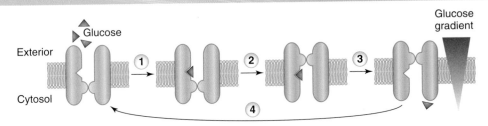

1 Glucose binds to outward-facing binding site

3 Glucose dissociates from inward-facing binding site

2 Conformational change generates inward-facing binding site

4 Reverse conformational change restores outward-facing binding site

3-2: Facilitated diffusion of glucose by the Na^+-independent glucose transporter (GLUT). In most cells, GLUT imports glucose delivered in the blood. The imported glucose is rapidly metabolized within cells, thereby maintaining the inward glucose gradient. However, all steps in the transport process are reversible. If the glucose gradient is reversed, GLUT can transport glucose from the cytosol to the extracellular space, as occurs in the liver during fasting.

D. Primary active transport

1. Pumps move molecules or ions against the concentration gradient with energy supplied by coupled ATP hydrolysis.
2. Pumps mediate unidirectional movement of each molecule transported.
3. Na^+/K^+-ATPase pump, located in the plasma membrane of every cell, maintains low intracellular Na^+ and high intracellular K^+ concentrations relative to the external environment.
   a. Hydrolysis of 1 ATP is coupled to the translocation of 3 Na^+ outward and 2 K^+ inward against their concentration gradients.
   b. Cardiotonic steroids, including digitalis and ouabain, specifically inhibit the Na^+/K^+ ATPase pump.
   c. Albuterol and insulin enhance the pump and drive K^+ from the extracellular compartment into the cell (i.e., hypokalemia).
   d. The β-blockers and succinylcholine inhibit the pump and drive K^+ from the intracellular compartment out into the interstitial space (i.e., hyperkalemia).
4. Ca^{2+}-ATPase pumps maintain low cytosolic Ca^{2+} concentration.
   a. Plasma membrane Ca^{2+}-ATPase, present in most cells, transports Ca^{2+} out of cells.
   b. Muscle Ca^{2+}-ATPase, located in the sarcoplasmic reticulum (SR) of skeletal muscle, transports Ca^{2+} from the cytosol to the SR lumen.
      (1) Release of stored Ca^{2+} from SR to cytosol triggers muscle contraction.
      (2) Rapid removal of Ca^{2+} by the ATPase pump and restoration of a low cytosolic level permits relaxation.
   c. In tissue hypoxia, the decrease in ATP production affects the Ca^{2+}-ATPase pump and allows Ca^{2+} into the cell, where it activates various enzymes (e.g., phospholipases, proteases, endonucleases, caspases [pro-apoptotic enzymes]), leading to irreversible cell damage.

E. Secondary active transport

1. Cotransport carrier proteins move one substance against its concentration gradient with energy supplied by the coupled movement of a second substance (usually Na^+ or H^+) down its gradient.
2. Na^+-linked symporters transport glucose and amino acids against a concentration gradient from the lumen into the epithelial cells lining the small intestine and renal tubules.
   a. Symporter in apical membrane couples movement of 1 or 2 Na^+ into the cell, down the concentration gradient with energetically unfavorable import of a second molecule (glucose or amino acid).
      (1) Absorption of glucose by epithelial cells of kidney tubules and intestine occurs against a steep glucose gradient by secondary active transport mediated by $Na^+/$ glucose symporter (SGLUT1).
      (2) For Na^+ to be reabsorbed in the small bowel, glucose must be present.
      (3) In patients with cholera, it is important to orally replenish Na^+.
   b. Na^+/K^+-ATPase pump in the basal membrane maintains a Na^+ gradient necessary for the operation of Na^+-linked symporters (Fig. 3-3).

Na^+/K^+-ATPase pump: Na^+ and K^+ in; inhibited by cardiotonic steroids, digitalis, and ouabain; lack of oxygen (hypoxia)

Albuterol, insulin: enhance Na^+/K^+-ATPase pump; hypokalemia

β-Blockers, succinylcholine: inhibit Na^+/K^+-ATPase pump (hyperkalemia)

Ca^{2+}-ATPase pumps Ca^{2+} out of cells

Tissue hypoxia: dysfunctional Ca^{2+}-ATPase pump; activation of intracellular enzymes

Secondary active transport: molecule moves against its concentration gradient with energy from movement of cotransported ion down its gradient

Glucose and amino acids transported by Na^+-linked symporters in gut and kidney

SGLUT1 symporter for Na^+/glucose is found in kidney and intestine.

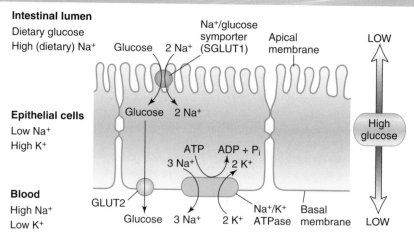

Intestinal lumen
Dietary glucose
High (dietary) Na⁺

Epithelial cells
Low Na⁺
High K⁺

Blood
High Na⁺
Low K⁺

3-3: Transport of glucose from the intestinal lumen to the blood. Three membrane transport proteins participate in this process. Glucose moves across the apical membrane into an epithelial cell against its gradient by means of an Na⁺/glucose symporter, also designated SGLUT1 (i.e., secondary active transport). Glucose exits from the basal surface by means of a GLUT2 uniporter (i.e., facilitated diffusion). Na⁺/K⁺-ATPase pumps Na⁺ out of cell (i.e., primary active transport), maintaining the low intracellular Na⁺ level needed for operation of the symporter. GLUT, glucose transporter.

Na⁺-linked Ca²⁺ antiporter in heart inhibited by digitalis; causes increase in cytosolic calcium and increased force of contraction

3. Na⁺-linked Ca²⁺ antiporter in the plasma membrane of cardiac muscle cells is primarily responsible for maintaining low cytosolic Ca²⁺.
   a. Coupled movement of 3 Na⁺ into the cell down the concentration gradient powers the export of 1 Ca²⁺.
   b. Operation of antiporter is indirectly inhibited by digitalis, accounting for its cardiotonic effect (Fig. 3-4).

F. **Hereditary defects**
   1. Hereditary defects in transport proteins cause diseases such as cystic fibrosis (Box 3-1).

Cystic fibrosis, cystinuria, and Hartnup's disease are caused by hereditary defects in transport proteins.

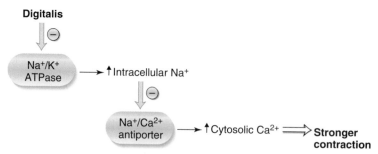

3-4: Mechanism of action of digitalis on cardiac muscle. The cardiotonic effect of digitalis stems from its inhibition of Na⁺/K⁺-ATPase, leading to an increase rise in the intracellular Na⁺ concentration and secondary inhibition of the Na⁺/Ca²⁺ antiporter. The increased cytosolic Ca²⁺ level results in an increase in cardiac muscle contraction. In skeletal muscle, control of the cytosolic Ca²⁺ level is effected by Na⁺-independent Ca²⁺ pumps; hence, digitalis does not affect skeletal muscle.

Box 3-1 DISORDERS CAUSED BY DEFECTIVE TRANSPORTERS

Cystic fibrosis is caused by an autosomal recessive defect in the cystic fibrosis transmembrane regulator gene (*CFTR*) on chromosome 7. The CFTR protein is a Cl⁻-ATPase pump in epithelial cells of the lungs, pancreas, intestines, and skin. The resulting dysfunction in exocrine glands leads to high Na⁺ and Cl⁻ concentrations in sweat (i.e., basis of the sweat test) and production of highly viscous mucus (dehydrated), which obstructs the airways and the pancreatic and bile ducts. Common symptoms include failure to thrive, malabsorption (i.e., atrophy of pancreatic exocrine glands), and recurrent respiratory infections due to *Pseudomonas aeruginosa*, which are the usual cause of death.

Cystinuria results from an autosomal recessive hereditary defect in the carrier protein that mediates reabsorption of dibasic amino acids (i.e., cystine, arginine, lysine, and ornithine) from renal tubules. Formation of cystine kidney stones and excessive urinary excretion of dibasic amino acids are common clinical features. Cystine is a hexagonal crystal in urine.

Hartnup's disease is caused by an autosomal recessive defect in the carrier protein that mediates intestinal and renal tubular absorption of neutral amino acids. Clinically, it is marked by pellagra-like symptoms (e.g., diarrhea, dermatitis, dementia) due to impaired absorption of tryptophan, which reduces the synthesis of niacin.

Familial hypercholesterolemia is an autosomal dominant disease characterized by a lack of functional receptors for low-density lipoprotein (LDL). The resulting high blood levels of cholesterol contribute to premature atherosclerosis and susceptibility to acute myocardial infarctions and stroke at an early age.

III. Receptors and Signal Transduction Cascades
A. Overview
1. Signal molecules such as hormones, growth factors, neurotransmitters, and cytokines bind to receptors to activate an intracellular signal pathway.
2. G protein–coupled receptors release a GTP-binding protein that activates membrane-bound adenylate cyclase leading to an increase in cAMP concentrations; produces a phosphorylation cascade.
3. Phosphoinositide coupled receptors release activated G_q protein that activates phospholipase C; releases inositol 1,4,5-triphosphate (IP_3) and diacylglycerol (DAG).
4. Calmodulin forms an active complex with Ca^{2+} that activates some protein kinases; DAG activates protein kinase C.
5. Receptor tyrosine kinases (RTKs) autophosphorylate during activation by hormone binding; activation of genes is related to growth.
6. Intracellular receptors function as transcription factors on hormone binding to regulate the expression of specific target genes.
7. Cell signaling is impaired by cholera and pertussis toxin, autoantibodies, gene mutation, and drugs.

B. Sequence of events in cell-cell signaling
1. Release of signal molecules normally occurs in response to a specific stimulus (e.g., increased blood glucose stimulates pancreatic β cells to release insulin).
   (a) Hormones, growth factors, neurotransmitters, and cytokines are the most common types of extracellular signals.
2. Binding of the signal to its specific receptor causes receptor activation.
3. Activated receptor-signal complex in turn functions as a signal, triggering an intracellular signal transduction cascade that ultimately leads to specific cellular responses (Fig. 3-5).

C. General properties of cell-surface receptors
1. Receptors located on the exterior surface of the cell bind peptide hormones and other hydrophilic extracellular signals.
   a. In contrast, steroid hormones, thyroxine, and retinoic acid are lipophilic and diffuse through the plasma membrane to receptors in the cytosol (steroid hormones) or nucleus (thyroxine, retinoic acid).
2. Binding interaction between the receptor and the hormone demonstrates reversibility (like enzyme-substrate interactions) and inhibition by antagonists (competitive or noncompetitive).
3. Cellular response to a hormone may be positive or negative (even in the same tissue), depending on the particular receptors present.
4. Signal amplification by a transduction cascade means that binding and activation of only a small fraction of receptors generates an effective response.
5. Receptor-hormone dissociation constants correlate with physiologic concentrations of hormones.

D. Common features of G protein–coupled receptors (GPCRs)
1. GPCRs are monomeric proteins (i.e., single polypeptide chain) containing seven transmembrane α-helices.
   a. Extracellular domain contains hormone-binding site

Margin notes:

G protein–coupled receptors: increased cAMP concentration, causing phosphorylation cascade

Phosphoinositide coupled receptors: release of IP_3 and DAG

Calmodulin complexed with Ca^{2+}: activates protein kinase C

Receptor tyrosine kinases: autophosphorylate

Intracellular receptors: transcription factors on hormone binding

The same receptor can have different target proteins in different cells.

A cell can respond only to those signal molecules whose specific receptor proteins it expresses.

Cell-cell signaling: release of signal molecule, binding of signal molecule to receptor, signal transduction (e.g., cascade), intracellular response

Extracellular signals: hydrophilic

Intracellular signals: lipophilic

Signal molecule: does not cross membrane; specific and reversible binding

Signal transduction cascade: amplifies small amount of signal molecule

---

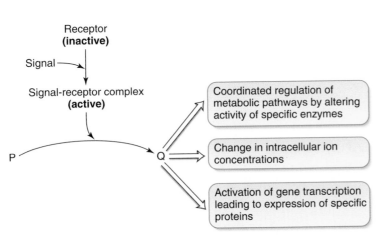

3-5: Signal transduction cascade. Binding of an extracellular signal to its receptor activates the receptor. The activated receptor transduces the signal by binding to a molecule within the cell (P) and converting it into another molecule (Q). Q can then act as a signal (often with intervening transducing molecules), leading to three major types of effects. Amplification of the signal occurs at every step after signal-receptor binding. For example, one active receptor molecule can interact with many molecules of P, yielding many Q molecules.

TABLE 3-3. **Major Trimeric G Proteins**

G_α TYPE	FUNCTION*	COUPLED RECEPTORS
G_s	Stimulates adenylate cyclase (↑ cAMP)	Dopamine (D_1), epinephrine (β_1, β_2), glucagon histamine (H_2), vasopressin (V_2)
G_i	Inhibits adenylate cyclase (↓ cAMP)	Dopamine (D_2), epinephrine (α_2)
G_q	Stimulates phospholipase C (↑ IP_3, DAG)	Angiotensin II, epinephrine (α_1), oxytocin, vasopressin (V_1)
G_t (transducin)	Stimulates cGMP phosphodiesterase (↑ cGMP)	Rhodopsin (light sensitive)

*In some signaling pathways, G_s and G_i are associated with ion channels, which open or close in response to hormone binding. DAG, diacylglycerol; IP_3, inositol triphosphate.

Activated G proteins bind GTP; they revert to the inactive state by hydrolyzing GTP to GDP.

G protein–coupled receptors transduce signals through second messengers: cyclic AMP (cAMP pathway); IP_3 and DAG (phosphoinositide pathway)

Accommodation: phosphorylation of β-adrenergic receptor–G protein complex prevents release of active G protein; requires increased epinephrine to overcome.

 b. Cytosolic domain interacts with trimeric G protein consisting of three subunits (α, β, and γ)

 2. Trimeric G proteins alternate between an active (dissociated) state with bound guanosine triphosphate (GTP) and an inactive (trimeric) state with bound guanosine diphosphate (GDP).

  a. In the active state, which is generated by the hormone binding to the coupled receptor, the α-subunit (G_α) binds to effector protein either to stimulate or inhibit an associated effector protein

 3. Multiple G proteins are coupled to different receptors and transduce signals to different effector proteins, leading to a wide range of responses (Table 3-3).

E. Cyclic AMP (cAMP) pathway

 1. Receptors for glucagon, epinephrine (β receptors), and other hormones coupled to G_s protein transmit a hormonal signal by means of the second messenger cAMP (Fig. 3-6).

 2. Hormone binding to the appropriate receptor causes a conformational change in the intracellular domain, allowing the receptor to interact with the G_s protein.

 3. Hormone-induced elevation of cAMP produces a variety of effects in different tissues (Table 3-4).

3-6: Cyclic adenosine monophosphate (cAMP) pathway. After hormone binding, coupled G protein exchanges bound guanosine diphosphate (GDP) for guanosine triphosphate (GTP). Active $G_{s\alpha}$-GTP diffuses in the membrane and binds to membrane-bound adenylate cyclase, stimulating it to produce cAMP. Binding of cAMP to the regulatory subunits (R) of protein kinase A releases the active catalytic (C) subunits, which mediate various cellular responses.

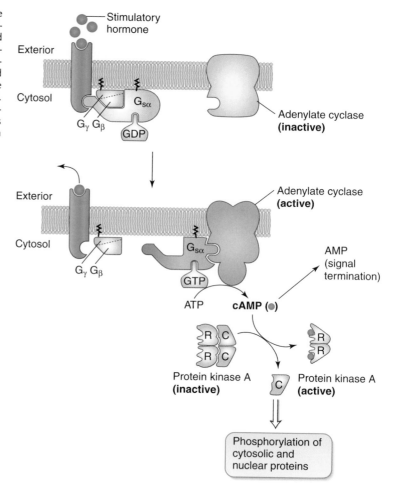

TABLE 3-4. **Effects of Elevated cAMP Levels in Various Tissues**

TISSUE OR CELL TYPE	HORMONE INCREASING cAMP	MAJOR CELLULAR RESPONSE
Adipose tissue	Epinephrine, adrenocorticotropic hormone (ACTH)	↑ Hydrolysis of triglycerides
Adrenal cortex	ACTH	Hormone secretion
Cardiac muscle	Epinephrine, norepinephrine	↑ Contraction rate
Intestinal mucosa	Vasoactive intestinal peptide (VIP), epinephrine	Secretion of water and electrolytes
Kidney tubules	Vasopressin (V_2 receptor)	Resorption of water
Liver	Glucagon, epinephrine	↑ Glycogen degradation
		↑ Glucose synthesis
Platelets	Prostacyclin (PGI_2)	Inhibition of aggregation
Skeletal muscle	Epinephrine	↑ Glycogen degradation
Smooth muscle (vascular and bronchial)	Epinephrine	Relaxation (bronchial) Contraction (arterioles)
Thyroid gland	Thyroid-stimulating hormone (TSH)	Synthesis and secretion of thyroxine

4. β-Adrenergic receptors undergo accommodation (reduction in physiologic response on repeated stimulation) when exposed to sustained, constant concentration of epinephrine (e.g., pheochromocytoma).
   a. Phosphorylation of receptor–G protein complex by β-adrenergic receptor kinase prevents the hormone-receptor complex from releasing activated G_s protein, attenuating the response to *unchanging* concentrations of epinephrine.
   b. Concentration of hormone must increase to generate new active hormone-receptor complexes.

F. **Phosphoinositide pathway**
   1. Receptors coupled to G_q protein transmit signals from hormones such as oxytocin, angiotensin II, and vasopressin (V_1 receptor) by means of several second messengers (Fig. 3-7, top).

Binding of epinephrine (β receptors) or glucagon leads to phosphorylation inside cell by the cAMP pathway; stored energy mobilized

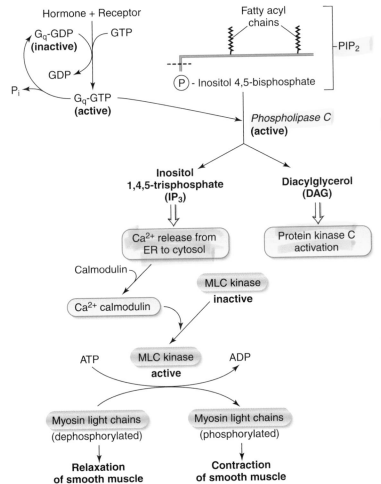

3-7: Phosphoinositide pathway linked to a G_q-coupled receptor. *Top,* The two fatty acyl chains of phosphatidylinositol 4,5-bisphosphate (PIP_2) are embedded in the plasma membrane with the polar phosphorylated inositol group extending into the cytosol. Hydrolysis of PIP_2 *(dashed line)* produces diacylglycerol (DAG), which remains associated with the membrane, and inositol triphosphate (IP_3), which is released into the cytosol. *Bottom,* Contraction of smooth muscle induced by hormones such as epinephrine (α_1 receptor), oxytocin, and vasopressin (V_1 receptor) results from the IP_3-stimulated increase in the level of cytosolic Ca^{2+}, which forms a Ca^{2+}-calmodulin complex that activates myosin light-chain (MLC) kinase. MLC kinase phosphorylates myosin light chains, leading to muscle contractions. *ER,* endoplasmic reticulum.

IP_3 increases cytosolic Ca^{2+} concentrations by stimulating release of stored Ca^{2+} from the ER.

DAG activates protein kinase C.

Extracellular calcium concentration is 10,000 times cytosolic calcium concentration.

Ca^{2+}-calmodulin complex activates protein kinases.

DAG is required for activation of protein kinase C.

RTKs undergo autophosphorylation.

2. Phospholipase C is stimulated by active $G_{q\alpha}$ subunit (similar to stimulation of adenylate cyclase by $G_{s\alpha}$) and cleaves phosphatidyl inositol 4,5-bisphosphate (PIP_2) to yield two second messengers:

   a. IP_3 can diffuse in the cytosol.

   b. DAG remains associated with the plasma membrane.

3. IP_3, a second messenger in the phosphoinositide pathway, causes a rapid release of Ca^{2+} from the endoplasmic reticulum (ER) by opening Ca^{2+} channels in the ER membrane.

   a. Ca^{2+} is a potent enzyme activator, and its access to the cytoplasm is tightly regulated.

      (1) Free Ca^{2+} concentrations in the cytosol are normally about 100 nM, whereas extracellular concentrations of calcium are 10,000-fold higher.

   b. Calmodulin binds cytosolic Ca^{2+}, forming the Ca^{2+}-calmodulin complex that activates Ca^{2+}-calmodulin-dependent protein kinases.

   c. Hormone-induced contraction of smooth muscle results from activation of myosin light-chain (MLC) kinase by Ca^{2+}-calmodulin (see Fig. 3-7, bottom).

4. DAG activates protein kinase C, which regulates various target proteins by phosphorylation.

   a. Elevated cytosolic Ca^{2+} promotes the interaction of inactive protein kinase C with the plasma membrane, where it can be activated by DAG.

G. Receptor tyrosine kinases (RTKs)

1. RTKs contain a single transmembrane α-helix, an extracellular hormone-binding domain, and a cytosolic domain with tyrosine kinase catalytic activity.

2. Hormone-binding (e.g., insulin) activates tyrosine kinase activity, leading to autophosphorylation of the receptor.

3. Insulin receptor is a disulfide-bonded tetrameric RTK that uses insulin receptor substrate 1 (IRS-1) to transduce the insulin signal by two pathways (Fig. 3-8).

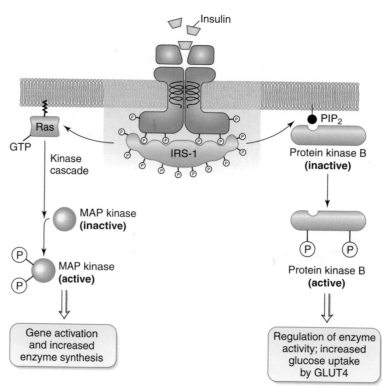

3-8: Signal transduction from an insulin receptor. Insulin binding induces autophosphorylation of the cytosolic domain. The insulin receptor substrate (IRS-1) then binds and is phosphorylated by the receptor's tyrosine kinase activity. Long-term effects of insulin, such as increased glucose uptake by muscle and adipocytes, are mediated through the RAS pathway, which is activated by mitogen-activated protein (MAP) kinase *(left)*. Two adapter proteins transmit the signal from IRS-1 to RAS, converting it to the active form. Short-term effects of insulin, such as increased synthesis of glucokinase in the liver, are mediated by the protein kinase B (PKB) pathway *(right)*. A kinase that binds to IRS-1 converts phosphatidylinositol in the membrane to phosphatidylinositol 4,5-bisphosphate (PIP_2), which binds cytosolic PKB and localizes it to the membrane. Membrane-bound kinases then phosphorylate and activate PKB.

a. RAS-dependent pathway, similar to that used by growth-factor RTKs, mediates the long-term effects of insulin (e.g., increased synthesis of glucokinase in liver).

b. RAS-independent pathway, which leads to activation of protein kinase B, mediates the short-term effects of insulin (e.g., increased glucose uptake by muscle and adipocytes, increased activity of glycogen synthase).

4. RAS, another type of G protein, functions in the signaling pathway from receptors for growth factors such as epidermal growth factor and platelet-derived growth factor (PDGF) receptor.

a. These monomeric receptors aggregate on binding of hormone, usually forming dimers.

b. The hormonal signal is transmitted from activated receptors by adapter proteins to membrane-bound RAS, converting it to the active GTP-bound form.

c. Kinase cascade triggered by RAS-GTP culminates in activation of mitogen-activated protein (MAP) kinase, which translocates to the nucleus and regulates the activity of transcription factors, leading to changes in gene expression.

(1) Signaling pathways that regulate gene expression take hours to days to produce cellular responses, whereas those that control the activity of existing proteins produce cellular responses much more quickly (seconds to minutes).

> Insulin receptor: RAS-dependent pathway (MAP kinase) has long-term effects, and the RAS-independent pathway (protein kinase B) has short-term effects.
>
> RAS: signaling pathway for growth factors

H. Intracellular receptors for lipophilic hormones (Fig. 3-9)

1. Lipophilic hormones, such as steroid hormones (e.g., cortisol), thyroid hormone, and retinoic acid, have receptors that are located in the cytosol or nucleus.

a. Steroid hormones bind to their receptors in the cytosol, and the hormone-receptor complexes move to the nucleus.

b. Thyroxine and retinoic acid have receptors only in the nucleus.

(1) These receptors contain a hormone-binding domain and DNA-binding domain.

2. Hormone-receptor complexes function as transcription factors, which regulate the expression of specific target genes.

> Steroid hormone receptors are located in the cytosol.
>
> Thyroxine and retinoic acid receptors are located in the nucleus.

I. Clinical aspects of cell-cell signaling

1. Cholera, enterotoxigenic *Escherichia coli*, and pertussis toxins catalyze ADP-ribosylation of the α-subunit of G proteins.

a. Cholera toxin produced by *Vibrio cholerae* and the toxin produced by enterotoxigenic *E. coli* permanently activate G_s protein (by ADP-ribosylation of G_s protein), which permanently activates adenylate cyclase.

(1) Resulting increase in cAMP in intestinal mucosa produces a massive secretory diarrhea with a loss of isotonic fluid (approximately the same tonicity as plasma).

b. Pertussis toxin, produced by *Bordetella pertussis*, permanently inactivates G_i protein (by ADP-ribosylation of G_i protein), which permanently activates adenylate cyclase.

(1) Resulting increase in cAMP causes increased mucus secretion in the respiratory tract (e.g., whooping cough).

> Cholera and enterotoxigenic *E. coli* toxin permanently activate G_s, and pertussis toxin permanently inactivates G_i by ADP-ribosylation.

2. Graves' disease results from inappropriate stimulation of thyroid-stimulating hormone (TSH) receptors by IgG autoantibodies.

a. Manifestations include thyromegaly, exophthalmos, and signs of hyperthyroidism, which include weight loss, fatigue, heat intolerance, diarrhea, and hand tremors.

b. Autoantibodies against receptors are type II hypersensitivity reactions.

> Graves' disease: IgG antibodies against TSH receptors produce hyperthyroidism.

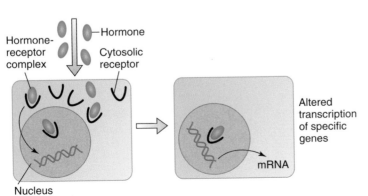

3-9: Signaling by hormones with intracellular receptors. Steroid hormones (e.g., cortisol) bind to their receptors in the cytosol, and the hormone-receptor complex moves to the nucleus. In contrast, the receptors for thyroid hormone and retinoic acid are located only in the nucleus. Binding of the hormone-receptor complex to regulatory sites in DNA activates gene transcription.

RAS protooncogene encodes a mutant RAS protein permanently in the active state that leads to cancer (e.g., pancreas, colon).

Ephedrine: α_1-adrenergic agonist; vasoconstriction

Albuterol, terbutaline: selective for α_2 receptors; bronchodilators

β-Nonselective and -selective blockers: decrease blood pressure; precipitate asthma

Losartan: blocks angiotensin II receptors; decreases blood pressure

Chlorpromazine, haloperidol: blocks dopamine D_2 receptors

3. Mutant RAS protein encoded by the *RAS* protooncogene, which has undergone a point mutation, has a very low GTPase activity and is permanently in the active state.
   a. Cell responds as if high levels of hormone were present, leading to increased cell proliferation.
   b. Mutation of *RAS* protooncogene has been implicated in the development of various types of cancers (i.e., pancreatic, colon, endometrial, and thyroid cancers).
4. Many drugs bind to receptors and either stimulate or inhibit intracellular signaling.
   a. Agonists activate receptors and mimic the action of normal signal molecules.
      (1) Ephedrine, a general agonist of α_1-adrenergic receptors; produces vasoconstriction (nasal decongestant)
      (2) Albuterol and terbutaline, which are relatively selective for α_2-receptors; produce smooth muscle relaxation (i.e., bronchodilator for asthma)
   b. Antagonists inhibit the action of normal signal molecules by blocking access to receptor. Examples:
      (1) β-Blockers, which may be nonselective (e.g., propranolol, timolol) or β_1-selective agents (e.g., metoprolol, atenolol), decrease blood pressure and may precipitate asthma.
      (2) Losartan, which blocks angiotensin II receptors, decreases blood pressure.
      (3) Antipsychotics, such as chlorpromazine and haloperidol, block dopamine D_2 receptors.

I. Key Dietary Terminology

A. Overview

1. The dietary reference intake (DRI) is a set of several reference values, including the recommended dietary allowance related to adequate intakes and upper levels of intakes.
2. The recommended dietary allowance (RDA) refers to intakes needed for optimal health; it is not a minimum daily requirement.
3. Daily energy expenditure depends most importantly on the basal metabolic rate (BMR) and on postprandial thermogenesis and the degree of physical activity.
4. The respiratory exchange rate (RER), also called the respiratory quotient, is the rate of oxygen consumption for different carbon sources.
5. The body mass index (BMI) is used to determine obesity, defined as a BMI greater than 30 (normal, 20 to 25).

B. DRI

1. DRI is a nutrient reference value intended to serve as a guide for good nutrition.
   a. Reference values are specified on the basis of age, gender, and life stage.
   b. Covers more than 40 vitamins and minerals
2. Basis of DRI calculations
   a. Estimated average requirement to satisfy the needs of 50% of the population in that gender and age group
   b. RDA, a daily dietary intake level based on the estimated average requirement of a nutrient
   c. Adequate intake (AI), when a recommended dietary allowance cannot be based on an estimated average requirement
   d. Tolerable upper intake levels (UL), above which risk of toxicity would increase (e.g., vitamin A)

C. RDA

1. RDA represents an optimal dietary intake of nutrients that under ordinary conditions can maintain a healthy general population.
2. RDA varies with sex, age, body weight, diet, and physiologic status.
   a. Example: RDAs for nutrients increase in childhood, in pregnancy, and during lactation.

D. Basal metabolic rate (BMR)

1. BMR accounts for about 60% of daily energy expenditure and refers to the energy consumption of an individual at rest.
   a. BMR reflects the energy involved in normal body functions (e.g., cardiac function, maintaining ion pumps, brown fat generation of heat) and primarily depends on body weight.
      (1) An estimate of BMR is obtained by multiplying the body weight in kilograms times 24 (1 kg = 2.2 lb):

$$\text{BMR} : 24 \times \text{body weight in kg}$$

   b. Other factors affecting BMR
      (1) Gender: males have higher BMRs than females
      (2) Age: children have higher BMRs than adults
      (3) Fever: BMR increased
      (4) Thyroid function: BMR increased in hyperthyroidism, decreased in hypothyroidism

---

DRI: reference values, including RDA and upper level (UL) intake

RDA: optimal dietary intake of nutrients

BMR: basal metabolic rate; energy for normal daily maintenance

BMR + postprandial thermogenesis + physical activity = daily energy expenditure

RER: respiratory quotient

BMI: normal = 20 to 25; overweight = 25 to 30; obese > 30

2. Postprandial thermogenesis
   a. Energy used in digestion, absorption, and distribution of nutrients
   b. Accounts for about 10% of daily energy expenditure
3. Physical activity is variable and is expressed as an activity factor, which when multiplied by the BMR equals the daily energy expenditure.
   a. The activity factor is 1.3 for a sedentary person, 1.5 for a moderately active person, and 1.7 for a very active person (e.g., marathon runner).
4. Sample calculations: calculate the BMR and daily energy expenditure for a 220-lb man (patient A) with sedentary habits and for a 110-lb woman (patient B), who runs 10 miles each day and is an aerobic exercise instructor at night.
   a. Patient A: 220 lb = 220/2.2 = 100 kg
      BMR: 24 × 100 = 2400 kcal/day
      Daily energy expenditure: 2400 × 1.3 = 3120 kcal/day
   b. Patient B: 110 lb = 110/2.2 = 50 kg
      BMR: 24 × 50 = 1200 kcal/day
      Daily energy expenditure: 1200 × 1.7 = 2040 kcal/day

E. **Respiratory exchange rate (RER)**
   1. The respiratory exchange rate (RER), also called the respiratory quotient, is the rate of oxygen consumption for different carbon sources.
   2. RER = V_{CO_2} (carbon dioxide production)/V_{O_2} (oxygen consumption)
      a. RER for carbohydrates = 1.0; RER for fats = 0.7

F. **BMI**
   1. The BMI is used to determine obesity, defined as a BMI greater than 30 (normal is 20 to 25).
   2. BMI = weight in kg/height in m^2
   3. To improve BMI, the total calories expended must be greater than the total intake of calories.
      a. When primarily drawing on adipose tissue to meet energy needs, to lose about 1 lb, a person must expend 3500 calories more than are consumed.
   4. Sample calculations using two patients, A and B:
      a. Patient A consumes 3600 kcal/day consisting of 168 g of fat, 108 g of protein, and 414 g of carbohydrates. Calculate the percentage of each of the nutrients. Is the patient gaining, maintaining, or losing weight?
         (1) Fat kcal: 168 g × 9 kcal/g = 1512 kcal/day; fat percent = 1512/3600 = 42% (exceedingly high)
         (2) Protein kcal: 108 g × 4 kcal/g = 432 kcal/day; protein percent = 432/3600 = 12% (normal)
         (3) Carbohydrate kcal: 414 g × 4 kcal/g = 1656 kcal/day; carbohydrate percent = 1656/3600 = 46% (slightly decreased)
         (4) Because the patient's calculated daily energy expenditure is 3120 kcal/day (see calculation 4.a (above)), the patient is consuming more calories (3600 kcal/day) than are being expended. A net gain of 480 kcal/day results in a gain of 1 lb in about 7 days (3500/480 = 7.3).
      b. Patient B consumes 2000 kcal/day consisting of 67 g of fat, 60 g of protein, and 290 g of carbohydrates. Calculate the percentage of each of the nutrients. Is the patient gaining, maintaining, or losing weight?
         (1) Fat kcal: 67 g × 9 kcal/g = 603 kcal/day; fat percent = 603/2000 = 30% (normal)
         (2) Protein kcal: 60 g × 4 kcal/g = 240 kcal/day; protein percent = 240/2000 = 12% (normal)
         (3) Carbohydrate kcal: 290 g × 4 kcal/g = 1160 kcal/day; carbohydrate percent = 1160/2000 = 58% (normal)
         (4) Because the patient's calculated daily energy expenditure is 2040 kcal/day (see 4.b (above)), the patient is consuming almost the same number of calories (2000 kcal/day) as are being expended, and the current weight will likely be maintained.

II. **Dietary Fuels**
   A. **Overview**
      1. Dietary carbohydrates with α-1,4 glycosidic linkages are digested to monosaccharides and transported directly to the liver through the hepatic portal vein.
      2. Dietary carbohydrates with β-1,4 glycosidic linkages are *not* digested but serve other functions in the gut such as reducing cholesterol absorption and softening the stool.
      3. Triacylglycerols are the major dietary lipids, although phospholipids and cholesterol are also consumed in the diet.
      4. Long-chain triacylglycerols and cholesterol are packaged in chylomicrons and bypass the liver by transport through the lymphatics to the subclavian vein.

5. Dietary proteins are digested to free amino acids for the synthesis of proteins and to supply carbon skeletons for the synthesis of glucose for energy.
6. Nitrogen balance is an indication of net synthesis (growth), loss (breakdown), or stability in bodily proteins.

B. **Dietary carbohydrates (see Chapter 1)**
1. Carbohydrates include the following:
   a. Polysaccharides (e.g., starch)
   b. Disaccharides (e.g., lactose, sucrose, maltose)
   c. Monosaccharides (e.g., glucose, galactose, fructose)
   d. Insoluble fiber (e.g., cellulose, lignin)
   e. Soluble fiber (e.g., pectins, hemicellulose)
2. α-Amylase, which is found in saliva in the mouth and in pancreatic secretions in the small intestine, cleaves the α-1,4 linkages in starch, producing smaller molecules (e.g., oligosaccharides and disaccharides).

   > Starch digestion by α-amylase

3. Intestinal brush border enzymes (e.g., lactase, sucrase, maltase) hydrolyze dietary disaccharides into the monosaccharides glucose, galactose, and fructose, which are reabsorbed into the portal circulation by carrier proteins in intestinal epithelial cells (see Chapter 3).
   a. Glucose is the predominant sugar in human blood.
   b. Glucose is stored as glycogen, which is primarily located in liver and muscle.
   c. Complete oxidation of carbohydrates to CO_2 and H_2O in the body produces 4 kcal/g.
   d. Lactose intolerance due to lactase deficiency is discussed in Box 4-1.

   > Lactose intolerance: inability to digest lactose provides growth medium for intestinal bacteria; converted into lactic acid and hydrogen gas; osmotic diarrhea
   >
   > Carbohydrate: 4 kcal/g

4. Insoluble and soluble dietary fiber has β-1,4 glycosidic linkages which cannot be hydrolyzed by amylase and supply no energy, but they serve several important functions in the body.
   a. Fiber increases intestinal motility, which results in less contact of bowel mucosa with potential carcinogens (e.g., lithocholic acid).
   b. Fiber reduces the risk for colorectal cancer by absorbing carcinogens and reducing transit time.
   c. Fiber softens the stool, which alleviates constipation and reduces the incidence of diverticulosis of the sigmoid colon.
   d. Fiber reduces absorption of cholesterol (decreasing blood cholesterol), fat-soluble vitamins, and some minerals (e.g., zinc).
   e. Soluble fiber (e.g., oat bran, psyllium seeds) has a greater cholesterol-lowering effect than insoluble fiber (e.g., wheat bran).
   f. β-Glucan from oat bran fosters growth of beneficial bacteria; prebiotic effect (bacteria are probiotic).

   > Insoluble fiber: increases transit time; decreases exposure to carcinogens
   >
   > Soluble fiber: increases favorable bacteria; decreases serum cholesterol level
   >
   > Fiber: softens stool; prevents sigmoid diverticulosis

C. **Dietary lipids**
1. Long-chain free fatty acids from triacylglycerols provide the major source of energy to cells, with the exception of RBCs and the brain.
2. Dietary fats also contain essential fatty acids and are required for the absorption of fat-soluble vitamins.
3. The composition of dietary triacylglycerols varies in plants and animals.
   a. Plants primarily contain unsaturated and saturated fats.
      (1) Monounsaturated fats (one double bond, long chain) are present in olive oil and canola oil.
      (2) Polyunsaturated fats (two or more double bonds, long chain) are present in soybean oil and corn oil.
      (3) Saturated fats (no double bonds, medium chain) are present in coconut oil and palm oil; acetic acid in vinegar is a short-chain saturated free fatty acid.

   > Plant fats from oils: monounsaturated and polyunsaturated, long chain; saturated, medium chain
   >
   > Animal fats from adipose: saturated, long chain
   >
   > Animal fats from muscle and organ tissues: polyunsaturated and monounsaturated

---

BOX 4-1 LACTOSE INTOLERANCE

Lactose intolerance, the most common type of digestive enzyme deficiency, is caused by insufficient lactase activity. It may result from an inherited decrease in lactase production or from damage to mucosal cells by drugs, diarrhea, or protein deficiency (e.g., kwashiorkor). The incidence of lactose intolerance is much higher (up to 90%) in those of Asian and African descent than in those of northern European descent (<10%).

The signs and symptoms of lactose intolerance result from the inability to digest lactose, a disaccharide that is present in dairy products. Unabsorbed lactose is osmotically active, causing retention of water in the gastrointestinal tract and production of a watery diarrhea. Bacterial degradation of lactose produces lactic acid and hydrogen gas, which causes abdominal bloating, cramps, and flatulence. The stool has an acid pH. Elimination of dairy products from the diet is the most effective treatment.

b. Animals also contain unsaturated and saturated fats.
   (1) The dominant (≈50% fat energy) fatty acids in the fat storage depots (adipocytes) of cattle are saturated fatty acids (long-chain SFAs, also includes cheese, butter).
   (2) The six major sources of SFAs in the United States diet are fatty meats, baked goods, cheese, milk, margarine, and butter.
   (3) The dominant fatty acids in muscle and all other organ tissues are long-chain polyunsaturated fatty acids (PUFAs) and long-chain monounsaturated fatty acids (MUFAs).
4. Essential fatty acids are required in the diet and include the polyunsaturated fatty acids linoleic (n-6) and linolenic (n-3) acids.
   a. Functions of essential fatty acids
      (1) Help maintain fluidity of cellular membranes
      (2) Precursors for arachidonic acid (linoleic acid) from which the eicosanoids (e.g., prostaglandins) are derived
      (3) Prevent platelet aggregation (linolenic acid), which reduces the incidence of strokes and myocardial infarctions
   b. Essential fatty acids are present in high concentration in fish oils, canola oil, and walnuts.
   c. Deficiency results in scaly dermatitis, poor wound healing, and hair loss.
5. Dietary triacylglycerols are digested primarily in the small intestine (Fig. 4-1).
   a. Pancreatic lipase (aided by colipase) degrades triacylglycerol into 2-monoacylglycerol and free fatty acids.
   b. Pancreatic cholesterol esterase hydrolyzes cholesteryl esters and releases free cholesterol.
   c. 2-Monoacylglycerol, free fatty acids, and cholesterol, along with fat-soluble vitamins and phospholipids, are emulsified by bile salts to form micelles that are absorbed into intestinal mucosal cells by passive diffusion.
      (1) Resynthesis of triacylglycerols and of cholesteryl esters occurs within mucosal cells.
      (2) Short- and medium-chain fatty acids are directly absorbed and released into the portal circulation; they also bypass the carnitine cycle and are used directly by the mitochondria.
   d. Nascent chylomicrons are assembled in mucosal cells and contain triacylglycerols (≈85%), cholesteryl esters (≈3%), phospholipids, the fat-soluble vitamins (i.e., A, D, E, and K), and apolipoprotein B-48, which is necessary for secretion of chylomicrons into the lymphatics.

Essential fatty acids: linoleic (ω-6) and linolenic (ω-3)

Higher (n-3)/(n-6) PUFA ratios in diet are healthier

Essential fatty acid sources: fish oils, walnuts, canola oil

Deficiency: dermatitis, slow healing, hair loss

Pancreatic lipase: converts triacylglycerols into 2-monoacylglycerol and free fatty acids

Bile salts: form micelles for reabsorption by small intestine villi

Triacylglycerols and cholesterol esters are resynthesized in the mucosal cells after absorption.

Short- and medium-chain fatty acids enter the portal circulation directly and are not reassembled into triacylglycerols.

ApoB-48: important in formation of chylomicrons and secretion into lymphatics

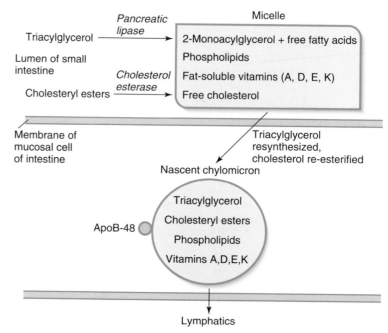

4-1: Digestion of dietary lipids and assembly of nascent chylomicrons. Pancreatic lipase acts on triacylglycerol to produce 2-monoacylglycerol and free fatty acids, and pancreatic cholesterol esterase hydrolyzes cholesteryl esters to free cholesterol. These degradation products, as well as phospholipids and fat-soluble vitamins, are micellarized by bile salts and absorbed into intestinal cells by passive diffusion. Triacylglycerols are resynthesized, cholesterol is re-esterified, and all components are packaged into nascent chylomicrons with apolipoprotein B-48 on their surfaces. ApoB-48 is required for secretion of chylomicrons into the lymphatics and bloodstream.

BOX 4-2   PATHOGENESIS OF MALABSORPTION

Malabsorption is a general term referring to increased fecal excretion of fat, called *steatorrhea*, with concurrent deficiencies of vitamins (particularly fat-soluble vitamins), minerals, carbohydrates, and proteins. The pathophysiology of malabsorption is classified as pancreatic insufficiency, bile salt deficiency, and small bowel disease.

Pancreatic insufficiency causes a maldigestion of fats due to diminished lipase activity, resulting in the presence of undigested neutral fats and fat droplets in stool. There is also maldigestion of proteins due to diminished trypsin, leading to undigested meat fibers in stool. Carbohydrate digestion is not affected because of the presence of amylase in salivary glands and disaccharidases in brush border enzymes. Chronic pancreatitis due to alcoholism is the most common cause of pancreatic insufficiency in adults; chronic pancreatitis due to cystic fibrosis is the most common cause in children.

Bile salt deficiency results in defective emulsification of fats, which is necessary for their absorption by small intestinal villi. Causes of bile salt deficiency include cirrhosis (i.e., inadequate production of bile salts and acids from cholesterol); intrahepatic or extrahepatic blockage of bile (e.g., calculus in common bile duct); bacterial overgrowth in the small bowel with destruction of bile salts; excess binding of bile salts (e.g., use of cholestyramine); and terminal ileal disease (i.e., inability to recycle bile salts and acids).

Small bowel disease associated with a loss of the villous surface leads to a malassimilation of fats, proteins, and carbohydrates. Celiac disease, or sprue, an autoimmune disease caused by antibodies directed against gluten in wheat, and Crohn's disease, an inflammatory bowel disease involving the terminal ileum, commonly produce malabsorption.

Characteristic clinical findings for malabsorption include weight loss, anemia, chronic diarrhea, and malnutrition. The signs and symptoms associated with multiple fat-soluble vitamin deficiencies are usually present. Night blindness (i.e., vitamin A deficiency), rickets (i.e., vitamin D deficiency), and a hemorrhagic diathesis with ecchymoses and gastrointestinal bleeding (i.e., vitamin K deficiency) are the usual findings.

     e. When discharged into the lymphatic vessels, lipoproteins rich in triacylglycerols ultimately enter the bloodstream and circulate to deliver fatty acids to tissues (see Chapter 7).

     f. Complete oxidation of fats to CO_2 and H_2O in the body produces 9 kcal/g.

  6. The pathophysiology of lipid malabsorption is discussed in Box 4-2.

> Dietary triacylglycerol: 9 kcal/g

D. Dietary proteins

  1. The biologic value of a dietary protein is determined by its content of essential amino acids (see Chapter 1).

    a. Plant proteins (e.g., rice, wheat, corn, beans) are of low biologic value unless they are combined to provide all of the essential amino acids.

      (1) Combining rice and beans or corn and beans provides higher biologic value than any one source.

    b. Animal proteins (e.g., eggs, meat, poultry, fish, and dairy products) are of high biologic value because they contain all of the essential amino acids.

    c. The dietary RDA for high-quality protein is 0.8 g/kg for men and women, which equals about 60 g/day for men and about 50 g/day for women.

> Biologic value of protein: determined by degree of representation by essential amino acids
>
> Biologic value for animal protein: higher than for plant protein

  2. Digestion of dietary proteins begins in the stomach, where the low pH denatures proteins, making them more susceptible to enzymatic hydrolysis.

    a. In the stomach, pepsinogen is secreted by gastric chief cells.

      (1) Acid converts pepsinogen to pepsin.

      (2) Pepsin cleaves proteins to smaller polypeptides.

    b. Pancreatic proteases (e.g., trypsin) and peptidases secreted by intestinal epithelial cells digest the polypeptides from the stomach to free amino acids.

    c. Uptake of amino acids from the intestinal lumen and their release into the portal circulation by carrier proteins in the intestinal epithelial cells are driven by the hydrolysis of ATP.

    d. Complete oxidation of proteins to CO_2 and H_2O produces 4 kcal/g.

> Achlorhydria: lack of acid leads to inability to properly digest protein (e.g., pernicious anemia)

  3. Proteins are in a constant state of degradation and resynthesis.

    a. When amino acids are oxidized, their nitrogen atoms are fed into the urea cycle in the liver and excreted as urea in the urine (primary route), feces, and sweat (see Chapter 8).

      (1) Other nitrogenous excretory products include uric acid, creatinine, and ammonia.

    b. Nitrogen balance is the difference in the amount of nitrogen (protein) consumed and the amount of nitrogen excreted in the urine, sweat, and feces.

      (1) Healthy adults are normally in zero nitrogen balance (i.e., nitrogen intake equals nitrogen loss).

> Proteins: 4 kcal/g

(2) A positive nitrogen balance (i.e., nitrogen intake is greater than nitrogen loss) indicates active synthesis of new protein. Examples include pregnancy and lactation, growth in children, and recovery from surgery, trauma, or extreme starvation.

(3) A negative nitrogen balance (i.e., nitrogen loss is greater than nitrogen intake) indicates breakdown of tissue proteins. Examples include diets containing protein of low biologic value, physiologic stress (e.g., third-degree burns).

Nitrogen balance: nitrogen consumed — nitrogen excreted

Positive for growth; negative for surgery recovery and burns, neutral for general health.

4. Carbohydrates have a protein-sparing effect.
   a. An adequate intake of carbohydrate provides the glucose that is necessary for maintaining normal blood sugar levels.
   b. An inadequate intake of carbohydrate (<150 g/day) causes degradation of skeletal muscle to provide carbon skeletons (e.g., pyruvate) for gluconeogenesis (see Chapter 9).

Carbohydrates are protein sparing.

Marasmus: total calorie deprivation

5. Protein-energy malnutrition results from inadequate intake of protein or calories.
   a. Marasmus is caused by a diet deficient in protein and calories (e.g., total calorie deprivation).
      (1) Marasmus is marked by extreme muscle wasting (i.e., broomstick extremities) due to breakdown of muscle protein for energy and by growth retardation.
      (2) It occurs primarily during the first year of life.
   b. Kwashiorkor is caused by a diet inadequate in protein in the presence of an adequate carbohydrate intake.
      (1) Marked by pitting edema and ascites (i.e., loss of the oncotic effect of albumin), enlarged fatty liver (i.e., decreased apolipoproteins), anemia, diarrhea (i.e., loss of brush border enzymes), and defects in cellular immunity
      (2) Less extreme muscle wasting than occurs in marasmus due to the protein-sparing effect of carbohydrates

Kwashiorkor: inadequate protein intake; protein sparing due to increased carbohydrates

III. Water-Soluble Vitamins
A. Overview (Fig. 4-2 and Table 4-1)
1. Thiamine functions in oxidative decarboxylation and pentose phosphate pathway; deficiency produces beriberi.
2. Riboflavin functions primarily in electron transport; deficiency produces glossitis, stomatitis.
3. Niacin functions in NAD^+ and $NADP^+$ as cofactors for redox reactions; deficiency produces pellagra.
4. Pantothenic acid functions in fat and carbohydrate metabolism as a component of acetyl CoA and fatty acid synthase; deficiency symptoms are unknown.
5. Pyridoxine functions in transamination reactions, heme synthesis, glycogenolysis, and various other amino acid conversions; deficiency produces sideroblastic anemia and peripheral neuropathy.
6. Cobalamin functions in single carbon metabolism; deficiency produces macrocytic anemia and pernicious anemia.

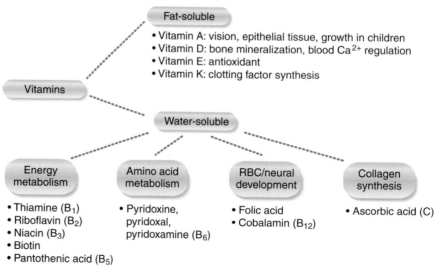

4-2: Classification and functions of the vitamins.

TABLE 4-1. Water-Soluble Vitamins: Signs and Symptoms of Deficiency

VITAMIN	SIGNS AND SYMPTOMS OF DEFICIENCY
Thiamine (vitamin B_1)	Wernicke-Korsakoff syndrome (confusion, ataxia, nystagmus, ophthalmoplegia, antegrade and retrograde amnesia, precipitated by giving thiamine with glucose in intravenous solution); peripheral neuropathy (dry beriberi); dilated cardiomyopathy (wet beriberi)
Riboflavin (vitamin B_2)	Corneal neovascularization; glossitis; cheilosis; angular stomatitis
Niacin (vitamin B_3)	Pellagra, with diarrhea, dermatitis, dementia
Pantothenic acid (vitamin B_5)	None identified
Pyridoxine (vitamin B_6)	Sideroblastic anemia; peripheral neuropathy; convulsions
Cobalamin (vitamin B_{12})	Macrocytic (megaloblastic) anemia; neutropenia and thrombocytopenia; hypersegmented neutrophils; glossitis; subacute combined degeneration (posterior column and lateral corticospinal tract demyelination); dementia; achlorhydria, atrophic gastritis body and fundus, increased serum gastrin (only in pernicious anemia); increased plasma homocysteine; increased urine methylmalonic acid
Folic acid	Same as vitamin B_{12} deficiency with the following exceptions: no neurologic dysfunction and normal urine methylmalonic acid level
Biotin	Dermatitis, alopecia, glossitis, lactic acidosis
Ascorbic acid (vitamin C)	Bleeding diathesis (ecchymoses, hemarthroses, bleeding gums, perifollicular hemorrhages, corkscrew hairs); loosened teeth; poor wound healing; glossitis

7. Folic acid functions in single carbon metabolism; deficiency produces macrocytic anemia.
8. Biotin functions in carboxylase reactions; deficiency produces dermatitis, alopecia, glossitis.
9. Ascorbic acid (vitamin C) functions in hydroxylation reactions; deficiency produces scurvy.

B. Thiamine (vitamin B_1)
  1. Sources of thiamine include enriched and whole-grain cereals, brewer's yeast, meats, legumes, and nuts.
  2. Thiamine pyrophosphate is the active form of the vitamin.
     a. Cofactor for dehydrogenases in oxidative decarboxylation of α-keto acids (e.g., pyruvate dehydrogenase conversion of pyruvate into acetyl CoA)
     b. Cofactor for transketolase (two-carbon transfer reactions) in the pentose phosphate pathway
  3. Thiamine deficiency most commonly occurs in alcoholics or malnourished individuals.
     a. Most clinical findings in thiamine deficiency reflect the loss of ATP from dysfunction of the pyruvate and α-ketoglutarate dehydrogenase reactions, which normally gain 2 NADH (6 ATP).
     b. Intravenous infusion of a glucose-containing fluid may precipitate acute thiamine deficiency in alcoholics (depleted by the pyruvate dehydrogenase reaction); deficiency is manifested by Wernicke-Korsakoff syndrome (WKS).
     c. WKS includes confusion, ataxia, nystagmus, eye muscle weakness, and retrograde and antegrade memory deficits.
     d. Clinical findings associated with thiamine deficiency (see Table 4-1)
C. Riboflavin (vitamin B_2)
  1. Sources of riboflavin include milk, eggs, meats, poultry, fish, and green leafy vegetables.
  2. Flavin adenine dinucleotide (FAD) and flavin mononucleotide (FMN) are the active forms of riboflavin.
     a. FAD is a cofactor associated with succinate dehydrogenase, which converts succinate to fumarate in the citric acid cycle.
     b. FMN is a component of the electron transport chain and accepts two hydrogen atoms (becomes $FMNH_2$) from NADH in a reaction catalyzed by NADH dehydrogenase.
  3. Riboflavin deficiency is usually seen in severely malnourished individuals or pure vegans, who lack intake of dairy products.
     a. Clinical findings associated with riboflavin deficiency (see Table 4-1)
D. Niacin (vitamin B_3 or nicotinic acid)
  1. Sources of niacin include meat, enriched and whole-grain cereals, and synthesis from tryptophan-containing foods, such as milk and eggs.
     a. Excess tryptophan is metabolized to niacin and supplies about 10% of the niacin RDA.

Water-soluble vitamins are readily excreted in the urine and rarely reach toxic levels.

Thiamine: cofactor for oxidative decarboxylation and transketolase

Thiamine deficiency due to alcoholism and undernourishment

Alcoholics: give intravenous thiamine before infusion fluids with glucose.

Thiamine (B_1) deficiency causes beriberi and Wernicke-Korsakoff syndrome.

Riboflavin's active forms: FAD and FMN

Riboflavin deficiency: lack of dairy products or general malnourishment

Riboflavin deficiency symptoms: glossitis, stomatitis, cheilosis

Niacin functions in redox reactions as NAD^+ and $NADP^+$.

Niacin deficiency causes: malnourishment, Hartnup's disease, carcinoid syndrome

2. The two active forms of niacin are NAD^+ and $NADP^+$.
   a. NAD^+ reactions are primarily catabolic (e.g., glycolysis).
   b. $NADP^+$ reactions are primarily anabolic (e.g., fatty acid synthesis).
3. Nicotinic acid is a lipid-lowering agent that decreases cholesterol and triacylglycerol and increased high-density lipoprotein.
4. Niacin deficiency, also known as pellagra, primarily occurs in individuals whose diets are deficient in niacin and tryptophan or in conditions in which tryptophan is lost in urine and feces (e.g., Hartnup's disease) or excessively used (e.g., carcinoid syndrome).
   a. Individuals who consume corn-based diets are particularly prone to pellagra, because maize protein has a low tryptophan content and niacin is in a bound form that cannot be reabsorbed (i.e., treating corn with lime, or calcium carbonate, releases bound niacin).
   b. Hartnup's disease is an autosomal recessive disease with a defect in the intestinal and renal reabsorption of neutral amino acids (e.g., tryptophan).
   c. In carcinoid syndrome, tryptophan is used to synthesize serotonin, which produces the flushing and diarrhea associated with the syndrome.
      (1) Carcinoid tumor in the small intestine metastasizes to the liver.
      (2) Metastatic nodules secrete serotonin into hepatic vein tributaries producing carcinoid syndrome.
   d. Clinical findings associated with pellagra (see Table 4-1)
5. Excessive intake of niacin leads to flushing due to vasodilatation.
   a. Flushing can be prevented by taking aspirin 30 minutes before taking niacin.

E. Pantothenic acid (vitamin B_5)
1. Pantothenic acid is present in a wide variety of foods.
2. It is a component of coenzyme A (CoA) and the fatty acid synthase complex, which is involved in fatty acid synthesis.
3. Pantothenic acid deficiency is uncommon.

F. Pyridoxine (vitamin B_6)
1. Sources of pyridoxine include whole-grain cereals, eggs, meats, fish, soybeans, and nuts.
2. Pyridoxal phosphate is the active form of the vitamin.
3. Functions of pyridoxine
   a. Pyridoxine is involved in transamination reactions (reversible conversion of amino acids to α-ketoacids), which are catalyzed by the transaminases alanine aminotransferase (ALT) and aspartate aminotransferase (AST).
   b. It is a cofactor for δ-aminolevulinic acid (ALA) synthase, which catalyzes the rate-limiting reaction that converts succinyl CoA + glycine into δ-ALA in heme synthesis.
      (1) Deficiency of pyridoxine leads to a defect in heme synthesis and anemia (i.e., sideroblastic anemia).
   c. Pyridoxine is involved in the synthesis of neurotransmitters such as γ-aminobutyrate (GABA), serotonin, and norepinephrine.
   d. It is a cofactor in the following:
      (1) Decarboxylation reactions (e.g., conversion of histidine to histamine)
      (2) Glycogenolysis (e.g., glycogen phosphorylase)
      (3) Deamination reactions (e.g., conversion of serine to pyruvate and ammonia)
      (4) Conversion of tryptophan to niacin
4. Pyridoxine deficiency is most commonly seen in alcoholics and in patients receiving isoniazid therapy for tuberculosis.
   a. Unfortified goat's milk may also cause pyridoxine deficiency.
5. Clinical findings associated with pyridoxine deficiency (see Table 4-1)

G. Cobalamin (vitamin B_{12}; contains cobalt)
1. Sources of cobalamin include meats, shellfish, poultry, eggs, and dairy products only.
   a. Pure vegans lack vitamin B_{12}, whereas ovolactovegetarians obtain adequate sources of vitamin B_{12} from eggs and dairy products.
   b. Pure vegans who are pregnant or who are breast-feeding require vitamin B_{12} supplements to prevent anemia from developing in the infant.
2. Functions of cobalamin (Fig. 4-3)
   a. Cobalamin removes the methyl group from N^5-methyltetrahydrofolate (N^5-methyl-FH_4) to form tetrahydrofolate (FH_4), which is used to synthesize deoxythymidine monophosphate (dTMP) from deoxyuridine monophosphate (dUMP) for DNA synthesis.
   b. It transfers methyl groups to homocysteine to form methionine.

Margin notes:

Niacin (B_3) deficiency: causes pellagra, with diarrhea, dermatitis, dementia

Niacin deficiency: corn-based diets low in tryptophan and niacin

Hartnup's disease: defect in intestinal and renal uptake of neutral amino acids

Carcinoid syndrome: tryptophan used up in synthesizing serotonin; serotonin produces diarrhea and flushing.

Pantothenic acid: most common active form is coenzyme A; also in fatty acid synthase

Pyridoxine: cofactor for transaminase, ALA synthase, glycogen phosphorylase, and neurotransmitter synthesis

Pyridoxine deficiency: alcoholism and isoniazid therapy

Pyridoxine: cofactor for ALA synthase in heme synthesis; deficiency leads to defect in heme synthesis and anemia

Pyridoxine (B_6) deficiency: sideroblastic anemia, peripheral neuropathy, convulsions

Pure vegans: must take vitamin B_{12} to prevent anemia in their babies

Cobalamin: single-carbon metabolism

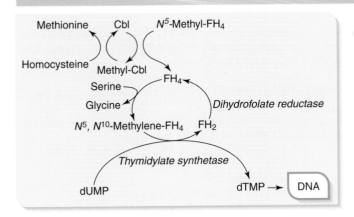

4-3: Vitamin B$_{12}$ (cobalamin) and folic acid in DNA synthesis. Cobalamin (Cbl) is important in the demethylation of N^5-methyltetrahydrofolate (N^5-methyl-FH$_4$) and methylation of homocysteine to form methionine. Tetrahydrofolate (FH$_4$) receives a methylene group (–CH$_2$–) from serine to produce N^5, N^{10}-methylene-FH$_4$, which transfers the methylene group to deoxyuridine monophosphate (dUMP) to produce deoxythymidine monophosphate (dTMP). Methyl-Cbl, methylcobalamin.

(1) A deficiency of vitamin B$_{12}$ or folic acid leads to an increase in plasma homocysteine levels, which damages vessels and poses a risk for vessel thrombosis.

(2) Folate deficiency is the most common cause of an increase in homocysteine in the United States.

c. Cobalamin is involved in odd-chain fatty acid metabolism.

(1) Propionyl CoA, the end product of odd-chain fatty acid metabolism, is converted into methylmalonyl CoA.

(2) Cobalamin is a cofactor for methylmalonyl CoA mutase, which converts methylmalonyl CoA into succinyl CoA (i.e., citric acid cycle).

(3) Vitamin B$_{12}$ deficiency results in an accumulation of methylmalonyl CoA (proximal to the enzyme block), which is converted into methylmalonic acid.

(4) Increase in propionyl CoA proximal to the block leads to demyelination of the spinal cord (i.e., subacute combined degeneration), peripheral nerves (i.e., peripheral neuropathy), and brain (i.e., dementia).

(5) Propionyl CoA replaces acetyl CoA in myelin synthesis.

3. Vitamin B$_{12}$ metabolism

a. Vitamin B$_{12}$ complexes with the R factor in saliva.

(1) R factor complex prevents degradation of vitamin B$_{12}$ by stomach acid.

b. Intrinsic factor (IF) is synthesized in parietal cells located in the body and fundus of the stomach.

(1) Autoantibodies in pernicious anemia destroy the parietal cells, causing deficiency of intrinsic factor leading to vitamin B$_{12}$ deficiency.

c. Pancreatic enzymes cleave off R factor, which allows vitamin B$_{12}$ to complex with IF in the duodenum.

(1) Chronic pancreatitis causes malabsorption of vitamin B$_{12}$.

(2) Bacterial overgrowth destroys vitamin B$_{12}$-IF complex.

(3) Fish tapeworm (*Diphyllobothrium latum*) uses vitamin B$_{12}$ in its metabolism.

d. The vitamin B$_{12}$-IF complex binds to intrinsic factor receptors in the terminal ileum and is absorbed.

(1) Terminal ileal disease (e.g., Crohn's disease) causes malabsorption of vitamin B$_{12}$.

e. After absorption, vitamin B$_{12}$ is bound to transcobalamin II in the plasma and is delivered to metabolically active cells or stored in the liver (6- to 9-year supply).

4. Vitamin B$_{12}$ deficiency is most commonly caused by autoimmune destruction of parietal cells (called pernicious anemia).

5. Summary of causes of vitamin B$_{12}$ deficiency

a. Pure vegan diet

b. Pernicious anemia: destruction of parietal cells

c. Chronic pancreatitis: cannot cleave off R factor

d. Bacterial overgrowth: vitamin B$_{12}$-IF complex is destroyed

e. Fish tapeworm (*Diphyllobothrium latum*)

f. Terminal ileal disease (e.g., Crohn's disease)

6. Clinical findings associated with vitamin B$_{12}$ deficiency (see Table 4-1)

H. Folic acid

1. Sources of folic acid include green leafy vegetables, liver, legumes, whole-grain cereals, and yeast.

Increased homocysteine is caused by decreases in folate or vitamin B$_{12}$ levels.

Cobalamin function: converts homocysteine to methionine; odd-chain fatty acid metabolism

Vitamin B$_{12}$ deficiency: increases urine methylmalonic acid levels

Increased propionyl CoA: produces demyelination in spinal cord, peripheral nerves, brain

Vitamin B$_{12}$ requires IF for absorption in the terminal ileum.

Vitamin B$_{12}$ (cobalamin) deficiency: caused by pernicious anemia, pure vegan diet, ileal disease, chronic pancreatitis

Cobalamin deficiency: macrocytic anemia, methylmalonic acidemia

Folate active form is tetrahydrofolate, needed for thymidylate synthase in DNA synthesis.

Fluorouracil inhibits thymidylate synthase.

2. Functions of folic acid (see Fig. 4-3)
  a. Tetrahydrofolate (FH_4) receives a methylene group ($-CH_2-$), an example of a one-carbon transfer reaction, from serine to produce N^5, N^{10}-methylene-tetrahydrofolate.
    (1) The methylene group is then transferred by thymidylate synthase to dUMP to produce dTMP for DNA synthesis.
    (2) Folic acid deficiency impairs DNA replication due to the shortage of purine nucleotides and thymine.
    (3) Fluorouracil, a chemotherapeutic drug, is converted to a compound that binds to thymidylate synthase, which irreversibly inhibits its function.
  b. Two hydrogens from FH_4 are used in the formation of dTMP, resulting in the formation of dihydrofolate (FH_2).
  c. FH_2 is reduced to FH_4 by dihydrofolate reductase.
    (1) Methotrexate and trimethoprim inhibit dihydrofolate reductase.
3. Folic acid metabolism
  a. Folic acid is ingested in a polyglutamate form.
  b. Polyglutamates are converted into monoglutamates in the jejunum by intestinal conjugase.
    (1) The drug phenytoin inhibits intestinal conjugase.
  c. Folate monoglutamate is absorbed in the jejunum.
    (1) Folate monoglutamate absorption is blocked by alcohol and by oral contraceptives, leading to folic acid deficiency.
  d. Folic acid circulates and is measured in the blood as methyltetrahydrofolate.
    (1) Deficiency of vitamin B_{12} traps N^5-methyl-FH_4 in its circulating form; may falsely increase the serum folate in 30% of cases.
  e. Only a 3- to 4-month supply of folic acid is stored in the liver.
4. Folic acid deficiency is most commonly caused by alcoholism and other causes:
  a. Diet lacking fruits and vegetables
  b. Drugs: 5-fluorouracil, methotrexate, trimethoprim, phenytoin, oral contraceptives
  c. Pregnancy and lactation: increased use of folate by the fetus
    (1) Women must have adequate levels of folate before becoming pregnant to prevent failure of the neural tube to close between the 23rd and 28th day of embryogenesis.
  d. Rapidly growing cancers (e.g., leukemia): malignant cells use folic acid
  e. Small bowel malabsorption (e.g., celiac disease)
  f. Unfortified goat's milk
5. Clinical and laboratory findings are similar to vitamin B_{12} deficiency except for the absence of neurologic deficits and normal levels of methylmalonic acid (see Table 4-1).

I. Biotin
1. Most of the daily requirement of biotin is supplied by bacterial synthesis in the intestine.
2. It is a cofactor in carboxylase reactions (e.g., pyruvate carboxylase, acetyl CoA carboxylase, propionyl CoA carboxylase).
3. Biotin deficiency is caused by eating raw eggs (egg whites contain avidin, which binds biotin) and by taking broad-spectrum antibiotics, which prevent bacterial synthesis of the vitamin.
4. Clinical findings associated with biotin deficiency (see Table 4-1)

J. Ascorbic acid (vitamin C)
1. Sources of vitamin C include citrus fruits, potatoes, green and red peppers, broccoli, tomatoes, spinach, and strawberries.
2. Functions of vitamin C
  a. Hydroxylation of lysine and proline residues during collagen synthesis
  b. Antioxidant activity (inactivates hydroxyl free radicals)
  c. Reduces nonheme iron (Fe^{3+}) from plants to the ferrous (Fe^{2+}) state for absorption in the duodenum
  d. Keeps tetrahydrofolate (FH_4) in its reduced form
  e. Cofactor in the conversion of dopamine to norepinephrine in catecholamine synthesis
3. Causes of vitamin C deficiency (scurvy) include diets lacking fruits and vegetables (i.e., tea and toast diet) and smoking cigarettes, because the vitamin is used up in neutralizing free radicals in cigarette smoke.
4. Excess intake of vitamin C may result in the formation of renal calculi.
5. Clinical findings associated with vitamin C deficiency (see Table 4-1)

Folic acid deficiency causes macrocytic anemia.

Folate absorption occurs in the jejunum.

Folic acid is stored in the liver, providing a 3- to 4-month supply.

Drugs causing folic acid deficiency: 5-fluorouracil, methotrexate, trimethoprim, alcohol, phenytoin, oral contraceptives, *metformin, sulfasalazine, triamterene*

Malignant cells compete for folate, cause deficiency.

Folic acid supplementation before pregnancy reduces risk of neural tube defects.

Biotin: supplied by intestinal bacteria; absorption blocked by avidin in raw egg white

Biotin deficiency: dermatitis, alopecia, glossitis, lactic acidosis

Ascorbate required for hydroxylation of lysine and proline during collagen synthesis.

Ascorbate is a cofactor in the conversion of dopamine to norepinephrine in catecholamine synthesis.

Ascorbic acid deficiency: insufficient fruits and vegetables; causes scurvy

Excessive ascorbate (>4 g/day) can produce oxalate stones.

IV. Fat-Soluble Vitamins

A. Overview

1. Vitamin A functions as a component of the visual pigments and in cell differentiation; deficiency produces night blindness, skin abnormalities, and growth abnormalities.
2. Vitamin D functions in calcium metabolism; deficiency produces rickets.
3. Vitamin E functions as an antioxidant; deficiency produces hemolytic anemia and peripheral neuropathy.
4. Vitamin K functions in the γ-carboxylation of clotting factors; deficiency produces prolonged prothrombin time and a bleeding diathesis.

B. Vitamin A (retinol)

1. Sources of vitamin A include cod liver oil, dairy products, and egg yolk.
2. Retinol (alcohol), retinal (aldehyde), and retinoic acid are the active forms of vitamin A (Fig. 4-4).
   a. β-Carotenes (i.e., provitamin A) found in dark green leafy and yellow vegetables (e.g., spinach and carrots) and retinol esters in the diet are converted into retinol, which is esterified (forming retinol esters) in the enterocytes of the small intestine.
      (1) An excess of β-carotenes in the diet turns the skin yellow, but the sclera remains white (unlike in jaundice).
   b. Retinol esters are packaged into chylomicrons and transported to the liver for storage.
   c. Retinol is released from the liver, complexes with retinol-binding protein (RBP), and is delivered to target tissues throughout the body (except heart and skeletal muscle).
   d. In the cytosol, retinol is irreversibly oxidized to retinoic acid.
   e. Retinoic acid (similar to steroid hormones and vitamin D) binds to nuclear receptors, forming a complex that activates gene transcription of protein products.
3. Functions of vitamin A
   a. A component of the visual pigments within rod and cone cells of the retina
   b. Important in normal cell differentiation and prevents epithelial cells from undergoing squamous metaplasia
   c. Important in normal bone and tooth development
   d. Supports spermatogenesis and placental development
   e. Drugs used in the treatment of skin disorders and acute promyelocytic leukemia
      (1) Topical tretinoin (all-*trans*-retinoic acid) is used in the treatment of psoriasis and mild acne.
      (2) Oral isotretinoin is used to treat severe cystic acne; however, because it is teratogenic, women must have a pregnancy test before the drug is prescribed.
      (3) All-*trans*-retinoic acid is used to treat acute promyelocytic leukemia (hypergranular M3) and is thought to induce maturation of the leukemic cells.
4. Causes of vitamin A deficiency include a diet poor in dark green leafy and yellow vegetables and diseases such as fat malabsorption (e.g., celiac disease).
5. Causes of vitamin A excess include eating polar bear liver and isotretinoin therapy.
6. Clinical findings associated with vitamin A deficiency and excess (Table 4-2)

The fat-soluble vitamins are absorbed with fats, transported in chylomicrons, and stored in the liver and adipose tissue.

Increased β-carotene levels: yellow skin but normal sclera

Active forms of vitamin A: retinol, retinal, retinoic acid

Retinol-binding protein delivers retinol to target tissues.

Vitamin A functions in cell differentiation and spermatogenesis.

Oral isotretinoin: severe cystic acne; teratogenic

*All-*trans*-retinoic acid: treatment for acute promyelocytic leukemia*

Vitamin A deficiency produces night blindness, poor wound healing, and follicular hyperkeratosis.

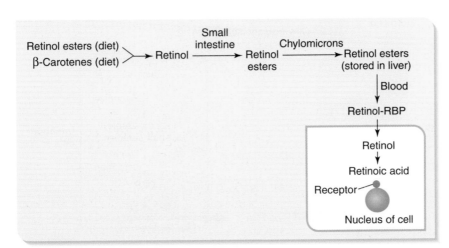

4-4: Vitamin A absorption and transport. Ingested retinol esters and β-carotenes are converted to retinol, the key absorption and transport form of vitamin A. In the small intestine, retinol is converted to retinol esters, the key storage form of vitamin A. When needed, retinol is released from the liver into the bloodstream, where it complexes with retinol-binding protein (RBP). Within cells, retinol is irreversibly oxidized to retinoic acid, which binds to nuclear receptors and activates gene transcription.

TABLE 4-2. **Fat-Soluble Vitamins: Signs and Symptoms of Deficiency and Excess**

VITAMIN	SIGNS AND SYMPTOMS OF DEFICIENCY	SIGNS AND SYMPTOMS OF EXCESS
Vitamin A	Night blindness; eye abnormalities (dry eyes, blindness); skin abnormalities (follicular hyperkeratosis, dry skin); lung abnormalities (bronchitis, pneumonia, possibly lung cancer); growth retardation; poor wound healing	Increased intracranial pressure (papilledema, convulsions); liver toxicity; bone pain
Vitamin D	Rickets in children and osteomalacia in adults; findings in both conditions: pathologic fractures, excess osteoid, bowed legs; findings in rickets only: craniotabes (soft bone), skeletal deformities, rachitic rosary (excess osteoid in epiphysis), defective epiphyseal plates with growth retardation	Hypercalcemia, renal calculi
Vitamin E	Hemolytic anemia; peripheral neuropathy; posterior column degeneration (poor joint sensation and absent vibratory sensation); retinal degeneration; myopathy *(rare)*	Decreased synthesis of vitamin K–dependent coagulation factors in the liver (enhances anticoagulation effect of coumarin derivatives)
Vitamin K	Bleeding diathesis (gastrointestinal bleeding, ecchymoses); prolonged prothrombin time	Hemolytic anemia and jaundice in newborns if mother receives excess vitamin K

C. Vitamin D

1. Sources of vitamin D include liver, egg yolk, saltwater fish, and vitamin D–fortified foods.
2. Synthesis of calcitriol (1,25-dihydroxycholecalciferol), the active form of vitamin D, occurs in the following sequence (Fig. 4-5).

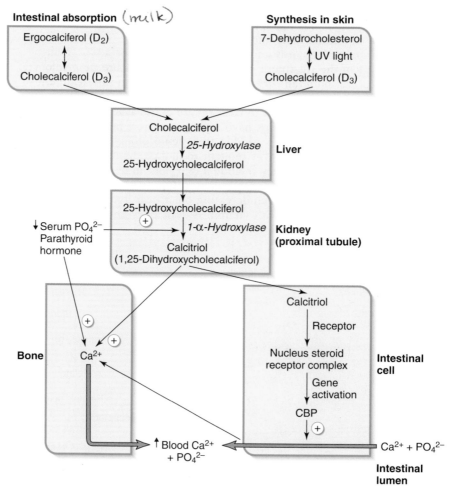

4-5: Formation of calcitriol, the active form of vitamin D, and its action in calcium homeostasis. The key functions of calcitriol are to mineralize bone using calcium and phosphorus and, in combination with parathyroid hormone (PTH), to maintain serum calcium levels. CBP, calcium-binding protein.

a. Preformed vitamin D in the diet consists of cholecalciferol (vitamin D_3) and ergocalciferol (vitamin D_2, found in plants), which is interconvertible with vitamin D_3.

b. Endogenous vitamin D is produced by photoconversion of 7-dehydrocholesterol to vitamin D_3 in sun-exposed skin (most important source of vitamin D).

c. The first hydroxylation, which produces 25-hydroxycholecalciferol, occurs in the liver within the cytochrome P450 system.

d. The second hydroxylation, by 1α-hydroxylase, produces 1,25-dihydroxycholecalciferol (calcitriol) and occurs in the proximal tubules of the kidneys.

   (1) Parathyroid hormone (PTH) increases the synthesis of 1α-hydroxylase in the proximal tubule cells of the kidneys.

e. Receptors for vitamin D are located in the intestine, kidneys, and on osteoblasts in bone.

3. Functions of vitamin D

  a. Vitamin D increases intestinal absorption of calcium and phosphorus and renal absorption of calcium.

   (1) Absorption of calcium and phosphorus provides an adequate solubility product (calcium × phosphorus) for vitamin D to mineralize bone.

   (2) When vitamin D interacts with its receptors on osteoblasts, alkaline phosphatase is released, leading to bone mineralization.

    (a) Alkaline phosphatase hydrolyzes pyrophosphate, an inhibitor of bone mineralization.

    (b) Alkaline phosphatase is three to five times higher in children than in adults because of their increased bone mineralization.

  b. In combination with PTH, vitamin D has the following effects:

   (1) It increases the mobilization of calcium from bone by stimulating the conversion of monocyte stem cells in the bone marrow to osteoclasts.

   (2) It maintains the serum calcium concentration.

4. Causes of vitamin D deficiency

  a. Inadequate exposure to sunlight: decreased synthesis of cholecalciferol (vitamin D_3)

  b. Fat malabsorption: cannot reabsorb fat-soluble vitamins (e.g., celiac disease)

  c. Chronic liver disease: cannot carry out the first hydroxylation of vitamin D_3

  d. Enhanced liver cytochrome P450 system (e.g., alcohol, phenytoin, barbiturates): increased conversion of 25-hydroxycholecalciferol into an inactive metabolite

  e. Renal failure (most common cause): due to deficiency of 1α-hydroxylase enzyme

  f. Primary hypoparathyroidism: PTH is required for enhancing the activity of 1α-hydroxylase

  g. Type I vitamin D–dependent rickets: deficiency of 1α-hydroxylase

  h. Type II vitamin D–dependent rickets: deficiency of vitamin D receptors in target tissue

5. Patients taking megadoses of vitamin D may develop vitamin D toxicity.

6. Clinical findings associated with vitamin D deficiency and excess (see Table 4-2)

D. Vitamin E

1. α-Tocopherol has the highest biologic activity of the naturally occurring tocopherols that constitute vitamin E and is abundant in fruits, vegetables, and grains.

2. Vitamin E is an antioxidant and scavenger of free radicals that protects polyunsaturated fats and fatty acids in cell membranes from lipid peroxidation and protects low-density lipoprotein (LDL) from oxidation.

  a. Oxidized LDL produces foam cells (macrophages) during atherosclerosis.

  b. Vitamin E protects erythrocytes from oxidative damage that leads to hemolysis.

3. Vitamin E deficiency is uncommon and primarily occurs in children with malabsorption caused by cystic fibrosis and in abetalipoproteinemia (see Chapter 7).

4. Patients taking megadoses of vitamin E may develop vitamin E toxicity.

  a. Inability to synthesize the vitamin K–dependent coagulation factors in the liver

5. Clinical findings associated with vitamin E deficiency and excess (see Table 4-2)

E. Vitamin K

1. Sources of vitamin K include green leafy vegetables (which supply vitamin K_1, phylloquinone) and bacterial synthesis in the colon (which supplies vitamin K_2, or menaquinone).

  a. After absorption in the bowel, vitamin K is oxidized to form an inactive epoxide.

  b. Most vitamin K comes from bacterial synthesis of the vitamin by colonic bacteria.

   (1) To be catalytically active, vitamin K synthesized by colonic bacteria must be reduced by epoxide reductase in the liver.

Margin notes:

Active form of vitamin D is a steroid hormone, calcitriol.

Skin: cholecalciferol

Liver: 25-hydroxycholecalciferol

Kidney: 1,25-dihydroxycholecalciferol (calcitriol)

Calcitriol functions in absorption of calcium from the intestine and kidneys; in combination with PTH, it mobilizes calcium from bone.

Hydroxyapatite, $Ca_{10}(PO_4)_6(OH)_2$, is the crystalline salt deposited in bone.

Alkaline phosphatase hydrolyzes pyrophosphate, an inhibitor of bone mineralization.

PTH and vitamin D: maintain ionized calcium level

Renal failure: most common cause of vitamin D deficiency

Vitamin E protects erythrocytes from oxidative damage leading to hemolysis.

Vitamin E deficiency produces hemolytic anemia and peripheral neuropathy.

Vitamin K is absorbed from gut as menaquinone, the active form.

Vitamin K is activated by epoxide reductase in the liver.

Vitamin K γ-carboxylates liver-derived vitamin K–dependent coagulation factors.

2. Vitamin K γ-carboxylates glutamate residues in the vitamin K–dependent coagulation factors, which are factors II (prothrombin), VII, IX, and X and proteins C and S.
   a. The vitamin K–dependent coagulation factors are all synthesized in the liver in a nonfunctional state.
   b. When γ-carboxylated in the liver by vitamin K and released into the circulation, these coagulation factors are able to bind to calcium, which is essential to the formation of a fibrin clot.
   c. The prothrombin time (PT) is a coagulation test that evaluates all of the vitamin K–dependent factors, except factor IX and proteins C and S.
   d. Coumarin derivatives act as anticoagulants by inhibiting the activity of epoxide reductase; hence, the vitamin K–dependent coagulation factors are rendered nonfunctional by their inability to bind to calcium.
3. Vitamin K also functions in bone calcification; γ-carboxylates glutamate residues in osteocalcin.
4. Vitamin K deficiency is rare, but can be caused by the use of broad-spectrum antibiotics, which destroy colonic bacterial synthesis of the vitamin.
5. Other causes of vitamin K deficiency
   a. Therapy with coumarin (warfarin) derivatives: inhibits hepatic epoxide reductase
   b. Fat malabsorption: cannot reabsorb fat-soluble vitamins (e.g., celiac disease)
   c. Newborns: lack bacterial colonization of the bowel and must receive an intramuscular vitamin K injection at birth to prevent hemorrhagic disease of the newborn
6. Excessive intake of vitamin K leading to toxicity is uncommon.
7. Clinical findings associated with vitamin K deficiency and excess (see Table 4-2)

V. Minerals and Electrolytes

A. Overview

1. Calcium functions in bone formation, nerve conduction, muscle contraction, blood clotting, and cell signaling hypocalcemia produces tetany.
2. Magnesium functions in energy metabolism, PTH synthesis, bone formation, nerve conduction, and muscle contraction.
3. Sodium functions in acid-base balance, osmotic pressure, muscle and nerve excitability, active transport, and membrane potential; deficiency produces abnormalities in mental status and convulsions.
4. Potassium functions in acid-base balance, osmotic pressure, muscle and nerve excitability, and insulin secretion; deficiency produces muscle weakness and polyuria.
5. Phosphate functions in bone formation, nucleotide structure, metabolic intermediates, metabolic regulation, vitamin function, and acid-base balance; deficiency produces muscle weakness, rhabdomyolysis, and hemolytic anemia.
6. Chloride functions in acid-base balance, osmotic pressure, and nerve and muscle excitability; deficiency symptoms are undefined.

B. Calcium

1. Sources of calcium include dairy products, leafy green vegetables, legumes, nuts, and whole grains.
2. Functions of calcium
   a. Bone formation and teeth
   b. Nerve conduction
   c. Skeletal, cardiac, smooth muscle contraction
   d. Binds to vitamin K–dependent coagulation factors and activates factor XIII to cross-link fibrin strands
   e. Calcium-calmodulin complex activates many enzymes.
3. Regulation of calcium
   a. Parathyroid hormone increases reabsorption in the early distal tubule of the kidneys and mobilizes calcium from bone.
   b. Vitamin D (calcitriol) increases calcium absorption in the intestine and kidneys.
   c. Calcitonin, which is synthesized by C cells in the thyroid gland, inhibits osteoclasts, thereby inhibiting the release of calcium from bones.
   d. Approximately 40% of calcium is bound to albumin; 13% is bound to phosphorus and citrates; and 47% circulates as free, ionized calcium, which is metabolically active.
      (1) Alkalotic conditions decrease ionized calcium levels by increasing the amount of calcium bound to albumin.
      (2) Alkalotic state increases calcium bound to albumin at the expense of the ionized calcium level.
      (3) Hypoalbuminemia decreases the total serum calcium *without* altering the ionized calcium level.

Coumarin derivatives are present in rat poison, which is a common cause of poisoning in children.

Coumarin (warfarin): anticoagulant that inhibits hepatic epoxide reductase

Vitamin K: γ-carboxylation of osteocalcin

Breast milk: inadequate supply of vitamin K

RDAs: minerals (>100 mg/day); trace elements (<100 mg/day)

Calcium: bone and teeth, nerve conduction, muscle contraction, clotting cascade, enzyme regulation

PTH: increased tubular reabsorption of calcium; increased mobilization of calcium from bone

Calcitonin: decreased calcium mobilization from bone

Alkalotic state lowers ionized calcium levels, causing tetany.

TABLE 4-3. Minerals: Signs and Symptoms of Deficiency and Excess

MINERAL	SIGNS AND SYMPTOMS OF DEFICIENCY	SIGNS AND SYMPTOMS OF EXCESS
Calcium	Tetany (signs of tetany: carpopedal spasm, Chvostek's sign, muscle twitching); osteoporosis	Renal calculi; metastatic calcification; polyuria
Phosphorus (phosphate)	Muscle weakness; rhabdomyolysis with myoglobinuria (due to low ATP); hemolytic anemia (due to low ATP)	Drives calcium into normal tissue (metastatic calcification) causing hypocalcemia
Sodium	Mental status abnormalities (cerebral edema); convulsions	Mental status abnormalities; convulsions
Chloride	No specific signs and symptoms	No specific signs and symptoms
Potassium	Muscle weakness; polyuria: U waves on an electrocardiogram	Heart stops in diastole; peaked T waves on an electrocardiogram
Magnesium	Hypocalcemia with tetany; tachycardia	Neuromuscular depression; bradycardia

4. Causes of hypocalcemia
   a. Hypoalbuminemia: most common nonpathologic cause
   b. Hypomagnesemia: most common pathologic cause
   c. Vitamin D deficiency
   d. Primary hypoparathyroidism
5. Causes of hypercalcemia
   a. Primary hyperparathyroidism: most common cause in the ambulatory population
   b. Malignancy: most common cause in a hospitalized patient *(b/c tumor secretes PTH)*
   c. Sarcoidosis: granuloma synthesis of vitamin D
6. Clinical findings associated with hypocalcemia and hypercalcemia (Table 4-3)

C. Magnesium
1. Sources of magnesium include green vegetables, nuts, and legumes.
2. Functions of magnesium
   a. Calcium metabolism
      (1) Cofactor of adenylate cyclase involved in the activation of PTH
      (2) Increases PTH synthesis and release
   b. Component of bone
   c. Muscle contraction: modulates the vasoconstrictive effects of intracellular calcium
   d. Nerve impulse propagation
   e. Cofactor in ATPases (e.g., Na^{2+}/K^+-ATPase, Ca^{2+}-ATPase)
3. Causes of hypomagnesemia
   a. Alcoholism: most common cause; increased renal loss
   b. Diuretics: increased renal loss
   c. Drugs (e.g., aminoglycosides, cisplatinum): increased renal loss
   d. Diarrhea: lost in the stool
4. Causes of hypermagnesemia
   a. Renal failure: decreased excretion
   b. Treatment of eclampsia with magnesium sulfate
5. Clinical findings associated with hypomagnesemia and hypermagnesemia (see Table 4-3)

D. Sodium
1. Major source is table salt.
2. Functions of sodium
   a. Regulation of pH, osmotic pressure, and water movement in body fluids
   b. Maintains muscle and nerve excitability
   c. Active transport of glucose, galactose, and amino acids in the small intestine
   d. Maintains the diffusion potential of membranes
3. Control of sodium
   a. Diet
   b. Aldosterone: controls renal reabsorption (when present) and excretion (when absent)
   c. Atrial natriuretic peptide (ANP): decreases reabsorption in the kidneys
4. Causes of hyponatremia
   a. Thiazide and loop diuretics: most common cause; increase renal excretion
   b. Inappropriate secretion of antidiuretic hormone: dilutional effect in plasma of excess water reabsorption from the collecting tubules of the kidneys   *SIADH*
   c. Congestive heart failure and chronic liver disease: dilutional effect in plasma of excess water reabsorbed from the kidneys

Hypocalcemia: hypoalbuminemia, hypomagnesemia, vitamin D deficiency, hypoparathyroidism

Hypercalcemia: hyperparathyroidism, malignancy, sarcoidosis (synthesis of vitamin D)

Magnesium: important in function of PTH

Magnesium: bone formation, muscle contraction, nerve conduction, cofactor in ATPases

Hypomagnesemia: alcoholism, diuretics, diarrhea, drugs producing renal loss

Hypermagnesemia: renal failure, magnesium treatment of eclampsia

Sodium is the most abundant cation in extracellular fluid.

Sodium: acid-base balance, osmotic pressure, muscle and nerve excitability, active transport, and membrane potential

Sodium controls water movement between extracellular and intracellular fluid compartments.

Hyponatremia: loop diuretics, excess ADH, congestive heart failure

Hypernatremia produces hypertension.

5. Causes of hypernatremia
   a. Osmotic diuresis: most common cause; loss of a hypotonic salt solution in the kidneys due to glucosuria, excess urea, or mannitol (i.e., treatment of cerebral edema)
   b. Diabetes insipidus: loss of water due to deficiency or dysfunction of antidiuretic hormone
6. Clinical findings associated with hyponatremia and hypernatremia (see Table 4-3)

E. Potassium

1. Sources of potassium include meats, vegetables, fruits, nuts, and legumes.
2. Functions of potassium
   a. Regulation of pH and osmotic pressure
   b. Regulation of neuromuscular excitability and muscle contraction
   c. Regulates insulin secretion: hypokalemia inhibits insulin; hyperkalemia stimulates insulin secretion
3. Control of potassium
   a. Aldosterone: controls renal reabsorption (when absent) and excretion (when present)
   b. Arterial pH
      (1) Alkalotic conditions cause hydrogen ions to move out of the cell (provides protons) and potassium to move into the cell (leads to hypokalemia) to maintain electroneutrality.
      (2) Acidotic conditions cause hydrogen ions to move into cells (for buffering) in exchange for potassium (leads to hyperkalemia).
4. Causes of hypokalemia
   a. Thiazide and loop diuretics: increase renal excretion
   b. Vomiting and diarrhea: lost in the body fluids
   c. Aldosterone excess: increased renal excretion
5. Causes of hyperkalemia
   a. Renal failure: decreased excretion of potassium
   b. Addison's disease due to loss of aldosterone; decreased excretion
6. Clinical findings associated with hypokalemia and hyperkalemia (see Table 4-3)

F. Phosphate (phosphorus)

1. Sources of phosphate include most foods.
2. Functions of phosphate
   a. Mineralization of bones and teeth
   b. Component of DNA and RNA
   c. Component of phosphorylated vitamins (e.g., thiamine, pyridoxine) and ATP
   d. Traps monosaccharides in cells (e.g., phosphorylation of glucose in glycolysis)
   e. Activates enzymes (e.g., protein kinase) and deactivates enzymes (e.g., glycogen synthase)
   f. Maintains pH
      (1) Protons secreted into the renal tubule lumen react with dibasic phosphate (HPO_4^{2-}) to form monobasic phosphate ($H_2PO_4^-$), which is called titratable acidity.
3. Control of phosphate
   a. Parathyroid hormone has a phosphaturic effect.
   b. Vitamin D (calcitriol) increases absorption of phosphate in the small bowel.
4. Causes of hypophosphatemia
   a. Respiratory and metabolic alkalosis: most common cause; alkalosis enhances glycolysis and phosphorylation of glucose
   b. Hypovitaminosis D due to malabsorption: decreased intestinal absorption of phosphate
   c. Primary hyperparathyroidism: increased loss of phosphate in urine
5. Causes of hyperphosphatemia
   a. Renal failure: most common cause; decreased renal excretion
   b. Primary hypoparathyroidism
   c. Normal children: higher levels of phosphate help drive calcium into bone
6. Clinical findings associated with hypophosphatemia and hyperphosphatemia (see Table 4-3)

G. Chloride

1. It is primarily found in table salt.
2. Functions of chloride
   a. Regulation of pH and osmotic pressure
   b. Regulates neuromuscular excitability and muscle contraction

---

Potassium: most abundant intracellular cation

Potassium: acid-base regulation, osmotic pressure, neuromuscular excitability, muscle contraction, insulin secretion

Alkalosis: potassium moves into cells

Acidosis: potassium moves into serum

Hypokalemia: vomiting and diarrhea, loop diuretics, aldosterone excess

Hyperkalemia: renal failure, Addison's disease

Potassium: major intracellular cation

Phosphate is the most abundant intracellular anion.

Phosphate: component of ATP, traps carbohydrate intermediates in cell, activation of vitamins, nucleic acid structure, found in bones and teeth, acid-base buffer

Hypophosphatemia: alkalosis, hypovitaminosis D, hyperparathyroidism

Hyperphosphatemia: renal failure, hypoparathyroidism

Chloride is the most abundant anion in extracellular fluid.

Chloride: fluid and electrolyte balance, gastric fluid

3. Control of chloride
   a. Diet
   b. Aldosterone: controls renal reabsorption (when present) and excretion (when absent)
4. Causes of hypochloremia
   a. Thiazide and loop diuretics: increase renal excretion
   b. Vomiting
5. Causes of hyperchloremia
   a. Mineralocorticoid excess increases sodium and chloride reabsorption
   b. In renal tubular acidosis and diarrhea, the loss of bicarbonate causes an increase in chloride levels to offset the loss of negative charges.

VI. Trace Elements

A. Overview

1. Iron functions in oxygen transport, the electron transport chain, and as an enzyme cofactor; deficiency symptoms are anemia, Plummer-Vinson syndrome, and fatigue.
2. Zinc functions as a cofactor for metalloenzymes; deficiency produces poor wound healing, dysgeusia, anosmia, and growth retardation.
3. Copper functions as a cofactor for metalloenzymes and in cytochrome oxidase; deficiency produces microcytic anemia, aortic aneurysm, and poor wound healing.
4. Iodine functions in the synthesis of thyroid hormones; deficiency produces goiter (i.e., hypothyroidism).
5. Chromium is a component of the glucose tolerance factor; deficiency produces impaired glucose tolerance.
6. Selenium functions in antioxidant action as a component of glutathione peroxidase; deficiency produces muscle pain and weakness.
7. Fluoride functions in the structure of hydroxyapatite; deficiency produces dental caries.

B. Iron

1. Primary sources of iron include meat, eggs, vegetables, and fortified cereals.
2. Iron is the structural component of heme in hemoglobin, myoglobin, and the cytochrome oxidase system.

   Iron: important cofactor for enzymes (e.g., catalase)

   a. Meat contains heme iron, which is ferrous (Fe^{2+}) and available for absorption in the duodenum.
   (1) After it is reabsorbed by duodenal enterocytes, heme is enzymatically degraded to release iron.
   (2) Most of the iron is diverted to storage as apoferritin in the enterocyte, but a small amount is delivered to plasma transferrin, the circulating binding protein of iron.

   Iron: component of heme, iron-sulfur proteins

   b. Plants contain nonheme iron, which is in the ferric state (Fe^{3+}); hence, absorption of nonheme iron is more complex and involves a number of different binding proteins *before* it is transferred to transferrin.

   Ferritin: iron storage in intestinal mucosa, liver, spleen, and bone marrow

   c. Ferritin, a soluble iron-protein complex, is the storage form of iron in the intestinal mucosa, liver, spleen, and bone marrow.
   (1) It is synthesized by macrophages.
   (2) Serum ferritin levels reflect iron stores in the bone marrow.
   (3) Serum ferritin is the best screening test for iron deficiency and iron overload disorders (e.g., hemochromatosis).

   Serum ferritin: low in iron deficiency, high in hemochromatosis

   d. Hemosiderin is an insoluble storage product of ferritin degradation.
   (1) Hemosiderosis is an acquired accumulation of hemosiderin in macrophages in tissues throughout the body.
   (2) Hemochromatosis is an autosomal recessive disease characterized by unrestricted absorption of iron from the duodenum, leading to an accumulation of iron in liver, heart, pancreas, skin, and other tissues.
   (3) Hemosiderosis is an acquired iron overload disease (e.g., excessive blood transfusions, alcoholism).

   Hemosiderin: ferritin degradation product

   Patients who require ongoing transfusions are at risk for hemosiderosis.

   e. Transferrin, the primary binding protein for iron, is synthesized in the liver and transports iron to macrophages in the bone marrow for storage or to the developing RBCs for hemoglobin synthesis.
   (1) When iron stores in the bone marrow macrophages are decreased (e.g., iron deficiency), liver synthesis of transferrin increases, which increases total iron-binding capacity.
   (2) When iron stores in the bone marrow macrophages are increased (e.g., hemochromatosis), transferrin synthesis is decreased, which decreases the total iron-binding capacity.

   Hemochromatosis: cirrhosis of the liver, bronze skin color, diabetes mellitus, malabsorption, and heart failure

   Transferrin: functions in iron transport

   Low iron stores: transferrin increased

   Iron stores high: transferrin reduced

TABLE 4-4. **Trace Elements: Signs and Symptoms of Deficiency**

TRACE ELEMENT	SIGNS AND SYMPTOMS OF DEFICIENCY
Iron	Microcytic anemia; low serum ferritin level, low serum iron level, high total iron-binding capacity (correlates with decreased transferrin synthesis); Plummer-Vinson syndrome (esophageal webs, glossitis, spoon nails, achlorhydria); excessive fatigue
Zinc	Poor wound healing; dysgeusia (inability to taste); anosmia (inability to smell); perioral rash; hypogonadism; growth retardation
Copper	Microcytic anemia (decreased ferroxidase activity); dissecting aortic aneurysm; poor wound healing
Iodine	Goiter (due to relative or absolute deficiency of thyroid hormones)
Chromium	Impaired glucose tolerance, peripheral neuropathy
Selenium	Muscle pain and weakness, cardiomyopathy
Fluoride	Dental caries

3. Causes of iron deficiency vary by age.
   a. Newborn/infants: nutritional (most common)
   b. Child: bleeding Meckel's diverticulum; nutritional
   c. Woman <50 years old: menorrhagia
   d. Man <50 years old: peptic ulcer disease
   e. Men and women >50 years old: colon polyps or cancer
4. Causes of excess serum iron
   a. Iron poisoning
      (1) Common in children
      (2) Causes hemorrhagic gastritis and liver necrosis
   b. Iron overload diseases: hemochromatosis; hemosiderosis; sideroblastic anemia (due to pyridoxine deficiency, lead poisoning, alcoholism)
      (1) Sideroblastic anemias are associated with excess iron accumulation in mitochondria resulting from difficulties in heme synthesis.
      (2) Excess iron in mitochondria produces ringed sideroblasts (mitochondria are located around the nucleus of immature RBCs).
5. Clinical findings associated with iron deficiency (Table 4-4)

C. **Zinc**
   1. Primary sources of zinc include meat, liver, eggs, and oysters.
   2. Zinc primarily serves as a cofactor for metalloenzymes.
      a. Superoxide dismutase
      b. Collagenase
         (1) Important in remodeling of a wound and replacing type III collagen with type I collagen to increase tensile strength
      c. Alcohol dehydrogenase: converts alcohol into acetaldehyde
      d. Alkaline phosphatase
         (1) Important in bone mineralization
         (2) Marker of obstruction to bile flow in the liver or common bile duct
   3. Zinc is also important in spermatogenesis and in growth in children.
   4. Causes of zinc deficiency
      a. Various diseases
         (1) Alcoholism, rheumatoid arthritis, acute and chronic inflammatory diseases, chronic diarrhea
      b. Acrodermatitis enteropathica
         (1) Autosomal recessive disease associated with dermatitis, diarrhea, growth retardation in children, decreased spermatogenesis, and poor wound healing
   5. Clinical findings associated with zinc deficiency (see Table 4-4)

D. **Copper**
   1. Sources of copper include shellfish, organ meats, poultry, cereal, fruits, and dried beans.
   2. Copper primarily serves as a cofactor for metalloenzymes.
      a. Ferroxidase: binds iron to transferrin; causes iron deficiency if deficient
      b. Lysyl oxidase: cross-linking of collagen and elastic tissue
      c. Superoxide dismutase: neutralizes superoxide, an O_2 free radical
      d. Tyrosinase: important in melanin synthesis; deficient in albinism
      e. Cytochrome *c* oxidase: component of the electron transport chain
   3. Ceruloplasmin
      a. It is an enzyme that is synthesized in the liver.

Iron deficiency in women: menorrhagia, colon polyps

Iron deficiency in men: peptic ulcer

Iron deficiency in newborns, infants: nutritional

Iron overload: ringed sideroblasts

Zinc: metalloenzyme cofactor, superoxide dismutase, alcohol dehydrogenase, wound healing (strengthening)

Zinc deficiency: caused by alcoholism, diarrhea, inflammatory diseases

Acrodermatitis enteropathica: autosomal recessive zinc deficiency, dermatitis, diarrhea, growth retardation, poor wound healing, decreased spermatogenesis

Zinc deficiency: poor wound healing, loss of taste and smell

Copper: cofactor for metalloenzymes, cytochrome oxidase

Ceruloplasmin: copper-transport plasma protein

b. It contains 6 atoms of copper in its structure.

c. It is secreted into the plasma, where it represents 90% to 95% of the total serum copper concentration.

4. Copper deficiency (hypocupremia) is most often due to total parenteral nutrition (TPN).

5. An excess of free copper (i.e., hypercupremia) is present in Wilson's disease, an autosomal recessive disease associated with a defect in secreting copper into bile and in incorporating copper into ceruloplasmin.

a. Wilson's disease: chronic liver disease, deposition of free copper into the eye (Kayser-Fleischer ring) and lenticular nuclei (dementia, movement disorder), low serum ceruloplasmin level, and high free copper levels in blood and urine

6. Clinical findings associated with copper deficiency (see Table 4-4)

E. Iodine

1. Sources of iodine include iodized table salt and seafood.

2. Iodine is used in the synthesis of thyroid hormones.

3. Iodine deficiency is due primarily to an inadequate intake of seafood or iodized table salt.

4. Clinical findings associated with iodine deficiency (see Table 4-4)

F. Chromium

1. Sources of chromium include wheat germ, liver, and brewer's yeast.

2. Chromium is a component of glucose tolerance factor, which facilitates insulin action through post-receptor effects.

3. Chromium deficiency primarily occurs in patients receiving TPN.

4. Clinical findings associated with chromium deficiency (see Table 4-4)

G. Selenium

1. Sources of selenium include seafood and liver.

2. Selenium is a component of glutathione peroxidase, which converts oxidized glutathione (see Chapter 6) into reduced glutathione in the pentose phosphate pathway.

3. Selenium deficiency occurs primarily in patients receiving TPN.

4. Clinical findings associated with selenium deficiency (see Table 4-4)

H. Fluoride

1. Sources of fluoride include tea and fluoridated water.

2. Fluoride is a structural component of calcium hydroxyapatite in bone and teeth.

3. Deficiency of fluoride is primarily due to inadequate intake of fluoridated water.

4. An excess in fluoride primarily results from an excess of fluoride in drinking water.

5. Clinical findings associated with fluoride deficiency (see Table 4-4)

Copper deficiency: result of TPN

Iodine is needed for thyroid hormone synthesis.

Iodine deficiency produces goiter.

Chromium is a component of glucose tolerance factor.

Chromium deficiency can result from TPN.

Glutathione: antioxidant that neutralizes peroxide and peroxide free radicals

Selenium: component of glutathione peroxidase

Fluoride: structural component of hydroxyapatite in bone and teeth

Fluorosis: chalky deposits on the teeth, calcification of ligaments, an increased risk for bone fractures

Fluoride deficiency: dental caries

CHAPTER 5
GENERATION OF ENERGY FROM DIETARY FUELS

I. Energetics of Metabolic Pathways

A. Overview
1. Free energy change (ΔG) for a biochemical reaction indicates its tendency to proceed and the amount of free energy it will release or require.
2. Coupled reactions share a common intermediate.
3. ATP-ADP cycle is the most common mechanism of energy exchange in biologic systems.
4. Redox coenzymes serve as carrier molecules for hydrogen and electrons during biologic oxidation-reduction reactions

B. Change in free energy (ΔG)
1. A decrease in free energy for a biochemical reaction, or sequence of reactions, indicates its tendency to proceed.
2. The amount of decrease in free energy in kcal/mole is the amount of free energy a reaction will release or require.
3. A reaction that requires a free energy input must be coupled to another reaction that releases at least that much energy.

C. Coupled reactions
1. When two reactions share a common intermediate, they are considered to be coupled,
   a. For example, the common intermediate C couples these two reactions:

$$A + B \rightleftharpoons C + D$$

$$C \rightleftharpoons X$$

   b. Overall coupled reaction ($A + B \rightleftharpoons X$) proceeds spontaneously in a forward direction if the sum of ΔG values of the individual reactions is negative.
2. Metabolic pathways consist of a series of coupled reactions linked by common intermediates (Fig. 5-1).
   a. The ΔG values are additive for all pathway reactions.
      (1) An energetically favorable reaction (e.g., hydrolysis of ATP) drives an energetically unfavorable coupled reaction in the forward direction.
   b. If ΔG for consecutive reactions is negative, the reactions operate spontaneously in the forward direction.

D. ATP-ADP cycle (Fig. 5-2)
1. Hydrolysis of high-energy bonds in ATP has a ΔG of about 7 kcal/mol.
2. Hydrolysis of other nucleoside triphosphates and other high-energy compounds provides energy for a variety of metabolic processes. For example:
   a. GTP during ribosomal steps of protein synthesis
   b. CTP during lipid synthesis
   c. UTP during polysaccharide synthesis
   d. Phosphocreatine in muscle replenishes ATP.

E. Redox coenzymes (Fig. 5-3)
1. Nicotinamide adenine dinucleotide (NAD^+) and flavin adenine dinucleotide (FAD) are the major electron acceptors in the catabolism of fuel molecules.
   a. Reduced forms of NAD^+ and FAD (i.e., NADH and $FADH_2$) transfer electrons to the electron transport chain.
   b. Electrons are ultimately transferred to O_2 by the electron transport chain with the coupled formation of ATP (i.e., oxidative phosphorylation).
2. NADPH (a phosphorylated derivative of NADH) is the primary electron donor in reductive biosynthetic reactions (e.g., synthesis of fatty acids, cholesterol, and steroids).

Free energy change: tendency to react, energy released or used

Coupled reactions: common intermediate required

Negative ΔG: allows coupled reactions to proceed spontaneously in a forward direction

ATP-ADP cycle: most common mechanism of energy exchange in biologic systems

ATP: large negative ΔG values can drive many energy-requiring reactions

Redox coenzymes: carrier molecules for hydrogen and electrons during biologic oxidation-reduction reactions

NAD^+ and FAD: electron and hydrogen carriers

NAD^+/NADH: catabolic reactions

$NADP^+$/NADPH: anabolic reactions

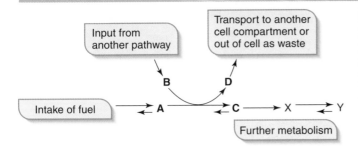

5-1: Processes that affect flow through metabolic pathways. In the absence of such processes, individual reversible reactions eventually reach equilibrium, and the flow of metabolites through a pathway ceases. For example, a genetic defect or inhibitor that reduces production of B also decreases operation of the pathway from fuel → A → Y.

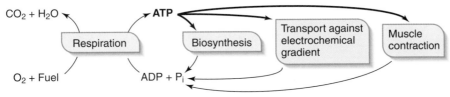

5-2: The ATP-ADP cycle. Energy stored in reduced fuel molecules is extracted during oxidative metabolism (i.e., respiration) and converted to ATP. Hydrolysis of ATP releases energy that cells use for major types of energy-requiring processes (*thick arrows*).

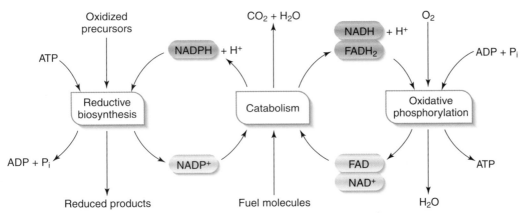

5-3: Role of redox coenzymes as electron carriers. Oxidized coenzymes (*light shading*) are reduced during catabolism. Reduced coenzymes (*dark shading*) are used in reductive biosynthesis (*left*) or in oxidative phosphorylation to generate ATP (*right*).

II. Introduction to Intermediary Metabolism
A. Overview
1. Catabolism of foodstuffs to generate ATP occurs in three stages: digestion to small components, formation of acetyl CoA, and oxidation of acetyl CoA in the citric acid cycle (CAC).
2. Metabolic pathways are localized to particular cellular sites.
3. There are five common aspects of metabolic pathways: reaction steps, regulated steps, unique characteristics, pathway interfaces, and clinical relevance.

B. Catabolic stages (Fig. 5-4).
1. Breakdown of large dietary constituents (i.e., carbohydrates, fats, and proteins) to their small building blocks by digestive enzymes
2. Formation of acetyl CoA by degradation of the products of digestion (e.g., glucose, fatty acids, glycerol, and amino acids)
3. Oxidation of acetyl CoA in the CAC and the flow of electrons through the electron transport chain with coupled formation of ATP (i.e., oxidative phosphorylation)
   a. Acetyl CoA is produced by fat oxidation and glucose oxidation.
   b. Acetyl CoA is used in fat synthesis, cholesterol synthesis, ketone body synthesis, and formation of acetylated molecules.

C. Compartmentation of metabolic pathways
1. Mitochondria
   a. Matrix
      (1) Production of acetyl CoA, CAC, β-oxidation of fatty acids, ketogenesis
   b. Inner membrane: oxidative phosphorylation

Acetyl CoA: product of fat and glucose oxidation

Acetyl CoA: a focal point in metabolism

5-4: Three stages in catabolism of the energy-yielding major nutrients. Notice the central role of pyruvate and acetyl CoA. Oxidative phosphorylation, the formation of ATP coupled to the flow of electrons (e⁻) through the electron transport chain (ETC), generates most of the ATP resulting from catabolism.

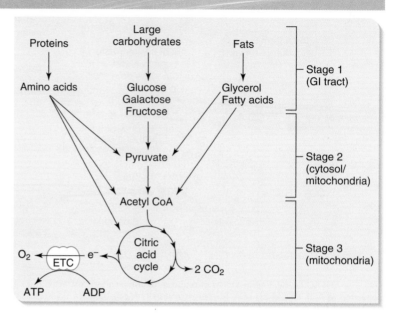

Metabolic pathways occur in specific cellular sites (e.g., glycolysis in cytosol, CAC oxidative phosphorylation in mitochondria).

    2. Cytosol
        a. Glycolysis, glycogenesis, glycogenolysis, pentose phosphate shunt, fatty acid synthesis, steroid synthesis (i.e., smooth endoplasmic reticulum), protein synthesis (i.e., rough endoplasmic reticulum)
    3. Mitochondria and cytosol
        a. Gluconeogenesis, urea cycle, heme synthesis
    4. Nucleus and mitochondria
        a. DNA and RNA synthesis

D. Five common perspectives for many metabolic pathways
    1. Pathway reaction steps
        a. Each reaction in a pathway has unique characteristics regarding substrates, products, enzymes, cofactors, and inhibitors.
    2. Regulated steps
        a. At least one step in a metabolic pathway is generally regulated by hormones or metabolites that restrict or accelerate the flow of metabolites through the pathway.
    3. Unique characteristics
        a. Each metabolic pathway has features that describe unique aspects of its function.
        b. This provides insight into its contribution to metabolism.
    4. Interface with other pathways
        a. Many intermediates within one pathway are substrates for other pathways, providing a means for different pathways to interact.
    5. Clinical relevance
        a. Significant reduction in the activity of an enzyme catalyzing one step in a pathway, caused by a genetic defect or inhibitor, leads to accumulation of some metabolites and reduced levels of others, often with pathologic consequences.
        b. Increased flow through a pathway results in toxic buildup (e.g., lactic acidosis).

Organize learning of metabolic pathways around four categories.

    6. These perspectives apply most easily to carbohydrate and lipid metabolism.
        a. Pathway reaction steps
        b. Regulated steps
        c. Unique characteristics
        d. Interface with other pathways
        e. Clinical relevance

CAC: acetyl CoA oxidized to CO_2

CAC located in matrix and inner membrane of mitochondria

Red blood cells and platelets lack mitochondria; cannot oxidize pyruvate or acetyl CoA

III. Citric Acid Cycle (CAC)
    A. Overview (Fig. 5-5)
        1. Acetyl CoA is oxidized to CO_2 in the CAC, producing reduced coenzymes and GTP.
        2. Regulation of citrate synthase, isocitrate dehydrogenase, and α-ketoglutarate dehydrogenase controls flow of metabolites in and out of the CAC.
        3. Many amino acids are interconverted with CAC intermediates through transamination.
        4. CAC activity is poisoned by fluoroacetate and impaired by dietary deficiencies of its vitamin cofactors.

Outer mitochondrial membrane
• Freely permeable to metabolites

Inner mitochondrial membrane
• Impermeable to metabolites
• Electron transport chain
• ATP synthase
• Succinate dehydrogenase
• ATP-ADP translocase

Mitochondrial matrix
• Low [H⁺] relative to intermembrane space
• Citric acid cycle enzymes except succinate dehydrogenase
• Pyruvate dehydrogenase
• Fatty acyl CoA dehydrogenase
• Ketogenesis enzymes (liver)
• Enzymes for β-oxidation of fatty acids

Intermembrane space
• High [H⁺] relative to matrix

5-5: Schematic diagram of a mitochondrion shows the location of key enzymes.

B. CAC: pathway reaction steps (Fig. 5-6)

1. Condensation of acetyl CoA with oxaloacetate to form 6-carbon citrate begins the cycle.
2. Two oxidative decarboxylation reactions release acetyl carbons as CO_2 and produce 2 NADH (6 ATP).
3. Conversion of succinyl CoA to malate occurs in three steps that also yield 1 GTP (energetically equivalent to 1 ATP) and 1 $FADH_2$ (2 ATP).
4. Regeneration of oxaloacetate produces a third molecule of NADH (3 ATP).
5. Total of 24 ATP per glucose molecule is generated by the CAC.

C. CAC: regulated steps

1. Three regulated enzymes in the cycle control the flow of substrates largely in response to the cell's need for ATP (see Fig. 5-6)
   a. Citrate synthase, isocitrate dehydrogenase, and α-ketoglutarate dehydrogenase.

CAC revolution produces 3 NADH (9 ATP), 1 GTP (1 ATP), and 1 FADH₂ (2 ATP).

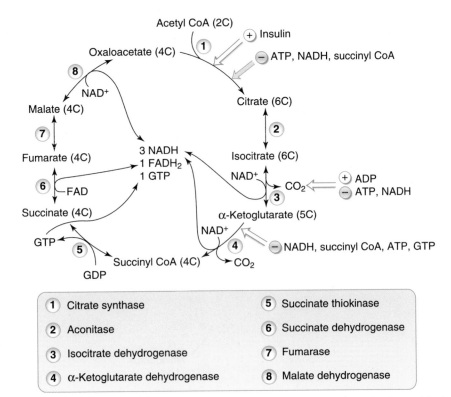

1. Citrate synthase
2. Aconitase
3. Isocitrate dehydrogenase
4. α-Ketoglutarate dehydrogenase
5. Succinate thiokinase
6. Succinate dehydrogenase
7. Fumarase
8. Malate dehydrogenase

5-6: The citric acid cycle is regulated primarily by three enzymes (i.e., citrate synthase, isocitrate dehydrogenase, and α-ketoglutarate dehydrogenase), which are inhibited (−) or activated (+) by the indicated metabolites.

2. Citrate synthase is inhibited by ATP, NADH, and succinyl CoA, and it is stimulated by insulin.

3. Isocitrate dehydrogenase, the cycle's pacemaker, is stimulated by ADP and inhibited by ATP and NADH.

4. α-Ketoglutarate dehydrogenase is inhibited by two of its products, NADH and succinyl CoA, as well as by ATP and GTP.

D. **CAC: unique characteristics**

1. Energy yield of the CAC is equivalent to 12 ATP per acetyl CoA oxidized.

2. Replenishing reactions restore cycle intermediates that are drained off to biosynthetic pathways or needed in greater amounts.

   a. Pyruvate carboxylase, which forms oxaloacetate by carboxylation of pyruvate, is allosterically activated by acetyl CoA.

   b. Various amino acids are converted to acetyl CoA, α-ketoglutarate, succinyl CoA, fumarate, or oxaloacetate (Fig. 5-7).

      (1) Example: glutamate is converted to α-ketoglutarate, methionine is converted to succinyl CoA, and aspartate is converted to oxaloacetate.

E. **CAC: interface with other pathways (see Fig. 5-7)**

1. The CAC accepts metabolites generated in catabolic pathways and contributes cycle intermediates for use in anabolic pathways, including gluconeogenesis.

2. Carbons entering the cycle as acetyl CoA are always oxidized to CO_2 and never contribute carbons to gluconeogenesis.

   a. Acetyl CoA is *not* a substrate for gluconeogenesis.

3. Carbons from glucogenic amino acids and pyruvate enter the cycle as intermediates that are convertible to malate, which is exported to the cytosol and converted to oxaloacetate to enter the gluconeogenic pathway (see Chapter 6).

F. **CAC: clinical relevance**

1. Fluoroacetate, a potent poison, is converted to fluorocitrate, which inhibits aconitase, blocking operation of the CAC and its energy output.

2. Five vitamins are precursors of coenzymes that participate in the CAC.

3. A deficiency of any of these vitamins negatively impacts operation of the cycle and impairs energy production.

   a. Niacin → NAD^+ (see Fig. 5-6, steps 3, 4, and 8)

   b. Riboflavin (vitamin B_2) → FAD (see Fig. 5-6, step 6)

   c. Thiamine (vitamin B_1) → thiamine pyrophosphate (see Fig. 5-6, step 4)

   d. Pantothenate → coenzyme A

   e. Vitamin B_{12} (cobalamin) → succinyl CoA by odd-chain fatty acid metabolism

Anaplerosis: replenishment of oxaloacetate by pyruvate carboxylation

Transamination: interconversion of CAC intermediates with amino acids

Fluoroacetate: CAC poison

Coenzyme-vitamin precursor relationships in citric acid cycle: NAD^+-niacin; FAD-riboflavin (B_2); thiamine pyrophosphate-thiamine (B_1); CoA-pantothenate; succinyl CoA-vitamin B_{12}

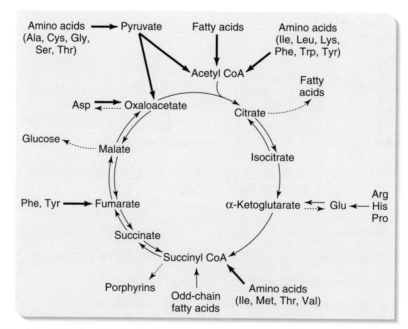

5-7: Interface of the citric acid cycle with other pathways. Many amino acids produce cycle intermediates *(thick arrows)*. Cycle intermediates also take part in synthetic pathways leading to glucose, fatty acids, porphyrins, and amino acids *(dashed arrows)*.

IV. Electron Transport and Oxidative Phosphorylation

A. Overview

1. The electron transport chain consists of a series of four large multiprotein complexes (I through IV) located in the inner mitochondrial membrane.
2. Oxidative phosphorylation is the synthesis of ATP coupled to the stepwise flow of electrons from NADH (donates 2 hydrogen atoms [$2e^- + 2H^+$]) and $FADH_2$ to O_2 in the electron transport chain.
3. ATP formation is driven by a proton gradient across the inner mitochondrial membrane, a process called chemiosmotic coupling.
4. ADP level controls rate of oxidative phosphorylation.
5. Mutations in mitochondrial DNA affect electron transport and ATP synthesis.

B. Electron transport chain (ETC)

1. The ETC consists of a series of four large multiprotein complexes (I through IV) located in the inner mitochondrial membrane that serve to transfer electrons from NADH and $FADH_2$ to oxygen (Fig. 5-8).
2. Coenzyme Q (CoQ) and cytochrome c, two smaller carriers, shuttle electrons between the large complexes.
3. Prosthetic groups in each complex reversibly accept and release electrons.
   a. FMN and FAD (riboflavin derivative): complexes I and II
   b. Heme groups: complexes III and IV
   c. Copper ions: complex IV
      (1) Cytochrome oxidase (complex IV) is inhibited by carbon monoxide (CO) and cyanide (CN^-), which stop electron transport.
   d. O_2 is an electron acceptor.
      (1) Hypoxia, an inadequate concentration of O_2, has its main effect on oxidative phosphorylation and the synthesis of ATP.
      (2) In tissue hypoxia, cells use anaerobic glycolysis to obtain ATP (2 ATP per glucose molecule).
4. Decrease in free energy as electrons move through electron-transport complexes favors electron flow from NADH and $FADH_2$ to O_2.

Electron transport: a conduit for electrons, not a metabolic pathway

Prosthetic groups: heme, FMN, FAD, and copper accept and transfer electrons and are bound to multiprotein complexes.

Cytochrome oxidase: inhibited by CO and CN^-

Anaerobic glycolysis: an outcome of tissue hypoxia; 2 ATP per glucose molecule

O_2: electron acceptor in oxidative phosphorylation; hypoxia stops ATP synthesis in the mitochondria

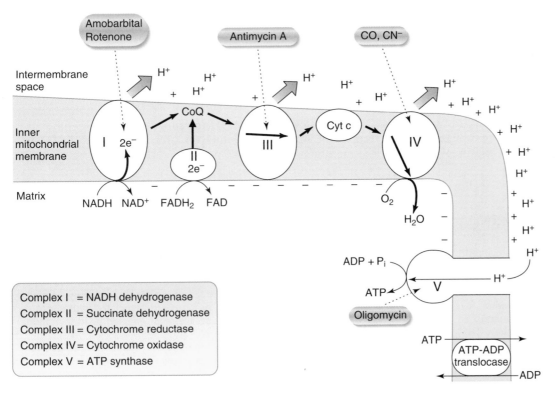

Complex I   = NADH dehydrogenase
Complex II = Succinate dehydrogenase
Complex III = Cytochrome reductase
Complex IV = Cytochrome oxidase
Complex V = ATP synthase

5-8: Overview of oxidative phosphorylation in the inner mitochondrial membrane. Electron flow *(thick arrows)* through complexes I, III, and IV provides energy to pump H^+ ions from the matrix to the intermembrane space *(thick arrows)* against the proton electrochemical gradient. The downhill movement of H^+ ions back into the matrix provides the energy for ATP synthesis by means of complex V. Preferential export of ATP from the matrix by ATP-ADP translocase (an antiport) maintains a high ADP/ATP ratio in the matrix. Inhibitors block electron flow through the indicated complexes *(dashed arrows)*; as a result, ATP synthesis also ceases. CN^-, cyanide; CO, carbon monoxide.

C. **Oxidative phosphorylation**
1. Most of the ATP generated by aerobic metabolism of fuel molecules is produced by oxidative phosphorylation.
   a. The rest of the ATP production is from substrate level phosphorylation in glycolysis (see Chapter 6).
2. The synthesis of ATP is coupled to the stepwise decreases in free energy as electrons move from one multiprotein complex to the next.
   a. The last step donates electrons and hydrogen to O_2 to form water.

D. **ATP synthase and chemiosmotic coupling (see Fig. 5-8)**
1. Proton gradient is established by complexes I, III, and IV, which pump H^+ ions from the mitochondrial matrix to the intermembrane space.
2. ATP synthase is inhibited by oligomycin.
3. The ATP yield from oxidative phosphorylation accounts for about 85% of the maximal 38 ATP generated in complete oxidation of glucose to CO_2.
   a. 1 NADH → 3 ATP
   b. 1 $FADH_2$ → 2 ATP
   c. Compare with anaerobic glycolysis → 2 ATP for each glucose metabolized to lactate.

E. **Respiratory control**
1. Tight coupling of electron flow and ATP synthesis in mitochondria ensures that O_2 consumption depends on availability of ADP, a phenomenon called respiratory control.
2. Low ADP (high ATP) decreases the flow of electrons, which decreases O_2 consumption.
3. High ADP (low ATP) increases the flow of electrons, which increases O_2 consumption.

F. **Mitochondrial DNA mutations (Box 5-1)**
1. Mutations in mitochondrial DNA can block the ETC and reduce ATP synthesis.

V. **Mitochondrial Transport Systems**

A. **Overview**
1. ATP-ADP translocase in the inner mitochondrial membrane carries out the tightly coupled exchange of ATP and ADP.
2. Shuttle mechanisms transport electrons from cytosolic NADH into the mitochondria.

B. **ATP-ADP translocase (see Fig. 5-8)**
1. The outer mitochondrial membrane is permeable to most small metabolites, whereas the inner membrane is not.

C. **NADH shuttle mechanisms**
1. NADH produced during catabolism of glucose to pyruvate (glycolysis) in the cytosol *cannot* cross the inner mitochondrial membrane.
2. Shuttle mechanisms permit regeneration of cytosolic NAD^+, which is necessary for glycolysis to continue.
3. Glycerol phosphate shuttle uses cytosolic NADH to reduce dihydroxyacetone phosphate (DHAP) to glycerol 3-phosphate, which transfers electrons to FAD^+ in the inner mitochondrial membrane (Fig. 5-9).

Margin notes:

Electrons donated to ETC by NADH and $FADH_2$

Chemiosmotic: creation of a proton gradient by pumping protons across membrane

ATP synthase: inhibited by oligomycin

ATP yield oxidative phosphorylation: maximum is 38 ATP

Respiratory control: rate of ETC controlled by ADP concentration

Increased ADP → decreased ETC

Decreased ADP → increased ETC

Mitochondrial DNA disorders: maternal transmission to all children; no paternal transmission

Outer mitochondrial membrane: permeable

Inner mitochondrial membrane: highly selective

Cytosolic NADH cannot cross inner membrane: requires shuttle

Glycerol phosphate shuttle: 2 NADH produce 4 ATP

Malate-aspartate shuttle: 2 NADH produce 6 ATP

BOX 5-1   HEREDITARY MITOCHONDRIAL DISEASES

Most mitochondrial diseases are caused by mutations in mitochondrial DNA (mtDNA), which has a higher mutation rate than nuclear DNA. Because all mitochondria in the zygote come from the ovum, these diseases exhibit maternal inheritance, by which affected mothers transmit the disease to all of their children. Affected males do not transmit mitochondrial diseases because the tail of the sperm, which contains the mitochondria, falls off after fertilization. Because mtDNA-encoded proteins are associated with electron transport and ATP synthesis, tissues with a very high oxygen demand are most affected by mitochondrial dysfunction.

- Leber's hereditary optic neuropathy: progressive loss of central vision and eventual blindness due to degeneration of the optic nerve. It is caused by a defect in NADH dehydrogenase (complex I). This disease affects more males than females, with onset most common in the third decade.

Other defects in mtDNA produce several syndromes known as mitochondrial myopathies:

- Kearns-Sayre syndrome: degeneration of retinal pigments, ophthalmoplegia, pain in the eyes, and cardiac conduction defects, which may cause death. Muscle biopsy reveals ragged red fibers marked by an irregular contour and structurally abnormal mitochondria that stain red. Onset occurs before age 20 years.
- MELAS syndrome: mitochondrial encephalomyopathy, lactic acidosis, and stroke-like episodes.
- MERRF syndrome: myoclonus epilepsy with ragged red fibers.

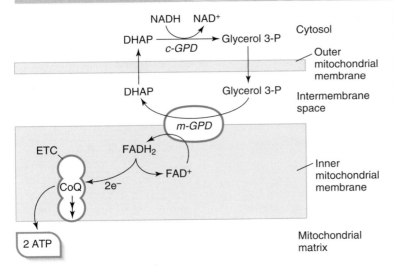

5-9: Glycerol phosphate shuttle for transporting electrons from NADH in the cytosol across the inner mitochondrial membrane. The glycerol phosphate shuttle operates only in one direction to import electrons from the cytosol to the mitochondrial matrix. Each NADH produces 2 ATP. CoQ, coenzyme Q in the electron transport chain (ETC); DHAP, dihydroxyacetone phosphate; GPD, glycerol-3-phosphate dehydrogenase in cytosolic (c) and mitochondrial (m) forms.

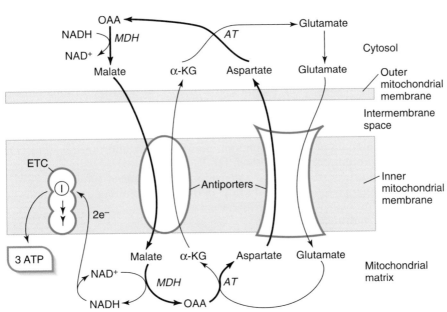

5-10: Malate-aspartate shuttle for transporting electrons from NADH in the cytosol across the inner mitochondrial membrane. The malate-aspartate shuttle is fully reversible and operates to import electrons into the mitochondrial matrix or to export electrons to the cytosol. Each NADH produces 3 ATP. The *thick arrows* trace the path of malate and aspartate. AT, aminotransferase; I, complex I in the electron transport chain (ETC); α-KG, α-ketoglutarate; MDH, malate dehydrogenase; OAA, oxaloacetate.

    4. Malate-aspartate shuttle uses cytosolic NADH to reduce oxaloacetate to malate, which enters the mitochondria by means of an antiporter (Fig. 5-10).

        a. Two cytosolic NADH by this shuttle produce 6 ATP.

  D. Specialized inner membrane transporters

    1. Transport of numerous other metabolites (e.g., PO_4^{3-}, Ca^{2+}, succinate, glutamate) in or out of mitochondria depends on specific transporters in the inner membrane.

VI. Inhibitors of Mitochondrial ATP Synthesis (Table 5-1)

  A. Overview

    1. Most inhibitors of mitochondrial ATP synthesis act to block electron flow or to uncouple electron flow from ATP synthesis.

    2. Electron-transport blockers: carriers upstream of the block become highly reduced and carriers downstream become oxidized.

    3. Uncouplers: electron flow proceeds in the absence of ATP synthesis, eliminating respiratory control.

Uncouplers of ETC destroy the proton gradient: pentachlorophenol, dinitrophenol, thermogenin (in brown fat)

Inner membrane impermeable to: NADH, acetyl CoA, oxaloacetate

Most inhibitors of mitochondrial ATP synthesis block electron flow or uncouple electron flow from ATP synthesis.

TABLE 5-1. Inhibition of Mitochondrial ATP Synthesis

INHIBITOR	MODE OF ACTION	SITE OF INHIBITION
Amobarbital (Amytal), rotenone	Blocks electron transport	Complex I (NADH dehydrogenase)
Antimycin A	Blocks electron transport	Complex III (cytochrome reductase)
Carbon monoxide (CO), cyanide (CN⁻)	Blocks electron transport	Complex IV (cytochrome oxidase)
Dinitrophenol, thermogenin	Acts as proton channel to reduce proton gradient, uncoupling ATP synthesis from electron transport	Throughout inner mitochondrial membrane
Oligomycin	Prevents ATP synthesis directly	Complex V (ATP synthase)

4. Nucleoside analogues (e.g., cytosine arabinoside, AZT): inhibition of mitochondrial DNA synthesis leads to a decrease in total number of mitochondria and reduced NADH oxidation in cells.

B. Electron-transport blockers

1. Electron-transport blockers prevent maintenance of the proton gradient so that ATP synthesis stops.
2. Electron-transport blockers include cyanide, carbon monoxide, amobarbital (Amytal), rotenone, and antimycin A (see Fig. 5-8).

C. Uncouplers

1. Uncouplers carry protons through the inner mitochondrial membrane *without* generating ATP.
2. Uncouplers short-circuit the proton gradient by transporting H^+ ions from the intermembrane space to the matrix, thereby abolishing the gradient.
3. Uncouplers dissipate energy from electron flow as heat (risk of hyperthermia).
4. Uncouplers increase O_2 consumption.
5. Uncouplers include pentachlorophenol, dinitrophenol, and thermogenin, a natural uncoupling protein present in brown fat.
   a. Unlike adults, newborns have a considerable amount of brown fat.
   b. The heat generated by electron flow in the mitochondria of brown fat is important in maintaining neonatal body temperature.

D. Mitochondrial malfunction

1. Nucleoside analogues (e.g., cytosine arabinoside, AZT)
   a. Inhibition of mitochondrial DNA synthesis leads to a decrease in total number of mitochondria and reduced NADH oxidation in cells.
2. Reduced NADH oxidation produces:
   a. Increased cytosolic NADH
   b. Shunting of pyruvate to lactate (i.e., lactic acidosis)
   c. Hyperglycemia due to increased gluconeogenesis
   d. Fatty liver due to inability to oxidize fatty acids

ETC carriers upstream of the block become highly reduced and carriers downstream become oxidized.

Uncouplers: risk for hyperthermia

Electron flow proceeds without ATP synthesis, eliminating respiratory control.

Dinitrophenol: used in making dynamite; was used for weight reduction

Thermogenin: uncoupler in newborn brown fat

Inhibition of mitochondrial DNA synthesis: reduced NADH oxidation

Reduced NADH oxidation: lactic acidosis, gluconeogenesis inhibition, fatty liver

CHAPTER 6
CARBOHYDRATE METABOLISM

I. Glycolysis and the Fate of Pyruvate

A. Overview

1. Glycolysis is composed of five reactions that consume ATP and five reactions that produce ATP.
2. Pyruvate dehydrogenase catalyzes three reactions within the same multienzyme complex.
3. Glycolysis is regulated at three points: hexokinase, phosphofructokinase (PFK), and pyruvate kinase.
4. Pyruvate dehydrogenase is regulated by covalent modification with phosphorylation.
5. Glycolysis interfaces with glycogen metabolism, the pentose phosphate pathway, the formation of amino sugars, triglyceride synthesis (by means of glycerol 3-phosphate), the production of lactate (a dead-end reaction), and transamination with alanine.
6. Pyruvate dehydrogenase interfaces with other pathways such as the citric acid cycle or fat synthesis through its product, acetyl CoA.
7. Lactic acidosis is caused by overproduction of pyruvate or NADH.
8. Deficiencies in any of the pyruvate dehydrogenase enzymes produce lactic acidosis.

B. Glycolysis and pyruvate oxidation: pathway reaction steps (Fig. 6-1)

1. Step 1
   a. Phosphorylation of glucose to glucose 6-phosphate, the first regulated step in glycolysis, is irreversible and traps glucose inside the cell.
   b. The reaction uses 1 ATP.
   c. The two enzymes that catalyze this step, hexokinase and glucokinase, are specialized to function under different conditions (Table 6-1).
   d. Hexokinase, present in all tissues, is active at low glucose concentrations (low K_m) and cannot rapidly phosphorylate large amounts of glucose (low V_{max}).
   e. Glucokinase, present in the liver and pancreatic β cells, is highly active only at high glucose concentrations (high K_m) and rapidly phosphorylates large amounts of glucose (high V_{max}).

2. Step 2
   a. Reversible reaction involving the conversion of glucose 6-phosphate to fructose 6-phosphate by phosphoglucose isomerase

3. Step 3
   a. Irreversible reaction involving the conversion of fructose 6-phosphate to fructose 1,6-bisphosphate by phosphofructokinase 1 (PFK-1), which is the rate-limiting enzyme of glycolysis
   b. The reaction uses 1 ATP.

4. Step 4
   a. Reversible conversion of fructose 1,6-bisphosphate to two 3-carbon intermediates by aldolase A
   b. Triose phosphate isomerase reversibly converts glyceraldehyde 3-phosphate to dihydroxyacetone phosphate (DHAP).
   c. DHAP is reversibly converted to glycerol 3-phosphate by glycerol 3-phosphate dehydrogenase, which uses NADH as a cofactor.
      (1) Glycerol 3-phosphate is the primary substrate for synthesis of triacylglycerol in the liver in the fed state (see Chapter 7).
      (2) Glycerol 3-phosphate is used to shuttle NADH to the electron transport chain (ETC) in the inner mitochondrial membrane (see Chapter 5).

Phosphorylation: traps glucose in the cell

Hexokinase: phosphorylates glucose; low K_m, low V_{max}

Hexokinase: low K_m ensures glucose uptake at fasting levels

Glucokinase: phosphorylates glucose; high K_m, high V_{max}

DHAP converted to glycerol 3-phosphate: triacylglycerol synthesis, gluconeogenesis, ETC shuttle

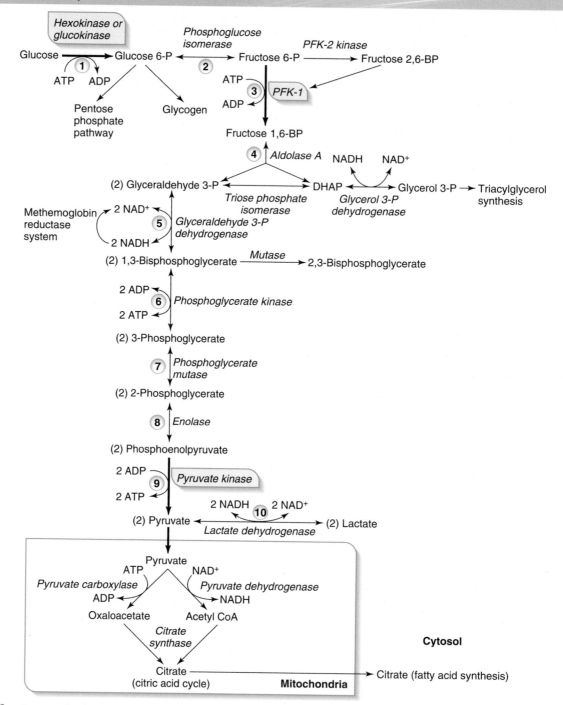

6-1: Steps in glycolysis. The regulated steps in glycolysis are indicated by *one-way arrows* and *boxed enzymes*. Reversible reactions are identified by two-way arrows. Interfaces with other pathways are also identified. BP, bisphosphate; MetHb, methemoglobin; P, phosphate; PFK, phosphofructokinase.

TABLE 6-1. Comparison of Hexokinase and Glucokinase

CHARACTERISTIC	HEXOKINASE	GLUCOKINASE
Tissue location	All	Liver and pancreatic β cells
K_m	Low (high substrate affinity)	High (low substrate affinity)
V_{max}	Low	High
Inhibited by glucose 6-phosphate	Yes	No
Inducible by insulin	No	Yes
Substrate specificity	Glucose, fructose, galactose	Glucose only
Physiologic role	Provides cells with basal level of glucose 6-phosphate needed for energy production	Permits accumulation of intracellular glucose for conversion to glycogen or triacylglycerols

        (3) In the fasting state, glycerol 3-phosphate is converted to DHAP, which is used as a substrate for gluconeogenesis.

  5. Step 5

    a. Reversible conversion of glyceraldehyde 3-phosphate to 1,3-bisphosphoglycerate (1,3-BPG) by glyceraldehyde 3-phosphate dehydrogenase using NAD^+ as a cofactor

    b. NAD^+ must be replenished for glycolysis to continue.

    c. NADH is shuttled into the ETC by the glycerol phosphate shuttle or the malate-aspartate shuttle (see Chapter 5).

    d. The methemoglobin reductase system and the pathway for synthesizing 2,3-bisphosphoglycerate (2,3-BPG) are ancillary pathways that emanate from this reaction.

  6. Step 6

    a. Reversible conversion of 1,3-BPG to 3-phosphoglycerate by phosphoglycerate kinase

    b. This reaction generates 2 ATP per glucose molecule, which replace the 1 ATP used in step 1 and the 1 ATP used in step 3.

  7. Step 7

    a. Reversible conversion of 3-phosphoglycerate to 2-phosphoglycerate by phosphoglycerate mutase

  8. Step 8

    a. Reversible conversion of 2-phosphoglycerate to phosphoenolpyruvate (PEP) by enolase

  9. Step 9

    a. Regulated, irreversible reaction involving the conversion of PEP to pyruvate by pyruvate kinase

    b. There is a net gain of 2 ATP per glucose molecule in this reaction.

  10. Step 10

    a. Reversible conversion of pyruvate to lactate by lactate dehydrogenase using NADH as a cofactor

    b. This reaction occurs in anaerobic glycolysis associated with shock and extreme exercise.

    c. This reaction also occurs in alcoholics because of an increase in NADH from alcohol metabolism (see Chapter 9).

  11. Pyruvate dehydrogenase, a large multienzyme complex located in the mitochondrial matrix, acts on pyruvate to produce acetyl CoA and 2 NADH.

    a. Five coenzymes are required for the above reaction.

      (1) Coenzymes derived from vitamins include thiamine pyrophosphate, FAD, NAD^+, and coenzyme A.

      (2) Lipoic acid is a coenzyme produced from octanoic acid.

    b. α-Ketoglutarate dehydrogenase, which carries out an analogous reaction to form succinyl CoA in the citric acid cycle, is also a multienzyme complex and requires the same coenzymes (see Chapter 5).

  12. Maximal ATP yield from oxidation of glucose is 36 to 38 ATP (Fig. 6-2).

    a. The maximum yield of ATP per glucose molecule depends on coupling of glycolysis with the citric acid cycle (see Chapter 5) by means of pyruvate dehydrogenase.

C. Glycolysis and pyruvate oxidation: regulated steps

  1. All the kinase reactions are irreversible and serve a regulatory role in glycolysis.

  2. Hexokinase versus glucokinase

    a. Hexokinase is inhibited by glucose 6-phosphate, the reaction product; glucokinase is not inhibited by glucose 6-phosphate.

    b. Glucokinase induction by insulin and lack of inhibition by glucose 6-phosphate promote clearance of blood glucose by the liver in the fed state.

  3. PFK-1

    a. Inhibitors include ATP (indicates that energy stores are high) and citrate (indicates that the cell is actively making ATP).

    b. Activators of PFK-1 include AMP (indicates that energy stores are depleted) and fructose 2,6-bisphosphate, which is formed by phosphofructokinase 2 (PFK-2).

    c. The fructose 2,6-bisphosphate level is controlled by the insulin-to-glucagon ratio by its effect on PFK-2, which is a bifunctional enzyme that has different activities in the fed and fasting states (Fig. 6-3).

Marginal notes:

Glycerol 3-phosphate fasting state: converted to DHAP and used as substrate for gluconeogenesis

NAD^+: must be replenished for glycolysis to continue

Ancillary pathways: methemoglobin reductase system; synthesis of 2,3-BPG

PEP conversion to pyruvate: net gain of 2 ATP

Pyruvate to lactate: anaerobic glycolysis; alcoholics

Lactic acidosis: often present in alcoholics

Anaerobic glycolysis: 2 ATP per glucose molecule

Aerobic glycolysis: 36 to 38 ATP per glucose molecule

Kinase reactions in glycolysis: irreversible; serve regulatory role

Hexokinase inhibited by glucose 6-phosphate

ATP and citrate inhibit glycolysis

AMP, fructose 2,6-bisphosphate stimulate glycolysis

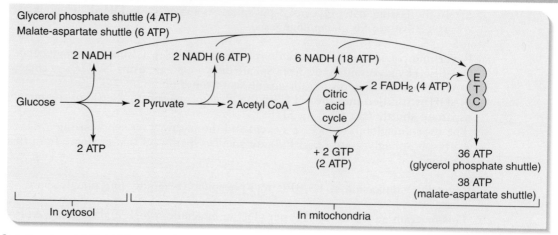

6-2: Overview of ATP yield from complete oxidation of glucose. Substrate-level phosphorylation generates 2 ATP per glucose molecule in the cytosol; however, the bulk of the energy output is derived from electron flow through the electron transport chain (ETC) and coupled oxidative phosphorylation. Electrons from cytosolic NADH move into mitochondria by the malate-aspartate shuttle to produce 38 ATP or by the glycerol phosphate shuttle, which results in a slightly lower ATP yield (i.e., 36 ATP).

6-3: Regulation of phosphofructokinase 1 (PFK-1). Fructose 2,6-bisphosphate (fructose 2,6-BP), the most potent activator of PFK-1, is formed by a separate enzyme, phosphofructokinase 2 (PFK-2). PFK-2 is a bifunctional enzyme that acts as a kinase at high insulin levels (i.e., fed state) and converts fructose 6-phosphate (fructose 6-P) into fructose 2,6-BP. In the fasting state, when glucagon levels are high, PFK-2 acts as a phosphatase, causing fructose 2,6-BP to convert back to fructose 6-P.

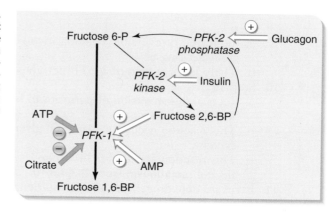

(1) In the fed state (i.e., high insulin, low glucagon), insulin dephosphorylates PFK-2, giving it kinase activity, which converts fructose 6-phosphate to fructose 2,6-bisphosphate, leading to activation of PFK-1 and enhanced glycolysis.

(2) In the fasting state (i.e., low insulin, high glucagon), protein kinase A (activated by glucagon) phosphorylates PFK-2, giving it phosphatase activity, resulting in conversion of fructose 2,6-bisphosphate back to fructose 6-phosphate, hence favoring gluconeogenesis.

4. Pyruvate kinase
   a. During glycolysis, pyruvate kinase is stimulated by fructose 1,6-bisphosphate, the product of the PFK-1 reaction.
   b. During gluconeogenesis, pyruvate kinase is inhibited by high levels of ATP, alanine, and active protein kinase A, thereby driving glucose formation.
   (1) During the fasting state, inactivation of liver pyruvate kinase through phosphorylation by active protein kinase further ensures enhancement of gluconeogenesis.

5. Pyruvate dehydrogenase
   a. Cycling between inactive and active forms of pyruvate dehydrogenase is catalyzed by a specific kinase (phosphorylates the enzyme) and phosphatase (dephosphorylates the enzyme).
   (1) Active pyruvate dehydrogenase (nonphosphorylated state) is favored by high insulin (fed state) and by high ADP (indicates the need for energy).
   (2) Inactive pyruvate dehydrogenase (phosphorylated state) is favored by increased acetyl CoA and NADH from fatty acid oxidation (fasting state).
   b. Acetyl CoA is a positive effector for pyruvate carboxylase, which favors generation of oxaloacetate as a substrate for gluconeogenesis.

Fed state: PFK-2 increases fructose 2,6-bisphosphate; activates of PFK-1

Fasting state: PFK-2 reduces fructose 2,6-bisphosphate; no activation of PFK-1

High insulin, low glucagon levels: favor glycolysis

Pyruvate kinase stimulated during glycolysis; inhibited during gluconeogenesis

Pyruvate dehydrogenase: stimulated by insulin and ADP in fed state

Pyruvate dehydrogenase: inhibited by acetyl CoA and NADH from fat oxidation in fasting state

D. Glycolysis and pyruvate oxidation: unique characteristics

1. Comparison of aerobic and anaerobic glycolysis
   a. NADH produced in the glycolytic pathway must be oxidized to regenerate NAD^+ for glycolysis to continue.
   b. In aerobic glycolysis, electrons from NADH are transferred to the ETC by the glycerol phosphate shuttle or the malate-aspartate shuttle, and NAD^+ is returned to the cytosol (see Figs. 5-9 and 5-10 in Chapter 5).
      (1) There is a net gain of 2 ATP and 2 NADH per glucose molecule in aerobic glycolysis.
      (2) 2 NADH produce a total of 4 ATP (glycerol phosphate shuttle) or 6 ATP (malate-aspartate shuttle) per glucose molecule.
2. In anaerobic glycolysis, reduction of pyruvate to lactate by lactate dehydrogenase regenerates NAD^+ (see Fig. 6-1).
   a. There is a net gain of 2 ATP and no NADH per glucose molecule in anaerobic glycolysis.
   b. Lactate is converted back to pyruvate in the liver or excreted in the urine.
   c. Mature red blood cells (RBCs) lack mitochondria and rely completely on anaerobic metabolism for generation of ATP.

E. Glycolysis and pyruvate oxidation: interface with other pathways (Fig. 6-4)

1. Glucose 6-phosphate, the first product formed in glycolysis, connects the glycolytic pathway to the pentose phosphate pathway and to glycogen synthesis, galactose metabolism, and the uronic acid pathway.
2. Fructose 6-phosphate is the precursor for synthesis of amino sugars (e.g., glucosamine and galactosamine), which ultimately are incorporated into glycoproteins and glycosaminoglycans (e.g., chondroitin sulfate, hyaluronic acid).
   a. Fructose 6-phosphate is also used to synthesize mannose, which is involved in the synthesis of glycoproteins and lysosomal enzymes.
3. Glycerol 3-phosphate, derived from dihydroxyacetone phosphate (DHAP), is used in the synthesis of triacylglycerols in the liver.
4. 1,3-BPG is converted by a mutase to 2,3-BPG in RBCs, which is the key substrate for shifting the O_2-binding curve to the right (decreasing O_2 affinity) (see Fig. 6-1).
5. In the RBC the reaction that converts glyceraldehyde 3-phosphate to 1,3-BPG, donates some of the NADH to be used by the methemoglobin reductase pathway to reduce methemoglobin (Fe^{3+}), which *cannot* bind O_2, to hemoglobin (Fe^{2+}), which binds O_2 to its heme iron (see Fig. 6-1, step 5).
6. Pyruvate, the end product of aerobic glycolysis, is a key metabolic intermediate whose fate differs in the fed and fasting states (Fig. 6-5).
   a. In the fed state, when glucose is plentiful, pyruvate is oxidized to acetyl CoA by pyruvate dehydrogenase.

Aerobic glycolysis: net gain of 2 ATP and 2 NADH molecules

Anaerobic glycolysis: net gain of 2 ATP molecules; NAD^+ replenished

Mature RBCs: lack mitochondria; use anaerobic glycolysis for ATP synthesis

Glycerol 3-phosphate: substrate for triacylglycerol in the liver

1,3-BPG: converted to 2,3-BPH by mutase

Methemoglobin reductase pathway: methemoglobin (Fe^{3+}) reduced to hemoglobin (Fe^{2+}), which can bind O_2

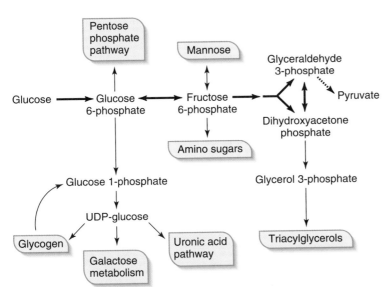

6-4: Interface of glycolytic intermediates *(boldface type)* with other pathways. Notice that the key glycolytic intermediates that branch off into other pathways are glucose 6-phosphate, fructose 6-phosphate, and dihydroxyacetone phosphate (DHAP).

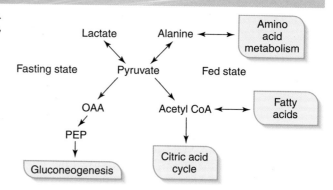

6-5: Fate of pyruvate in the fed and fasting states. Notice the importance of pyruvate in carbohydrate, protein, and fat metabolism.

Acetyl CoA fed state: enter CAC or used for fatty acid synthesis

Pyruvate fed state: transaminated to produce alanine

Fed state: pyruvate oxidized to acetyl CoA

Fasting state: pyruvate shunted to gluconeogenesis; derives from alanine

Pyruvate fasting state: used for gluconeogenesis

Acetyl CoA: precursor for fat synthesis; oxidation product of fat, ethanol, and ketone bodies

Pyruvate dehydrogenase deficiency: lactic acidosis, decreased acetyl CoA, early death

Pyruvate dehydrogenase deficiency: lactic acidosis, decreased acetyl CoA, early death

(1) Acetyl CoA enters the citric acid cycle (see Chapter 5) or is used for fatty acid synthesis (see Chapter 7).
(2) In the fed state, pyruvate, an α-ketoacid, can be converted by alanine aminotransferase to the amino acid alanine, which is used for protein synthesis.
(3) Excessive carbohydrate intake increases the amount of acetyl CoA available for fatty acid synthesis, which in turn increases the amount available for storage in fat.
   b. In the fasting state, when glucose is in short supply, pyruvate is carboxylated to oxaloacetate, providing carbon skeletons for gluconeogenesis (see Fig. 6-1).
7. Acetyl CoA is a precursor for fatty acid synthesis and a product of fatty acid oxidation.
   a. Acetyl CoA is also a product of ethanol metabolism and ketone body oxidation.
F. Glycolysis and pyruvate oxidation: clinical relevance
1. Lactic acidosis (increased lactate in blood)
   a. In aerobic glycolysis, electrons from NADH are transferred to the ETC by the glycerol phosphate shuttle or the malate-aspartate shuttle, and NAD^+ is returned to the cytosol (see Figs. 5-9 and 5-10 in Chapter 5).
   b. In anaerobic glycolysis, reduction of pyruvate to lactate by lactate dehydrogenase regenerates NAD^+ (see Fig. 6-1).
      (1) Hypoxia causes intramitochondrial NADH to reverse the malate shuttle producing an increase in cytosolic NADH.
      (2) Other conditions such as pyruvate dehydrogenase deficiency (increased pyruvate) and ethanol metabolism (increased NADH) produce excess lactate.
   c. Lactate is converted back to pyruvate in the liver or excreted in the urine.
      (1) Mature RBCs lack mitochondria and rely completely on anaerobic metabolism for generation of ATP.
2. Glycolytic and pyruvate dehydrogenase enzyme defects (e.g., pyruvate kinase deficiency, pyruvate dehydrogenase deficiency) are summarized in Table 6-2.

II. Gluconeogenesis
   A. Overview
   1. Four key enzymes in the gluconeogenic pathway bypass irreversible steps in glycolysis
      a. Pyruvate carboxylase, phosphoenolpyruvate (PEP) carboxykinase, fructose 1,6-bisphosphatase, glucose 6-phosphatase.
   2. Reciprocal regulation by the insulin-to-glucagon ratio ensures that glycolysis and gluconeogenesis are *not* stimulated simultaneously.
   3. Gluconeogenesis is an energy-requiring pathway.
      a. The energy is supplied by fatty acid oxidation.
   4. Gluconeogenesis is a carbon skeleton–requiring pathway.
      a. The carbon skeletons are supplied by amino acids from skeletal muscle and lactate from muscle and RBCs.
   5. Only the liver can use the free glycerol from fatty acid mobilization for gluconeogenesis.
   6. Gluconeogenic enzyme deficiencies result in fasting hypoglycemia.
   B. Gluconeogenesis: pathway reaction steps (Fig. 6-6)
   1. Enzymes that are required to bypass the three irreversible steps in glycolysis are discussed in Box 6-1.
      a. Pyruvate carboxylase
      b. Phosphoenolpyruvate (PEP) carboxykinase
      c. Fructose 1,6-bisphosphatase (rate-limiting reaction)
      d. Glucose 6-phosphatase

Gluconeogenic enzyme deficiency: hypoglycemia, such as glucose 6-phosphatase deficiency in von Gierke's disease

TABLE 6-2. **Hereditary Defects in Catabolism of Sugars**

DISEASE	METABOLIC EFFECT	CLINICAL FEATURES
Glucose and Pyruvate Metabolism		
Pyruvate kinase deficiency (AR, most common enzyme deficiency in the glycolytic pathway)	Inadequate ATP for maintaining ion pumps in RBC membrane results in loss of H_2O and membrane damage (produces RBCs with spikes)	Hemolytic anemia and jaundice begin at birth; anemia is somewhat offset by an increase in 2,3-BPG, which shifts the O_2-binding curve to the right (decreased O_2 affinity)
	Damaged RBCs, subject to macrophage destruction in the spleen, lead to increase in unconjugated bilirubin	
	Increase in RBC 2,3-BPG proximal to the enzyme block	
Pyruvate dehydrogenase deficiency (AR)	Increase in pyruvate with concomitant increase in lactic acid and alanine (by transamination); decrease in production of acetyl CoA; severe reduction in ATP production	Lactic acidosis, neurologic defects, myopathy; usually fatal at early age
Galactose Metabolism		
Galactokinase deficiency (AR)	Increase in galactose and galactitol (sugar alcohol)	Cataracts
Galactosemia (AR)	Deficiency of GALT; increase in galactose (blood, urine), galactose 1-phosphate (very toxic), and galactitol (sugar alcohol)	Cirrhosis, mental retardation, cataracts, galactosuria
	Galactitol is osmotically active and damages tissues	Women with disorder can synthesize lactose in breast milk due to epimerase reaction.
Fructose Metabolism		
Essential fructosuria (AR)	Deficiency of fructokinase; decreased fructose in blood and urine	Benign condition marked by fructosuria
Hereditary fructose intolerance (AR)	Deficiency of aldolase B; increase in fructose and fructose 1-phosphate (toxic)	Toxic liver damage, renal disease
	Excess fructose traps phosphorus in cells, leading to hypophosphatemia, decrease in ATP, and increase in AMP	Severe fasting hypoglycemia, hypophosphatemia, increased uric acid level (metabolism of AMP)
Pentose Phosphate Pathway		
Glucose 6-phosphate dehydrogenase (G6PD) deficiency (sex-linked recessive)	Inadequate NADPH production; results in reduction in antioxidant activity of glutathione in mature RBCs	Hemolytic anemia often induced by infections, oxidant drugs (e.g., dapsone, primaquine), and fava beans

AR, autosomal recessive; 2,3-BPG, 2,3-bisphosphoglycerate; GALT, galactose 1-phosphate uridyltransferase; RBC, red blood cell.

2. Conversion of pyruvate to PEP, which bypasses the irreversible pyruvate kinase reaction, occurs in several steps.
   a. Pyruvate carboxylase, a biotin-containing mitochondrial enzyme, converts pyruvate to oxaloacetate (OAA) in an irreversible reaction that consumes ATP (see Fig. 6-6, step 1).
      (1) Biotin deficiency leads to a build up of pyruvate, which is converted to lactic acid and leads to lactic acidosis.
   b. OAA is reduced to malate (i.e., malate shuttle), which is transported to the cytosol and then reoxidized to OAA.
   c. PEP carboxykinase decarboxylates OAA to produce PEP in a reversible reaction that consumes GTP (step 2).
3. Conversion of PEP to fructose 1,6-bisphosphate occurs by simple reversal of six reactions in the glycolytic pathway.
4. Fructose 1,6-bisphosphatase (rate-limiting enzyme), which dephosphorylates fructose 1,6-bisphosphate to produce fructose 6-phosphate (step 3), bypasses the irreversible PFK-1 reaction.
5. Simple reversal of the phosphoglucose isomerase reaction converts fructose 6-phosphate to glucose 6-phosphate.
6. Glucose 6-phosphatase dephosphorylates glucose 6-phosphate to produce glucose (step 4), bypassing the irreversible hexokinase or glucokinase reaction.

Pyruvate kinase bypass: pyruvate carboxylase + PEP carboxykinase

OAA cannot exit the mitochondria and must be converted into malate.

Fructose 1,6-bisphosphatase: rate-limiting enzyme of gluconeogenesis; bypasses PFK

Glucose 6-phosphatase releases free glucose into blood; bypasses hexokinase

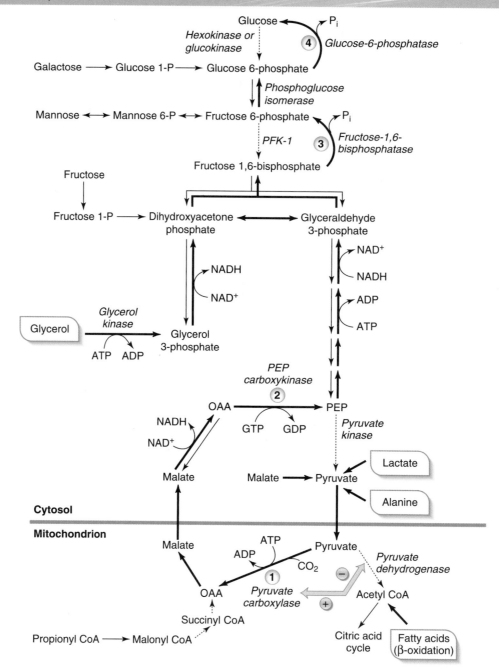

6-6: Overview of gluconeogenesis and its regulation, reading from bottom to top. Key enzymatic reactions that occur in the gluconeogenic pathway are numbered and labeled. *Thick arrows* indicate reactions leading to glucose synthesis. *Dashed arrows* indicate irreversible glycolytic reactions, which are blocked under conditions favoring gluconeogenesis in the fasting state. Lactate, alanine, and glycerol *(boxes)* are the primary sources of carbon skeletons for gluconeogenesis. Other fuels for gluconeogenesis that are depicted include malate (malate → pyruvate in fatty acid metabolism), succinyl CoA (propionyl metabolism), dihydroxyacetone phosphate + glyceraldehyde 3-phosphate (fructose metabolism), glucose 6-phosphate (galactose metabolism), and fructose 6-phosphate (mannose metabolism). OAA, oxaloacetate; PEP, phosphoenolpyruvate.

C. **Gluconeogenesis: regulated steps (see Fig. 6-6)**
1. Reciprocal regulation ensures that gluconeogenesis or glycolysis predominates, preventing futile cycling of glucose to pyruvate and back again to glucose.
2. Acetyl CoA, a product of fatty acid oxidation, is a positive allosteric effector of pyruvate carboxylase, which diverts pyruvate into the gluconeogenic pathway rather than the citric acid cycle.
3. The ratio of insulin to glucagon regulates pyruvate kinase (PEP conversion to pyruvate) and fructose 2,6-bisphosphate (see Fig. 6-3), which activates PFK-1.

Acetyl CoA: produced during gluconeogenesis from mobilized fatty acids

BOX 6-1 COMPARISON OF GLYCOLYSIS AND GLUCONEOGENESIS

Of the 10 reactions in the glycolytic pathway, 7 are reversible (see Fig. 6-1). During gluconeogenesis, when these reactions proceed in the reverse direction, the gluconeogenic reactions are catalyzed by the same enzymes that are used in glycolysis.

For the three irreversible reactions in glycolysis, four other enzymes are required for gluconeogenesis to occur (see Fig. 6-6). Notice that a pathway involving two enzymes, pyruvate carboxylase and phosphoenolpyruvate (PEP) carboxylase, is required in gluconeogenesis to reverse the effect of pyruvate kinase in glycolysis. In the following chart, a minus sign (−) indicates inhibition, and a plus sign (+) indicates activation.

Glycolysis	Gluconeogenesis
Hexokinase (or glucokinase)	Glucose 6-phosphatase
Glucose → glucose 6-phosphate (step 1 in Fig. 6-1)	Glucose 6-phosphate → glucose (step 4 in Fig. 6-6)
(−) Glucose 6-phosphate (hexokinase)	
Phosphofructokinase 1	Fructose 1,6-bisphosphatase
Fructose 6-phosphate → fructose 1,6-bisphosphate (step 3)	Fructose 1,6-bisphosphate → fructose 6-phosphate (step 3)
(+) AMP, fructose 2,6-bisphosphate	(−) AMP, fructose 2,6-bisphosphate
(−) ATP, citrate	(+) Citrate
Rate-limiting reaction	Rate-limiting reaction
Pyruvate kinase	Pyruvate carboxylase (in mitochondria)
PEP → pyruvate (step 9)	Pyruvate → oxaloacetate (OAA) (step 1)
(+) Fructose 1,6-bisphosphate	(+) Acetyl CoA
(−) ATP, alanine, glucagon	Requires biotin; OAA is converted to malate in the mitochondria, and malate is converted back to OAA in the cytosol
	PEP carboxykinase (in cytosol)
	OAA PEP (step 2)
	Only reversible reaction in gluconeogenesis

    a. High insulin and low glucagon levels (fed state)
       (1) Leads to increased pyruvate kinase activity and increased fructose 2,6-bisphosphate levels
       (2) Result: increased glycolysis (particularly in the liver) and decreased gluconeogenesis
    b. Low insulin and high glucagon levels (fasting state)
       (1) Leads to decreased pyruvate kinase activity and decreased fructose 2,6-bisphosphate levels
       (2) Result: decreased glycolysis (particularly in the liver) and increased gluconeogenesis (maintains blood glucose)
  4. Three allosteric effectors have opposite effects on fructose 1,6-bisphosphate and PFK-1.
    a. Fructose 2,6-bisphosphate and AMP
       (1) Stimulate PFK-1, which results in increased glycolysis
       (2) Inhibit fructose 1,6-bisphosphatase, which results in decreased gluconeogenesis
    b. Citrate
       (1) Inhibits PFK-1, which leads to decreased glycolysis
       (2) Stimulates fructose 1,6-bisphosphatase, which leads to increased gluconeogenesis

D. Gluconeogenesis: unique characteristics
  1. A total of 6 ATP per glucose molecule are consumed in gluconeogenesis.
  2. The liver is the most important site for gluconeogenesis, whereas the kidneys and the epithelium of the small intestine assume a less important role (mainly during starvation).
    a. In prolonged starvation, the kidneys assume a key role in gluconeogenesis.
  3. Gluconeogenesis is important in maintaining blood glucose levels in the fasting state for energy requirements in the brain, RBCs, exercising muscle, and the renal medulla.
  4. Gluconeogenesis occurs in part in the mitochondria (pyruvate carboxylase reaction) and in part in the cytosol.

High insulin and low glucagon levels: gluconeogenesis inhibited, glycolysis stimulated

Low insulin and high glucagon: gluconeogenesis stimulated, glycolysis inhibited

Fructose 2,6-bisphosphate and AMP: inhibition of gluconeogenesis

Citrate: stimulation of gluconeogenesis

Fat oxidation provides energy (6 ATP per glucose) for gluconeogenesis.

Gluconeogenesis: maintain glucose in fasting state

Gluconeogenesis sites: liver (major site), kidneys (starvation), epithelium of small intestine

E. **Gluconeogenesis: interface with other pathways**
1. Lactate provides approximately one third of the carbon skeletons used in gluconeogenesis.
   a. In the Cori cycle, lactate produced in exercising muscle and RBCs travels in the bloodstream to the liver for conversion to glucose, which then travels back to muscle and RBCs.
2. Glucogenic amino acids, derived from degradation of muscle protein, supply carbon atoms for gluconeogenesis by transamination to pyruvate (alanine), α-ketoglutarate (glutamate), and OAA (aspartate) (see Chapter 8).
3. Glycerol is an important source of carbon atoms for gluconeogenesis in fasting or starvation conditions, when triacylglycerols in adipose tissue are mobilized.
   a. Glycerol kinase, present only in the liver, converts glycerol to glycerol 3-phosphate, which is further converted to DHAP and used as a substrate for gluconeogenesis.

F. **Gluconeogenesis: clinical relevance**
1. Gluconeogenic enzyme deficiencies result in fasting hypoglycemia.
2. Example: In von Gierke's disease, a glycogen storage disease, the absence of glucose 6-phosphatase leads to a decrease in glucose synthesis.

III. **Glycogen Metabolism**
A. **Overview**
1. Glycogen is a highly branched glucose polymer found primarily in liver and skeletal muscle.

B. **Glycogen metabolism: pathway reaction steps**
1. Glycogenesis, the synthesis of glycogen, requires a preexisting fragment of glycogen or a primer glycoprotein (glycogenin) plus the activated glucose donor UDP-glucose (Fig. 6-7A).
   a. Phosphoglucomutase isomerizes glucose 6-phosphate to glucose 1-phosphate by a reversible reaction.
   b. Glucose 1-phosphate plus uridine triphosphate (UTP) produces uridine diphosphate-glucose (UDP-glucose), which is the activated form of glucose that is necessary for glycogen synthesis.
   c. Glycogen synthase (rate-limiting enzyme) forms an α-1,4-glycosidic bond between a glucose unit from UDP-glucose at the nonreducing end of an existing glycogen fragment or glycoprotein (glycogenin).
   d. The branching enzyme, glucosyl (4:6) transferase, removes a block of glucose units from the nonreducing end of a growing glycogen chain and reattaches glucose units in an α-1,6 linkage at a different site, creating a branch point (see Fig. 6-7B).

Cori cycle: liver conversion of lactate (muscle, RBCs) to glucose via gluconeogenesis; glucose returned to muscle

Alanine cycle: liver conversion of alanine (muscle) to glucose by gluconeogenesis; glucose returned to muscle

Glycerol kinase: only in hepatocytes; converts glycerol to glycerol 3-phosphate

Gluconeogenic enzyme deficiencies: fasting hypoglycemia

Glycogen constitutes a reserve supply of glucose that is rapidly mobilized in the fasting state.

Glycogen branches off of and re-enters glycolysis at glucose 6-phosphate

Glycogen synthase: rate-limiting enzyme of glycogen synthesis

Glucosyl (4:6) transferase: branching enzyme in glycogenesis

Branching provides more nonreducing ends for faster release of glucose

---

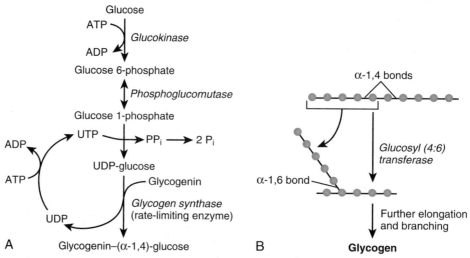

6-7: Glycogenesis. **A,** Formation of UDP-glucose and α-1,4 linkages in glycogen. Phosphoglucomutase isomerizes glucose 6-phosphate to glucose 1-phosphate in a reversible reaction. Glucose 1-phosphate is converted to UDP-glucose, which is the activated form of glucose required for glycogen synthesis. Glycogen synthase, the rate-limiting enzyme, adds one glucose unit at a time, in a 1,4 linkage, to the nonreducing end of glycogenin. **B,** Branch formation. After the main glycogen chain reaches a certain length, a block of α-1,4-linked residues is split from the end and reattached in an α-1,6 linkage by the branching enzyme, glucosyl 4:6 transferase.

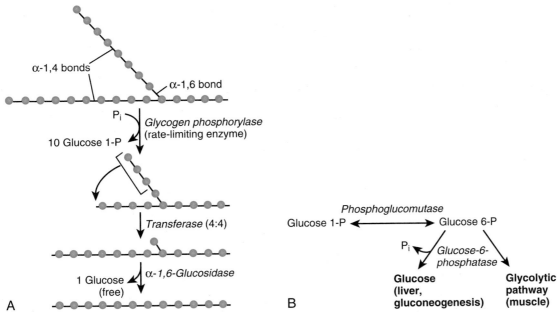

6-8: Glycogenolysis. **A,** Action of glycogen phosphorylase and the debranching enzyme, which catalyzes the transferase (4:4) and glucosidase (1:6) reactions. The ratio of glucose 1-phosphate to free glucose released depends on the number of branch points and length of the branches. **B,** Fate of glucose 1-phosphate in liver and muscle. Both tissues contain phosphoglucomutase, which catalyzes the reversible reaction between glucose 1-phosphate and glucose 6-phosphate. The liver contains glucose 6-phosphatase (i.e., gluconeogenic enzyme), and glucose 6-phosphate is primarily converted to glucose. Muscle oxidizes glucose 6-phosphate into ATP, which is used to power muscle contraction.

 (1) Branching increases the rate of glycogen synthesis and degradation by increasing the number of ends at which glucose units are added or removed.
2. Glycogenolysis entails the initial breakdown of glycogen to glucose 1-phosphate and free glucose in a ratio of about 10:1 (Fig. 6-8A).
 a. Shortening of chains is catalyzed by glycogen phosphorylase (rate-limiting enzyme), which cleaves α-1,4 bonds with inorganic phosphate (P_i) to produce glucose 1-phosphate.
  (1) Glycogen phosphorylase sequentially removes glucose units from the ends of all chains but stops four glucose units from each branch point.
 b. Branch removal of the four remaining glucose units is accomplished by a single debranching enzyme, which carries out two reactions.
  (1) Three glucose units remaining at one branch point are transferred to the nonreducing end of the linear glycogen chain by 4:4 transferase activity.
  (2) The last glucose unit at a branch point is split off by α-1,6-glucosidase activity, releasing a free glucose.
 c. Phosphoglucomutase reversibly converts glucose 1-phosphate to glucose 6-phosphate, and is therefore functional in glycogenesis and glycogenolysis (see Fig. 6-8B).
 d. The fate of glucose 6-phosphate derived from glycogenolysis differs in muscle and liver.
  (1) Liver contains glucose 6-phosphatase, a gluconeogenic enzyme that converts glucose 6-phosphate to free glucose, which helps maintain the blood glucose level in the fasting state.
  (2) Muscle glycogen is converted to glucose 6-phosphate, which is oxidized in the muscle to produce ATP.
 e. Some glycogen is degraded in lysosomes by α-1,4-glucosidase (acid maltase).
C. Glycogen metabolism: regulated steps
 1. Reciprocal regulation ensures that synthesis or degradation predominates, preventing the wasteful operation of both pathways simultaneously.
 2. Hormonal regulation involves cAMP-mediated signaling that controls cycling of glycogen synthase and glycogen phosphorylase between phosphorylated and nonphosphorylated forms, which differ in their activity (Fig. 6-9).
 a. High insulin (low glucagon), typical of the fed state, promotes glycogen synthase activity leading to glycogen synthesis.

Glycogen phosphorylase: rate-limiting enzyme of glycogen degradation

Debranching enzymes: release free glucose

Liver glycogenolysis: helps maintain blood glucose in fasting state

Muscle glycogenolysis: uses glucose for its own energy purposes

Glycogen metabolism is reciprocally regulated, as is gluconeogenesis.

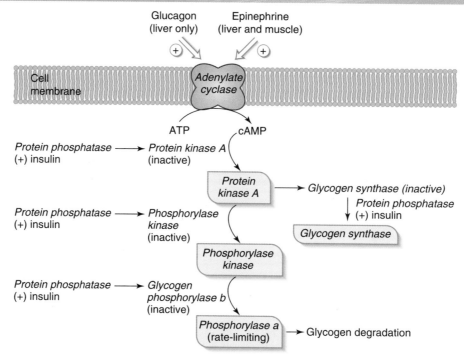

6-9: Hormonal regulation of glycogen synthesis and degradation involves cAMP signaling. *Shading* indicates activated forms of the enzymes. Active protein kinase A phosphorylates glycogen synthase, which inactivates the enzyme and prevents glycogenesis. Active protein kinase A activates phosphorylase kinase, which activates phosphorylase a, the rate-limiting reaction of glycogenolysis. Insulin activates hepatic protein phosphatase, which activates glycogen synthase and inactivates protein kinase A, phosphorylase kinase, and phosphorylase a.

Fed state: high insulin, low glucagon levels

Fasting state: low insulin, high glucagon levels

Glycogen mobilization by epinephrine: liver (α_1-adrenergic) and muscle (β-adrenergic)

Insulin (fed state) activates glycogen synthase; glucagon (fasting state) activates glycogen phosphorylase.

Activation of inactive form of glycogen phosphorylase by glucose 6-phosphate

Glucose 6-phosphatase: restricted to liver, kidney, and small intestine

Glycogen synthase: allosteric activation after a meal to trap glucose

    (1) Insulin activates hepatic protein phosphatase, which then removes phosphate groups from glycogen synthase (activating the enzyme), phosphorylase kinase (inactivating the enzyme), and glycogen phosphorylase (inactivating the enzyme).

  b. High glucagon (low insulin), typical of the fasting state, activates adenylate cyclase, leading to sequential activation of protein kinase A, phosphorylase kinase, and phosphorylase in the liver, and inactivates glycogen synthase (phosphorylation inhibits the enzyme), leading to glycogen degradation in the liver.

  c. Epinephrine, unlike glucagon, enhances glycogenolysis in muscle (β-adrenergic) and liver (α_1-adrenergic).

3. Allosteric regulation of enzymes increases glycogen synthesis or degradation more rapidly than hormone-induced activation of enzymes.

  a. Glucose 6-phosphate, which is elevated in the liver in the fed state, directly stimulates glycogen synthase b (the less active phosphorylated form), leads to an immediate increase in glycogen synthesis.

  b. Calcium, which is released from sarcoplasmic reticulum in contracting muscle, directly activates phosphorylase kinase, which leads to an immediate increase in glucose 6-phosphate from glycogen degradation and a concomitant increase in ATP production to power muscle contraction.

D. Glycogen metabolism: unique characteristics

1. The location of glucose 6-phosphatase in glucose-producing tissues (liver, kidney, and small intestine) ensures that glucose remains trapped in all other tissues for energy metabolism.

2. Glycogen is a highly branched glucose polymer for increased solubility and rapid degradation.

3. Glycogen synthesis consumes 2 ATP for each glucose molecule polymerized.

4. The ability to rapidly allosterically re-activate the inactive (phosphorylated) form of glycogen synthase by glucose 6-phosphate enables maximum storage of glucose immediately after a meal.

5. Because glycogen is polymerized by forming acetal bonds with the incoming glucose, the ring structure is maintained and the end molecule of glucose is nonreducing (reducing activity requires the open-chain form).

E. **Glycogen metabolism: interface with other pathways**
 1. Glucose 1-phosphate is the key metabolite linking glycogen synthesis to the glycolytic pathway (see Fig. 6-4).
 2. Excess glucose is shunted to glycogen synthesis in liver and muscle when the ATP supply is adequate.
 3. Mobilization of glucose from glycogen occurs in muscle when ATP is needed for contraction and in liver when blood glucose levels fall.
 4. UDP-glucose can also be converted to glucuronic acid, a precursor for conjugation with drugs and toxins in the liver (forms glucuronides).

F. **Glycogen metabolism: clinical relevance**
 1. The glycogenoses are autosomal recessive disorders that increase glycogen synthesis (e.g., von Gierke's disease) or prevent glycogenolysis (e.g., Pompe's disease, debranching enzyme deficiencies) and lead to an accumulation of structurally normal or abnormal glycogen within cells (Table 6-3).
 2. Branching and debranching enzyme disorders produce structurally abnormal glycogen, whereas the other types accumulate normal glycogen.
 3. Clinical manifestations depend on which tissues (e.g., muscle, liver, kidney) are affected by glycogen accumulation.
 4. Hypoglycemia occurs only in glycogenoses that interfere with gluconeogenesis (e.g., von Gierke's disease) or liver glycogenolysis (e.g., deficiency of liver phosphorylase).
 5. Muscle glycogenoses (e.g., McArdle's disease) do *not* result in hypoglycemia, because muscle uses its glycogen to supply glucose for generation of ATP.

IV. **Metabolism of Galactose and Fructose**
 A. **Overview**
  1. Galactose and fructose metabolism produce glycolytic intermediates.
  2. Neither galactose or fructose metabolism are regulated.
  3. Galactose and fructose metabolism allow the use of lactose from milk and sucrose from many dietary sources such as fruits, honey, and beets.
  4. Genetic deficiencies in enzymes from galactose and fructose lead to serious clinical problems such as cataracts and liver damage.
 B. **Galactose metabolism: pathway reaction steps (Fig. 6-10)**
  1. Galactose undergoes an exchange reaction with UDP-glucose to produce glucose 1-phosphate and UDP-galactose, using the rate-limiting enzyme galactose 1-phosphate uridyltransferase (GALT).
  2. Glucose 1-phosphate is converted to glucose 6-phosphate by phosphoglucomutase.
  3. UDP-galactose provides galactose units for lactose synthesis in breast tissue (epimerase reaction) and synthesis of glycoproteins, glycolipids, and glycosaminoglycans in other tissues.

Glucose 1-phosphate links glycogen synthesis and breakdown

Glycogenoses: increased glycogens synthesis or decreased glycogenolysis

Glycogenoses: genetic disorders due to reduction or absence of glycogen enzymes

Liver defects usually produce fasting hypoglycemia.

Muscle defects produce cramps during exertion.

Pompe's disease: deficiency of α-1,4-glucosidase (lysosomal enzyme).

Von Gierke's disease: deficiency of glucose 6-phosphatase (gluconeogenic enzyme)

McArdle's disease: deficiency of muscle glycogen phosphorylase

GALT: rate-limiting enzyme of galactose metabolism; deficient in galactosemia

TABLE 6-3. Glycogen Storage Diseases

DISEASE	DEFICIENT ENZYME	GLYCOGEN STRUCTURE	CLINICAL FEATURES
Von Gierke's (AR)	Glucose 6-phosphatase (liver and kidney)	Normal	Severe fasting hypoglycemia, ketosis, hyperlipidemia, lactic acidosis, enlarged liver and kidneys (hepatorenomegaly)
Pompe's (AR)	α-1,4-Glucosidase (lysosomes)	Normal	Infant form: mental retardation, hypotonia, cardiomegaly leading to death by age 2
			Adult form: gradual skeletal myopathy
Cori's (AR)	Debranching enzyme (muscle and liver), amylo-α-1,6-glucosidase	Abnormal: many short-branched chains (α-limit dextrins)	Mild hypoglycemia, hepatomegaly
			Decreased in free glucose after epinephrine challenge
Andersen's (AR)	Branching enzyme (liver and spleen), glucosyl 4,6-transferase	Few long chains with very few branches	Hepatosplenomegaly, cirrhosis, liver failure leading to death by age 2
McArdle's (AR)	Muscle glycogen phosphorylase	Normal	Muscle cramping, fatigue, and myoglobinuria with strenuous exercise; *no* increase in lactic acid after exercise
Hers' (AR)	Liver glycogen phosphorylase	Normal	Similar to von Gierke's disease but less severe

AR, autosomal recessive.

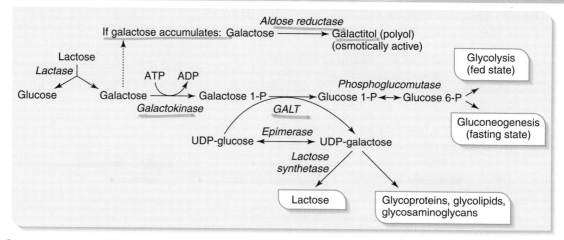

6-10: Galactose metabolism. Lactose is the primary source of galactose. Galactose 1-phosphate uridyltransferase (GALT) is the rate-limiting reaction. Glucose 1-phosphate is converted to glucose 6-phosphate, which is used as a substrate for glycolysis (fed state) or gluconeogenesis (fasting state). UDP-galactose is used in the synthesis of lactose and other compounds. Aldolase reductase converts excess galactose (e.g., galactokinase deficiency) into a sugar alcohol (galactitol), which is osmotically active.

C. **Galactose metabolism: regulated steps**
   1. There is no known regulation of the conversion of galactose to glycolytic intermediates.

D. **Galactose metabolism: unique characteristics**
   1. The major dietary source of galactose is the disaccharide lactose, which is present in milk and milk products.
   2. Lactase, a brush border disaccharidase enzyme located in the epithelium of the small intestine, converts lactose to glucose and galactose.
   3. There is no net synthesis of the UDP-glucose intermediate; the net output of the pathway is glucose 6-phosphate.

E. **Galactose metabolism: interface with other pathways**
   1. If galactose accumulates in tissues containing aldose reductase (e.g., lens, neural tissue), it is converted into a sugar alcohol (polyol) called galactitol, which is osmotically active.

F. **Galactose metabolism: clinical relevance**
   1. Galactokinase deficiency and galactosemia, which are caused by a deficiency of GALT, are summarized in Table 6-2.

G. **Fructose metabolism: pathway reaction steps (Fig. 6-11)**
   1. Aldolase B, the rate-limiting enzyme in fructose metabolism, is primarily located in the liver and, to a lesser extent, in the small intestine and proximal renal tubules.
   2. Aldolase B converts fructose 1-phosphate into the 3-carbon intermediates (trioses) DHAP and glyceraldehyde.
   3. Glyceraldehyde is further converted into glyceraldehyde 3-phosphate by a kinase.
   4. In the fed state, glyceraldehyde 3-phosphate and DHAP are used as intermediates in the glycolytic pathway; in the fasting state, they are converted to glucose 6-phosphate and used for glucose synthesis.

Lactase in epithelium of small intestine converts lactose to glucose + galactose.

The net output of galactose metabolism is glucose 6-phosphate.

Aldolase B: rate-limiting enzyme of fructose metabolism; deficient in hereditary fructose intolerance

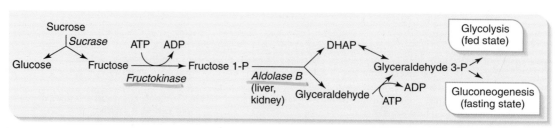

6-11: Fructose metabolism. Sucrose is the primary source of fructose. Aldolase B, the rate-limiting enzyme, converts fructose 1-phosphate to dihydroxyacetone phosphate (DHAP) and glyceraldehyde, which is further converted to glyceraldehyde 3-phosphate. DHAP and glyceraldehyde 3-phosphate are used as substrates for glycolysis (fed state) or gluconeogenesis (fasting state).

H. **Fructose metabolism: regulated steps**
   1. Similar to galactose metabolism, there is no known regulation of fructose metabolism.
   2. Because fructose metabolism bypasses PFK, it can be rapidly converted into acetyl CoA and into fat.
I. **Fructose metabolism: unique characteristics**
   1. The major dietary source of fructose is the disaccharide sucrose, which is present in table sugar, fruits, and honey.
      a. Sucrase, a brush border disaccharidase enzyme located in the epithelium of the small intestine, converts sucrose into glucose and fructose.
   2. Fructokinase, like galactokinase, is found primarily in the liver.
      a. It phosphorylates fructose at the first carbon (C1) position instead of the C6 position.
      b. Aldolase B can cleave both forms of fructose phosphate.
J. **Fructose metabolism: interface with other pathways**
   1. Fructose is a precursor for amino sugars in glycoproteins and glycolipids.
K. **Fructose metabolism: clinical relevance**
   1. Essential fructosuria and hereditary fructose intolerance, which are caused by a deficiency of aldolase B, are summarized in Table 6-2.

V. **Pentose Phosphate Pathway**
   A. **Overview**
      1. The oxidative branch consists of three irreversible reactions that convert glucose 6-phosphate to ribulose 5-phosphate with release of CO_2 and formation of NADPH.
      2. The nonoxidative branch consists of a series of reversible reactions that interconvert various sugars that produce ribose 5-phosphate and intermediates used in glycolysis or gluconeogenesis.
      3. Increased pathway activity occurs in tissues that consume NADPH in reductive biosynthetic pathways.
      4. Ribose 5-phosphate is used for RNA and DNA synthesis; NADPH is used for reductive biosynthesis and for maintaining glutathione (GSH) in the reduced state.
      5. G6PD deficiency produces hemolytic anemia under oxidative stress (e.g., primaquine).
   B. **Pentose phosphate pathway: pathway reaction steps**
      1. The oxidative branch consists of three irreversible reactions that convert glucose 6-phosphate to ribulose 5-phosphate with release of CO_2 and formation of NADPH (Fig. 6-12).
         a. Glucose 6-phosphate dehydrogenase (G6PD) is the rate-limiting enzyme that converts glucose 6-phosphate to 6-phosphogluconolactone, which is then converted via a series of intermediate reactions to ribulose 5-phosphate.
      2. The nonoxidative branch consists of a series of reversible reactions that interconvert various sugars that produce ribose 5-phosphate and intermediates used in glycolysis or gluconeogenesis.
         a. Transketolase reactions (thiamine-dependent) are responsible for two-carbon transfer reactions, while transaldolase reactions are involved in three-carbon transfer reactions.
         b. In the fed state, fructose 6-phosphate and glyceraldehyde 3-phosphate are used as substrates in the glycolytic pathway; in the fasting state, they are used as intermediates in gluconeogenesis.
   C. **Pentose phosphate pathway: regulated steps**
      1. Increased pathway activity occurs in tissues that consume NADPH in reductive biosynthetic pathways.
         a. Adipose tissue for fatty acid synthesis
         b. Gonads and adrenal cortex for steroid hormone synthesis
         c. Liver for fatty acid and cholesterol synthesis
      2. G6PD is competitively inhibited by its product, NADPH.
   D. **Pentose phosphate pathway: unique characteristics**
      1. The function of the pentose phosphate pathway is tailored to different cellular needs according to the relative needs of the cell
      2. Ribose 5-phosphate is used for RNA and DNA synthesis
      3. NADPH is used for reductive biosynthesis and for maintaining glutathione (GSH) in the reduced state.
         a. GSH is an antioxidant that neutralizes the oxidant activity of hydrogen peroxide (H_2O_2) by converting it to H_2O, using the enzyme glutathione peroxidase.

Pentose phosphate pathway: major source of NADPH (reductive biosynthesis); source of ribose 5-phosphate (RNA, DNA synthesis)

Oxidative pathway: production of ribulose 5-phosphate and NADPH

Glucose 6-phosphate dehydrogenase: rate-limiting enzyme of pentose phosphate pathway

Nonoxidative pathway: intermediates for glycolysis or gluconeogenesis

Transketolase reactions: thiamine-dependent

G6PD deficiency: decreased reduced glutathione; hemolytic anemia from oxidizing drugs (e.g., primaquine, dapsone) or infection

GSH: antioxidant; neutralizes hydrogen peroxide

Pentose phosphate pathway: major source of NADPH for reductive biosynthesis and of ribose 5-phosphate for nucleotide synthesis in all cells

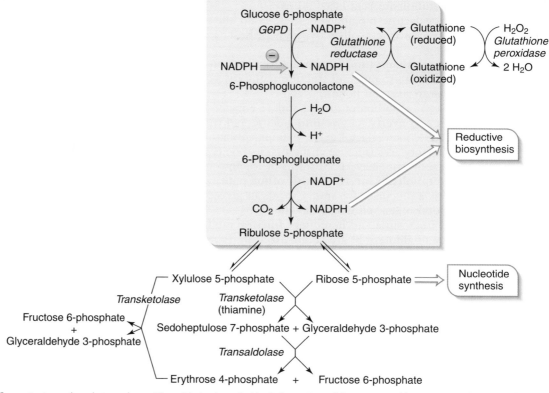

6-12: Pentose phosphate pathway. The oxidative branch *(shaded)* consists of three irreversible reactions. Glucose 6-phosphate dehydrogenase (G6PD), the rate-limiting enzyme, converts glucose 6-phosphate to 6-phosphogluconolactone, which is further converted in a series of reactions to ribulose 5-phosphate. The 2 NADPH formed are used for reductive biosynthesis and for maintaining the antioxidant glutathione in its reduced form. All of the reactions of the nonoxidative branch *(unshaded)* are reversible and produce ribose 5-phosphate for RNA and DNA synthesis. The intermediates fructose 6-phosphate and glyceraldehyde 3-phosphate are produced for glycolysis (fed state) or gluconeogenesis (fasting state).

4. The pentose phosphate pathway, also known as the hexose monophosphate pathway (HMP), occurs in the cytosol.
   E. **Pentose phosphate pathway: interface with other pathways**
      1. The nonoxidative branch produces glycolytic intermediates that allow the flow of carbon out of the pathway or into the pathway.
      2. The direction of the flow of carbon is determined by the relative use of NADPH and ribose 5-phosphate.
   F. **Pentose phosphate pathway: clinical relevance**
      1. G6PD deficiency is summarized in Table 6-2.
VI. **Glycoproteins and Proteoglycans**
   A. **Overview**
      1. Glycoproteins are short, branched oligosaccharides that function in blood group antigens, cell-cell adhesion, and coagulation factors.
      2. Proteoglycans are long, linear polysaccharides (glycosaminoglycans) attached to a protein core that function in the extracellular matrix.
   B. **Glycoproteins**
      1. Glycoproteins are proteins with attached short, branch-chained oligosaccharides, whose glycosidic linkages may be *N-* (contain dolichol phosphate) or *O*-linked.
      2. Clinically important glycoproteins
         a. Blood-group antigens on RBCs (Box 6-2) and other circulating blood proteins (e.g., coagulation factors, tumor markers such as prostate-specific antigen).
         b. Laminin, a glycoprotein in basement membranes, maintains the integrity of the basement membrane.
            (1) It also binds to integrin receptors producing a growth factor effect.
         c. Fibronectin, an adhesive glycoprotein, binds to cell-surface receptors (including the integrin receptors, as with laminin) and attaches cells to the extracellular matrix.
         d. Glycoproteins are components of lysosomal enzymes (e.g., mannose oligosaccharides) (Box 6-3).

Glycoproteins: short, branched oligosaccharides attached to protein

Proteoglycans: long, linear, unbranched polysaccharides attached to protein

Laminin and fibronectin bind integrin receptors; extracellular ground substance

BOX 6-2 ABO BLOOD GROUP ANTIGENS

ABO blood group antigens are genetically determined and are present on red blood cells (RBCs) and on epithelial cells located throughout the body. Individuals are identified as having O, A, B, or AB antigens on their RBCs. These antigens are produced by the H gene, which occurs in most individuals and codes for a glycosyltransferase that attaches fucose to a glycolipid to produce H antigen on RBCs. Individuals with the A gene, which codes for an N-acetylgalactosamine transferase that attaches N-acetylgalactosamine to the H antigen, produce A antigen. People with the B gene, which codes for a galactosyltransferase that attaches galactose to the H antigen, produce B antigen. In blood group AB individuals, A and B genes code for both transferases, and their RBCs contain A and B antigens. Individuals with the O gene cannot synthesize transferases, so the surfaces of their RBCs contain only H antigens.

After birth, individuals develop IgM antibodies (i.e., isohemagglutinins) against antigens they do *not* have on the surface of their RBCs (e.g., individuals with blood group A develop anti-B IgM antibodies). Individuals with blood group O develop anti-A IgM and anti-B IgM antibodies, as well as anti-A and anti-B IgG antibodies, which can cross the placenta and potentially attack and destroy fetal RBCs containing the A or B antigen. This is called ABO hemolytic disease of the newborn. The elderly often lose their isohemagglutinins. The ABO blood groups are summarized in the following chart.

Type	Antibodies	Comments
O	Anti-A IgM; anti-B IgM; anti-A,B IgG	Individuals with blood group O are universal donors; their RBCs can be transfused into all blood groups because there are no antigens on the surface of the RBCs to react with recipient isohemagglutinins. Blood group O individuals can receive only O blood. It is the most common blood group.
A	Anti-B IgM	Individuals have a predisposition for gastric carcinoma.
B	Anti-A IgM	
AB	None present	Individuals with blood group AB are universal recipients; they can be transfused with any blood group because they do not have isohemagglutinins to react against A or B antigens. It is the least common blood group.

BOX 6-3 SYNTHESIS OF LYSOSOMAL ENZYMES AND LYSOSOMAL DISEASES

Hydrolytic enzymes synthesized by the rough endoplasmic reticulum (RER) are transported to the Golgi apparatus for posttranslational modification. Modification involves attaching phosphate (by means of phosphotransferase) to mannose residues on hydrolytic enzymes to produce mannose 6-phosphate. The marked lysosomal enzymes attach to specific mannose 6-phosphate receptors on the Golgi membrane. Vesicles containing the receptor-bound lysosomal enzymes pinch off the Golgi to form primary lysosomes in the cytosol. Fusion of additional vesicles to the primary lysosome further increases their content of hydrolytic enzymes. Small vesicles containing only the receptors pinch off the primary lysosomes and return to the Golgi to bind more marked lysosomal enzymes, and the cycle repeats itself. Lysosomal functions include fusion with phagocytic vacuoles containing bacteria (i.e., lysosomes designated secondary or phagolysosomes), destruction of cell organelles, and degradation of complex substrates (e.g., sphingolipids, glycosaminoglycans).

Clinical Disorders Involving Lysosomes

Inclusion (I) cell disease is a rare, inherited condition in which there is a defect in posttranslational modification of lysosomal enzymes in the Golgi. Mannose residues on newly synthesized lysosomal enzymes coming from the rough endoplasmic reticulum are not phosphorylated because of a deficiency of phosphotransferase. Without mannose 6-phosphate to direct the enzymes to lysosomes, vesicles that pinch off the Golgi empty the unmarked enzymes into the extracellular space, where they are degraded in the bloodstream. Undigested substrates (e.g., carbohydrates, lipids, proteins) accumulate as large inclusions in the cytosol. Symptoms include psychomotor retardation and early death.

Deficiencies of degrading enzymes in lysosomes lead to the accumulation of complex substrates in lysosomes, producing lysosomal storage diseases. These diseases include genetic disorders with defects in the degradation of sphingolipids (e.g., Tay-Sachs disease) (see Chapter 7); defects in the degradation of glycosaminoglycans (e.g., mucopolysaccharidoses, such as Hurler's disease and Hunter's disease); and a single disorder with a defect in the degradation of glycogen by α-1,4-glucosidase in lysosomes, called Pompe's disease.

Hurler's disease is an autosomal recessive disease associated with a deficiency of α-L-iduronidase, which leads to lysosomal accumulation of dermatan sulfate and heparan sulfate. Clinical findings include severe mental retardation, coarse facial features, hepatosplenomegaly, corneal clouding, coronary artery disease (i.e., lipid accumulates in coronary vessels), and vacuoles in the lysosomes of peripheral blood leukocytes.

Hunter's disease is an X-linked recessive disease associated with a deficiency of iduronate sulfatase, leading to lysosomal accumulation of dermatan and heparan sulfate. It is milder than Hurler's disease.

3. The protein portion of glycoproteins is synthesized in rough endoplasmic reticulum (RER), and the carbohydrate portion (oligosaccharide) is attached to the protein in the lumen of the RER and in the Golgi apparatus.

   a. Carbohydrate precursors for glycoproteins are sugar nucleotides (e.g., UDP-galactose, UDP-glucose).

C. Proteoglycans

1. Proteoglycans are an important component of the extracellular matrix, where they interact with collagen and elastin, fibronectin, and laminin.

2. Proteoglycans consist of a core protein to which is attached numerous long, linear chains of glycosaminoglycans (GAGs).

   a. GAGs are complexes of unbranched, acidic polysaccharide chains containing repeating disaccharide units of amino sugars (e.g., glucosamine or galactosamine) and acid sugars (e.g., iduronic acid or glucuronic acid).

      (1) GAGs other than hyaluronic acid are also sulfated.

      (2) Sulfated GAGs attach covalently to a linear core of protein.

   b. GAGs are the major components of ground substance in the interstitial tissue and of mucins that compose mucus.

3. Clinically important GAGs

   a. Hyaluronic acid is the largest GAG (MW 10^7 D).

      (1) Binds large amount of water to form viscous solutions and gels

      (2) Major component of synovial fluid (joint lubricant)

      (3) Glucuronic acid: *N*-acetylglucosamine repeating disaccharide; only nonsulfated GAG

   b. Heparin is an anticoagulant that prevents excessive fibrin formation during inflammatory or allergic reactions

      (1) Released by mast cells and basophils

      (2) Enhances antithrombin III activity, leading to inactivation of activated serine protease coagulation factors (e.g., XII, XI, X).

      (3) Repeating disaccharide: Glucosamine—iduronic acid

   c. Heparan sulfate is located in the plasma membrane.

      (1) Imparts strong negative charge to basement membrane

      (2) Repels albumin in the glomerular basement membrane

      (3) Repeating disaccharide: Glucosamine—glucuronic acid (compare similarity with heparin, above)

      (4) Only important GAG in CNS

   d. Chondroitin sulfate, the most abundant GAG, is a major component of cartilage.

      (1) Chondroitin sulfate is lost in osteoarthritis because of the digestion of the proteoglycans by metalloproteinases.

      (2) Excess chondroitin sulfate and hyaluronic acid in interstitial tissue is called myxedema (e.g., pretibial myxedema in Graves' disease)

   e. Keratan sulfate is found in cartilage.

   f. Dermatan sulfate is primarily found in skin, blood vessels, tendon, and valvular tissue in the heart.

      (1) An increase in dermatan sulfate in heart valves (i.e., myxomatous degeneration) is associated with mitral valve prolapse (MVP).

      (2) I-cell disease is characterized by an inability to phosphorylate mannose residues of Golgi lysosomal enzymes, inclusion bodies in lysosomes, psychomotor retardation, and early death.

4. GAGs are degraded in lysosomes.

5. Lysosomal enzyme synthesis, I-cell disease, and lysosomal storage diseases, such as Hurler's disease and Hunter's disease, are discussed in Box 6-3.

Proteoglycans: core proteins with GAGs covalently attached

GAGs repeating disaccharide: amino sugars and acid sugars

Amino sugars: glucosamine, galactosamine

Acidic sugars: iduronic acid, glucuronic acid

Hyaluronic acid: largest GAG, viscous solutions and gels—vitreous body of eye, synovial fluid, Wharton jelly in umbilical cord

Heparin is an anticoagulant.

Heparan sulfate: integrity of extracellular matrix, especially in the CNS

Myxedema: increased chondroitin sulfate and hyaluronic acid levels in interstitial tissue

Chondroitin sulfate: most abundant GAG, component of cartilage

Osteoarthritis: metalloproteinase digestion of chondroitin containing proteoglycans

MVP: increased dermatan sulfate levels in heart valves (i.e., myxomatous degeneration)

Lysosomal storage diseases: deficiencies of lysosomal enzymes; complex substrates accumulate in lysosomes

CHAPTER 7
LIPID METABOLISM

I. Fatty Acid and Triacylglycerol Synthesis
A. Overview
1. Fatty acid and triacylglycerol synthesis occurs in the cytoplasm (oxidation occurs in the mitochondria) but its precursor, acetyl CoA, is formed in the mitochondrial matrix.
2. Fatty acid synthesis begins in the mitochondria with the formation of citrate as a 2-carbon transporter (acetyl CoA shuttle to cytoplasm).
3. Acetyl CoA carboxylase provides malonyl CoA to be used by the multienzyme complex, fatty acid synthase.
4. Regulation of fatty acid synthesis occurs at acetyl CoA carboxylase and is controlled by insulin, glucagon, and epinephrine.
5. Many phospholipids are derived from desaturated fatty acids, most of which are synthesized by the body.

B. Fatty acid and triacylglycerol synthesis: pathway reaction steps (Fig. 7-1)
1. Step 1
   a. The citrate shuttle transports acetyl CoA generated in the mitochondrion to the cytosol (see Fig. 7-1).
   b. Acetyl CoA *cannot* move across the mitochondrial membrane and must be converted into citrate.
   c. Acetyl CoA and oxaloacetate (OAA) undergo an irreversible condensation by citrate synthase to form citrate, which is transported across the mitochondrial membrane into the cytosol.
   d. Citrate remaining in the mitochondrion is used in the citric acid cycle.
2. Step 2
   a. Citrate is converted back to acetyl CoA and OAA by citrate lyase, an insulin-enhanced enzyme, in a reaction that requires ATP.
3. Step 3
   a. Acetyl CoA is converted to malonyl CoA (see Step 5 below for disposal of OAA), an important intermediate in fatty acid synthesis, by acetyl CoA carboxylase in an irreversible rate-limiting reaction that consumes ATP and requires biotin as a cofactor.
   b. Malonyl CoA inhibits carnitine acyltransferase I (see fatty acid oxidation below), preventing movement of newly synthesized fatty acids across the inner mitochondrial membrane into the matrix, where fatty acids undergo β-oxidation (futile cycling is thereby avoided).
4. Step 4
   a. Fatty acid synthase, a large multifunctional enzyme complex, initiates and elongates the fatty acid chain in a cyclical reaction sequence.
   b. Palmitate, a 16-carbon saturated fatty acid, is the final product of fatty acid synthesis.
   c. One glucose produces 2 acetyl CoA, and each acetyl CoA contains 2 carbons; therefore, 4 glucose molecules are required to produce the 16 carbons of palmitic acid.
5. Step 5
   a. OAA from citrate cleavage is converted to malate.
6. Step 6
   a. Malate is converted to pyruvate by malic enzyme, producing 1 NADPH.
   b. NADPH is required for synthesis of palmitate and elongation of fatty acids.
   c. NADPH is produced in the cytosol by malic enzyme and the pentose phosphate pathway, which is the primary source.

Acetyl CoA: converted into citrate to cross the mitochondrial membrane

Excess dietary carbohydrate is the major carbon source for fatty acid synthesis, which occurs primarily in the liver during the fed state.

Fatty acid synthesis: acetyl CoA carboxylase is rate-limiting enzyme; occurs in cytosol in fed state

Malonyl CoA: inhibits carnitine acyltransferase I

NADPH: produced by malic enzyme and by pentose phosphate pathway

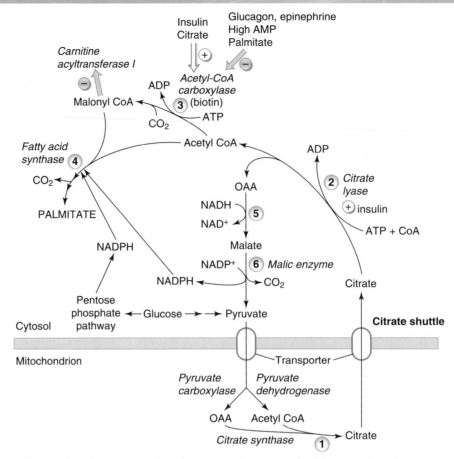

7-1: Overview of fatty acid synthesis. Fatty acid synthesis primarily occurs in the fed state and is enhanced by insulin. Palmitate, a 16-carbon saturated fat, is the end product of fatty acid synthesis. NADPH is required for synthesis of palmitate and elongation of the chain.

7. Conversion of fatty acids to triacylglycerols in liver and adipose tissue (Fig. 7-2)
    a. Step 1
       (1) In the fed state, fatty acids synthesized in the liver or released from chylomicrons and VLDL by capillary lipoprotein lipase, are used to synthesize triacylglycerol in liver and adipose tissue (see Fig. 7-2).
    b. Step 2
       (1) Glycerol 3-phosphate is derived from DHAP during glycolysis or from the conversion of glycerol into glycerol 3-phosphate by liver glycerol kinase.
       (2) Glycerol 3-phosphate is the carbohydrate intermediate that is used to synthesize triacylglycerol.
       (3) Decreasing the intake of carbohydrates is the most effective way of decreasing the serum concentration of triacylglycerol.
    c. Step 3
       (1) Newly synthesized fatty acids or those derived from hydrolysis of chylomicrons and VLDL are converted into fatty acyl CoAs by fatty acyl CoA synthetase.
    d. Step 4
       (1) Addition of 3 fatty acyl CoAs to glycerol 3-phosphate produces triacylglycerol (TG) in the liver.
    e. Step 5
       (1) Liver triacylglycerols are packaged into VLDL, which is stored in the liver and transports newly synthesized lipids through the bloodstream to peripheral tissues.
    f. Step 6
       (1) Synthesis and storage of triacylglycerol in adipose tissue require insulin-mediated uptake of glucose, leading to glycolysis and production of glycerol 3-phosphate, which is converted to triacylglycerol by the addition of 3 fatty acyl CoAs.

Only liver can capture glycerol; glycerol kinase only found in liver

Decrease triacylglycerol by decreasing carbohydrate intake.

Glycerol kinase: present only in liver, converts glycerol to glycerol 3-phosphate (precursor for triacylglycerol synthesis)

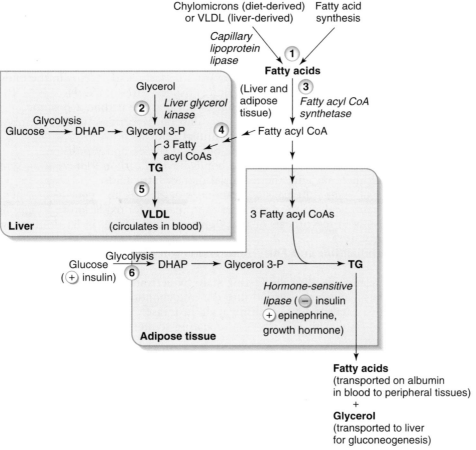

7-2: Triacylglycerol (TG) synthesis in liver and adipose tissue. Sources of fatty acids range from synthesis in the liver to hydrolysis of diet-derived chylomicrons and liver-derived very-low-density lipoprotein (VLDL) *(step 1)*. In the liver, glycerol 3-phosphate is derived from glycolysis or conversion of glycerol to glycerol 3-phosphate by liver glycerol kinase *(step 2)*. In adipose tissue, glycerol 3-phosphate is derived only from glycolysis *(step 6)*. DHAP, dihydroxyacetone phosphate.

> (2) Insulin inhibits hormone-sensitive lipase, which allows adipose cells to accumulate triacylglycerol for storage during the fed state.
> (3) Epinephrine and growth hormone activate hormone-sensitive lipase during the fasting state.

C. Fatty acid and triacylglycerol synthesis: regulated steps (see Fig. 7-1, step 3)
1. Formation of malonyl CoA from acetyl CoA, the irreversible regulated step in fatty acid synthesis, is controlled by two mechanisms.
   a. Allosteric regulation of acetyl CoA carboxylase
      (1) Stimulation by citrate ensures that fatty acid synthesis proceeds in the fed state.
      (2) End-product inhibition by palmitate downregulates synthesis when there is an excess of free fatty acids.
   b. Cycling between active and inactive forms of acetyl CoA carboxylase
      (1) High AMP level (low energy charge) inhibits fatty acid synthesis by phosphorylation of acetyl CoA carboxylase, which inactivates the enzyme.
      (2) Glucagon and epinephrine (fasting state) inhibit acetyl CoA carboxylase by phosphorylation (by protein kinase); insulin (fed state) activates the enzyme by dephosphorylation (by phosphatase).
2. Inhibition of acetyl CoA carboxylase enhances the oxidation of fatty acids, because malonyl CoA is no longer present to inhibit carnitine acyltransferase I.

D. Fatty acid and triacylglycerol synthesis: unique characteristics
1. Synthesis of longer-chain fatty acids and unsaturated fatty acids
   a. Chain-lengthening systems in the endoplasmic reticulum and mitochondria convert palmitate (16 carbons) to stearate (18 carbons) and other longer saturated fatty acids.
2. Compartmentation prevents competition between fat synthesis and fat oxidation.
   a. Synthesis in the cytosol ensures availability of NADPH from the pentose phosphate pathway.

Hormone-sensitive lipase: inhibited by insulin, prevents lipolysis

Palmitate is elongated in the endoplasmic reticulum and the mitochondrion; different elongation enzymes

   b. The product, palmitate, cannot undergo immediate oxidation without transport back into the matrix.
3. Adipose tissue does not contain glycerol kinase so the glycerol backbone of triacylglycerols must come from glycolysis.

E. Fatty acid synthesis: interface with other pathways
1. Desaturation of fatty acids to produce unsaturated fatty acids occurs in the endoplasmic reticulum in a complex process that requires oxygen either NADH or NADPH.
2. Unsaturated fatty acids are stored in triglycerides, at the carbon 2 position.
3. Unsaturated fatty acids are used in making phosphoglycerides for cell membranes.

F. Fatty acid and triacylglycerol synthesis: clinical relevance
1. Fatty acid desaturase introduces double bonds at the carbon 9 position.
   a. The desaturase *cannot* create double bonds beyond carbon 9 preventing synthesis of linoleic and linolenic acid, the essential dietary fatty acids.
   b. Deficiency of essential fatty acids produces dermatitis and poor wound healing.
2. An excess of fatty acids in the liver over the capacity for oxidation (e.g.. chronic alcoholics) results in resynthesis of triacylglycerol and storage in fat droplets, which produces a fatty liver.

II. Triacylglycerol Mobilization and Fatty Acid Oxidation (Fig. 7-3)
   A. Overview
   1. Fatty acids are mobilized in the fasting state by activating hormone-sensitive lipase.
   2. Long-chain fatty acids are shuttled into the mitochondrial matrix by formation of acyl-carnitine esters; catalyzed by carnitine acyltransferase.
   3. β-Oxidation of fatty acids consists of a repeating sequence of four enzymes to produce acetyl CoA.

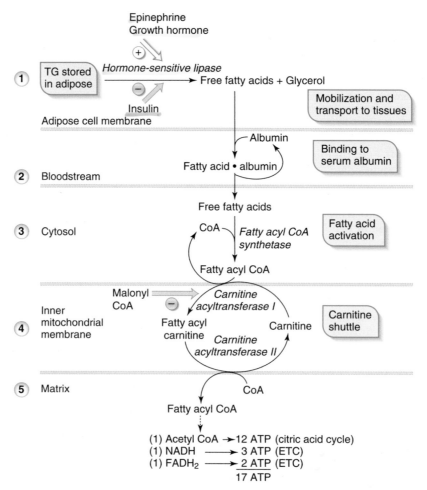

7-3: Overview of lipolysis and oxidation of long-chain fatty acids. Lipolysis occurs in the fasting state. Carnitine acyltransferase I is the rate-limiting reaction and is inhibited by malonyl CoA during the fed state. Oxidation of fatty acids yields the greatest amount of energy of all nutrients. ETC, electron transport chain; TG, triacylglycerol.

4. Fatty oxidation in the liver is unregulated; the only point of regulation of fat oxidation is hormone-sensitive lipase in the fat cell.
5. Odd-chain fatty acids undergo normal β-oxidation until propionyl CoA is produced; propionyl CoA is converted by normal β-oxidation to methylmalonyl CoA and then to succinyl CoA.
6. Unsaturated fatty acids enter the normal β-oxidation pathway at the trans-enoyl step.
7. Deficiencies in fatty acid oxidation often produce nonketotic hypoglycemia.

B. Triacylglycerol mobilization and fatty acid oxidation: pathway reaction steps
1. Step 1
   a. Mobilization of stored fatty acids from adipose tissue (lipolysis)
   b. Hormone-sensitive lipases in adipose tissue hydrolyze free fatty acids and glycerol from triacylglycerols stored in adipose tissue (see Fig. 7-3).
   c. Glycerol released during lipolysis is transported to the liver, phosphorylated into glycerol 3-phosphate by glycerol kinase, and used as a substrate for gluconeogenesis.
2. Step 2
   a. Free fatty acids released from adipose tissue are carried in the bloodstream bound to serum albumin.
3. Step 3
   a. The fatty acids are delivered to all tissues (e.g., liver, skeletal muscle, heart, kidney), except for brain and red blood cells.
   b. The fatty acids dissociate from the albumin and are transported into cells, where they are acetylated by fatty acyl CoA synthetase in the cytosol, forming fatty acyl CoAs.
4. Step 4
   a. The carnitine shuttle transports long-chain ($\geq$14-carbon) acetylated fatty acids across the inner mitochondrial membrane (see Fig. 7-3).
   b. Carnitine acyltransferase I (rate-limiting reaction) on the outer surface of the inner mitochondrial membrane removes the fatty acyl group from fatty acyl CoA and transfers it to carnitine to form fatty acyl carnitine.
   c. Carnitine acyltransferase II on the inner surface of the inner mitochondrial membrane restores fatty acyl CoA as fast as it is consumed.
   d. Medium-chain fatty acids are consumed directly by the mitochondria because they do not depend on the carnitine shuttle.
      (1) Medium-chain triglycerides are an effective dietary treatment for an infant with carnitine deficiency.
      (2) Medium-chain triglycerides spare glucose for the brain and red cells and serve as a fuel for all other tissues.
5. Step 5
   a. The oxidation system consists of four enzymes that act sequentially to yield a fatty acyl CoA that is two carbons shorter than the original and acetyl CoA, NADH, and FADH$_2$.
   b. Repetition of these four reactions eventually degrades even-numbered carbon chains entirely to acetyl CoA.
   c. Acetyl CoA enters the citric acid cycle, which is also in the matrix.

C. Triacylglycerol mobilization and fatty acid oxidation: regulated steps
1. Hormone-sensitive lipase is the only point in fat oxidation that is regulated by hormones.
   a. Epinephrine and norepinephrine (i.e., fasting, physical exercise states) activate lipolysis by converting hormone-sensitive lipase to an active phosphorylated form by their activation of protein kinase.
      (1) Perilipin coats the lipid droplets in adipose cells in the unstimulated state.
      (2) Phosphorylation of perilipin removes it from the lipid droplet so that the activated hormone-sensitive lipase can act to mobilize free fatty acids.
   b. Insulin (fed state) activates protein phosphatase, which inhibits lipolysis by converting hormone-sensitive lipase into an inactive dephosphorylated form.
   c. Glucocorticoids, growth hormone, and thyroid hormone induce the synthesis of hormone-sensitive lipase, which provides more enzyme available for activation (i.e., activation by these hormones is indirect).
2. Carnitine acyltransferase I is inhibited allosterically by malonyl CoA to prevent the unintended oxidation of newly synthesized palmitate.
   a. Malonyl CoA is the precursor used in fat synthesis, and its concentration reflects the active synthesis of palmitate.

Lipolysis occurs in the fasting state when fat is required for energy.

Hormone-sensitive lipase: activated by epinephrine and growth hormone, promotes lipolysis

β-Oxidation of fatty acids: occurs in mitochondrial matrix in fasting state

Fatty acids with 12 carbons or less enter the mitochondrion directly and are activated by mitochondrial synthetases.

Medium-chain fatty acids are consumed directly by the mitochondria; they spare glucose for the brain and red cells and serve as a fuel for all other tissues

Carnitine acyltransferase I: rate-limiting enzyme of fatty acid oxidation; shuttle for fatty acyl CoA

Acetyl CoA: end product of even-chain saturated fatty acids

Total energy yield from oxidation of long-chain fatty acids (e.g., palmitate, stearate) is more than 100 ATP per molecule.

Hormone-sensitive lipase is the only point in fat oxidation that is regulated by hormones.

TABLE 7-1. **Comparison of Fatty Acid Synthesis and Oxidation**

PROPERTY	SYNTHESIS	OXIDATION
Primary tissues	Liver	Muscle, liver
Subcellular site	Cytosol	Mitochondrial matrix
Carriers of acetyl and acyl groups	Citrate (mitochondria → cytosol)	Carnitine (cytosol → mitochondria)
Redox coenzyme	NADPH	NAD$^+$, FAD
Insulin effect	Stimulates	Inhibits
Epinephrine and growth hormone effect	Inhibits	Stimulates
Allosterically regulated enzyme	Acetyl CoA carboxylase (citrate stimulates; excess fatty acids inhibit)	Carnitine acyltransferase I (malonyl CoA inhibits)
Product of pathway	Palmitate	Acetyl CoA

Fatty acids are the major energy source (9 kcal/g) in human metabolism.

High insulin-to-glucagon ratio (fed state) leads to fatty acid synthesis; low insulin-to-glucagon ratio (fasting state) leads to fatty acid degradation.

Ketone bodies (acetone, acetoacetic acid, β-hydroxybutyric acid): fuel for muscle (fasting), brain (starvation), kidneys

The liver is the primary site for ketone body synthesis; HMG CoA synthase is the rate-limiting enzyme.

b. Malonyl CoA is absent in the fasting state when fatty acids are being actively oxidized.
3. Reciprocal regulation of fatty acid oxidation and synthesis is illustrated in Table 7-1.

D. Triacylglycerol mobilization and fatty acid oxidation: unique characteristics
1. Ketone body synthesis (Fig. 7-4) serves as an overflow pathway during excessive fatty acid supply (usually from accelerated mobilization)
2. Ketone body synthesis occurs in the mitochondrial matrix during the fasting state when excessive β-oxidation of fatty acids results in excess amounts of acetyl CoA.
   a. Ketone bodies (acetone, acetoacetate, and β-hydroxybutyrate) are used for fuel by muscle (skeletal and cardiac), the brain (starvation), and the kidneys.
   b. Ketone bodies spare blood glucose for use by the brain and red blood cells.
3. The sequence of biochemical reactions leading up to 3-hydroxy-3-methylglutaryl coenzyme A (HMG CoA) is similar to those in cholesterol synthesis; however, in ketone body synthesis, HMG CoA lyase (rather than HMG CoA reductase) is used (see Fig. 7-4).
4. Conditions associated with an excess production of ketone bodies include diabetic ketoacidosis, starvation, and pregnancy.
   a. An increase in the acetoacetate or β-hydroxybutyrate level produces an increased anion gap metabolic acidosis.
   b. The usual test for measuring ketone bodies in serum or urine (nitroprusside reaction) only detects acetoacetate and acetone, a spontaneous decomposition product of acetoacetate (see Fig. 7-4).

7-4: Ketone body synthesis. Synthesis of ketone bodies occurs primarily in the liver from leftover acetyl CoA. Ketone bodies are acetone, acetoacetate, and β-hydroxy-butyrate, and they are used as fuel by muscle (fasting), brain (starvation), and kidneys.

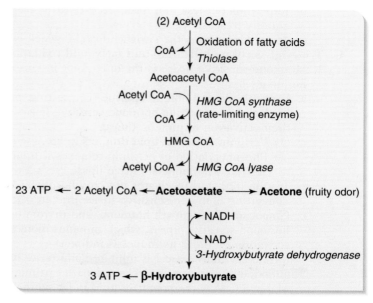

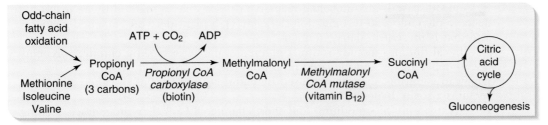

7-5: Sources of propionyl CoA (odd-chain fatty acid) and its conversion to succinyl CoA. Vitamin B$_{12}$ is a cofactor in odd-chain fatty acid metabolism, and succinyl CoA is used as a substrate for gluconeogenesis.

(1) Because of increased production of NADH in alcohol metabolism, the primary ketoacid that develops in alcoholics is β-hydroxybutyrate (NADH forces the reaction in the direction of β-hydroxybutyrate), which is not detected by standard laboratory tests.

   c. Acetone is a ketone with a fruity odor that can be detected in a patient undergoing a physical examination.

5. Degradation of ketone bodies in peripheral tissue (see Fig. 7-4) requires conversion of acetoacetate to acetyl CoA, which enters the citric acid cycle.

   a. Ketone bodies are short-chain fatty acids that do *not* require a special transport system for entry into the cell and into the mitochondria.

   b. Conversion of β-hydroxybutyrate back into acetoacetate generates NADH, which enters the electron transport chain.

   c. The liver cannot use ketones for fuel, because it lacks the enzyme succinyl CoA: acetoacetate CoA transferase, which is necessary to convert acetoacetate into acetyl CoA.

E. Triacylglycerol mobilization and fatty acid oxidation: interface with other pathways

1. Odd-numbered fatty acids undergo oxidation by the same pathway as saturated fatty acids, except that propionyl CoA (3 carbons) remains after the final cycle (Fig. 7-5).

   a. Propionyl CoA is converted first to methylmalonyl CoA and then to succinyl CoA, a citric acid cycle intermediate that enters the gluconeogenic pathway.

     (1) Vitamin B$_{12}$ is a cofactor for one of the enzymes (methylmalonyl CoA mutase) in this pathway.

     (2) A major difference between odd-chain fatty acid metabolism and even-chain fatty acid metabolism is that succinyl CoA is used as a substrate for gluconeogenesis, and acetyl CoA is not.

   b. Catabolism of methionine, isoleucine, and valine also produces propionyl CoA.

2. Unsaturated fatty acids are also degraded by entering β-oxidation at the trans-unsaturated intermediate with reduction or rearrangement of the unsaturated bond as needed.

3. Peroxisomal oxidation of very-long-chain fatty acids (20 to 26 carbons) is similar to mitochondrial oxidation but generates no ATP.

4. α-Oxidation of branched-chain fatty acids from plants occurs with release of terminal carboxyl as CO_2.

F. Triacylglycerol mobilization and fatty acid oxidation: clinical relevance

1. Carnitine deficiency or carnitine acyltransferase deficiency impairs the use of long-chain fatty acids by means of the carnitine shuttle for energy production.

   a. Clinical findings include muscle aches and fatigue following exercise, elevated free fatty acids in blood, and reduced ketone production in the liver during fasting (nonketotic hypoglycemia; acetyl CoA from β-oxidation is necessary for ketone production).

   b. Hypoglycemia occurs because all tissues are competing for glucose for energy.

2. Deficiency of medium-chain acyl CoA dehydrogenase (MCAD), the first enzyme in the oxidation sequence, is an autosomal recessive disorder.

   a. Clinical findings include recurring episodes of hypoglycemia (all tissues are competing for glucose), vomiting, lethargy, and minimal ketone production in the liver.

3. Adrenoleukodystrophy is an X-linked recessive disorder associated with defective peroxisomal oxidation of very-long-chain fatty acids.

   a. Clinical findings include adrenocortical insufficiency and diffuse abnormalities in the cerebral white matter, leading to neurologic disturbances such as progressive mental deterioration and spastic paralysis.

Side notes:

Ketone body, acetoacetate, and acetone measured with nitroprusside reaction; not β-hydroxybutyrate

Liver synthesizes ketone bodies but *cannot* use them for fuel; unidirectional flow from liver to peripheral tissues

Vitamin B$_{12}$: cofactor for mutase in odd-chain fatty acid metabolism

Odd-chain fatty acids: oxidized to propionyl CoA, then to methylmalonyl CoA, before formation of succinyl CoA

Carnitine deficiency: inability to metabolize long-chain free fatty acids; all tissues compete for glucose (hypoglycemia)

Medium-chain acyl CoA dehydrogenase deficiency: inability to fully metabolize long-chain fatty acids

Defective fatty acid catabolism: carnitine and MCAD deficiencies, adrenoleukodystrophy, Refsum's disease

Adrenoleukodystrophy: defective peroxisomal oxidation of fatty acids

4. Refsum's disease is an autosomal recessive disease that is marked by an inability to degrade phytanic acid (α-oxidation deficiency), a plant-derived branched-chain fatty acid that is present in dairy products.
   a. Clinical findings include retinitis pigmentosa; dry, scaly skin; chronic polyneuritis; cerebellar ataxia; and elevated protein in the cerebrospinal fluid.
5. Jamaican vomiting sickness is caused by eating unripe fruit of the akee tree that contains a toxin, hypoglycin.
   a. This toxin inhibits medium- and short-chain acyl CoA dehydrogenases, leading to nonketotic hypoglycemia.
6. Zellweger syndrome results from the absence of peroxisomes in the liver and kidneys.
   a. This results in the accumulation of very-long-chain fatty acids, especially in the brain.

IV. Cholesterol and Steroid Metabolism

A. Overview

1. Cholesterol, the most abundant steroid in human tissue, is important in cell membranes and is the precursor for bile acids and all the steroid hormones, including vitamin D, which is synthesized in the skin from 7-dehydrocholesterol.
2. Cholesterol synthesis occurs in the liver, and its rate of synthesis is determined by the activity of the rate-limiting enzyme HMG CoA reductase.
3. The bile acids are a major product of cholesterol synthesis and are converted into secondary forms by intestinal bacteria.
4. The steroid hormones are synthesized from cholesterol after it is converted to pregnenolone.
5. Deficiencies in the enzymes that convert progesterone to other steroid hormones produce the adrenogenital syndrome (congenital adrenal hyperplasia) due to disruption of normal hypothalamic-pituitary feedback.

B. Cholesterol synthesis and regulation (Fig. 7-6)

1. Step 1
   a. HMG CoA is formed by condensation of three molecules of acetyl CoA.
   b. In the liver, HMG CoA is also produced in the mitochondria matrix, where it serves as an intermediate in the synthesis of ketone bodies.

Although almost all tissues synthesize cholesterol, the liver, intestinal mucosa, adrenal cortex, testes, and ovaries are the major contributors to the body's cholesterol pool.

Cholesterol functions: cell membrane, bile acid synthesis, steroid hormone synthesis

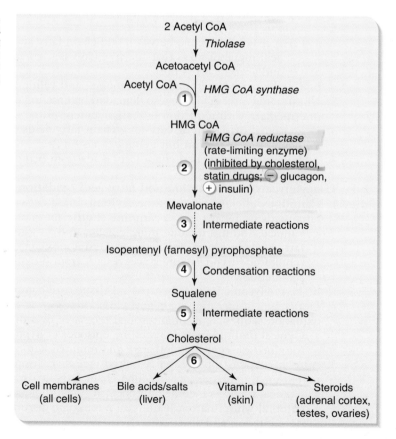

7-6: Overview of cholesterol synthesis. Hydroxy-3-methylglutaryl coenzyme A (HMG CoA) reductase is the rate-limiting enzyme, and it is inhibited by statin drugs and by cholesterol. Glucagon favors the inactive form of the enzyme; insulin favors the active form.

2. Step 2
   a. HMG CoA reductase conversion of HMG CoA to mevalonate is the rate-limiting step in cholesterol synthesis.
   b. Cholesterol is an allosteric inhibitor of HMG CoA reductase, and it also inhibits expression of the gene for HMG CoA reductase.
   c. Statin drugs, such as atorvastatin, simvastatin, and pravastatin, act as competitive inhibitors with mevalonate for binding to HMG CoA reductase.
   d. Hormones control cycling between the inactive and active forms of HMG CoA reductase by phosphorylation and dephosphorylation, respectively.
      (1) Glucagon favors the inactive form and leads to decreased cholesterol synthesis.
      (2) Insulin favors the active form and leads to increased cholesterol synthesis.
   e. Sterol-mediated decrease in expression of HMG CoA reductase provides long-term regulation.
      (1) Delivery of cholesterol to liver and other tissues by plasma lipoproteins, such as low-density lipoproteins (LDLs) and high-density lipoproteins (HDLs), leads to a reduction in de novo cholesterol synthesis and a decrease in the synthesis of LDL receptors.
3. Step 3
   a. Isopentenyl (farnesyl) pyrophosphate (IPP) is formed in several reactions from mevalonate and is the key five-carbon isoprenoid intermediate in cholesterol synthesis.
   b. Isopentenyl pyrophosphate (containing isoprene) is also a precursor in the synthesis of other cellular molecules:
      (1) The side chain of coenzyme Q (ubiquinone)
      (2) Dolichol, which functions in the synthesis of *N*-linked oligosaccharides in glycoproteins
      (3) The side chain of heme a
      (4) Geranylgeranyl and farnesyl groups that serve as highly hydrophobic membrane anchors for some membrane proteins
4. Step 4
   a. Squalene, a 30-carbon molecule, is formed by several condensation reactions involving isopentenyl pyrophosphate.
5. Step 5
   a. Conversion of squalene to cholesterol requires several reactions and requires NADPH.
6. Step 6
   a. Cholesterol is excreted in bile or used to synthesize bile acids and salts.
   b. The low solubility of cholesterol creates a tendency to form gallstones. Conditions in bile favoring gallstones are:
      (1) Excess cholesterol in bile
      (2) Low content of bile salts
      (3) Low content of lecithin (an emulsifying phospholipid)
7. Treatment of hypercholesterolemia
   a. Reduce cholesterol intake
      (1) A 50% reduction in intake only lowers serum cholesterol by about 5%.
   b. Decrease cholesterol synthesis by inhibiting HMG CoA reductase with statin drugs (most effective).
   c. Increase cholesterol excretion with bile acid–binding drugs (e.g., cholestyramine): leads to bile salt and acid deficiency and subsequent upregulation of LDL receptor synthesis in hepatocytes for synthesis of bile salts and acids by using cholesterol

C. Bile salts and bile acids

1. Bile salts are primarily used to emulsify fatty acids and monoacylglycerol and package them into micelles, along with fat-soluble vitamins, phospholipids, and cholesteryl esters, for reabsorption by villi in the small bowel (see Chapter 4).
2. Primary bile acids (e.g., cholic acid and chenodeoxycholic acid) are synthesized in the liver from cholesterol (Fig. 7-7).
   a. Primary bile acids are conjugated before secretion in the bile with taurine (taurochenodeoxycholic acid) or glycine (glycocholic acid).
   b. Bile acid synthesis is feedback inhibited by bile acids and stimulated by cholesterol at the gene transcription level; amount of 7α-hydroxylase (the committed step) is increased or decreased.

HMG CoA reductase: rate-limiting enzyme in cholesterol synthesis; blocked by statin drugs

Insulin stimulates cholesterol synthesis.

Statin drugs decrease synthesis of coenzyme Q, which may be responsible for muscle-related problems that occur when taking the drug.

Isoprene, an intermediate in cholesterol synthesis, also serves other functions in coenzyme Q and membrane anchoring of proteins.

Gallstones form from excess concentration of cholesterol and reduced concentration of bile acids and phospholipids in bile.

Treating hypercholesterolemia: ↓ cholesterol intake; ↓ cholesterol synthesis; ↑ cholesterol excretion

About 70% to 80% of cholesterol is converted to bile acids.

Primary bile salts from liver, secondary bile salts from intestinal bacteria

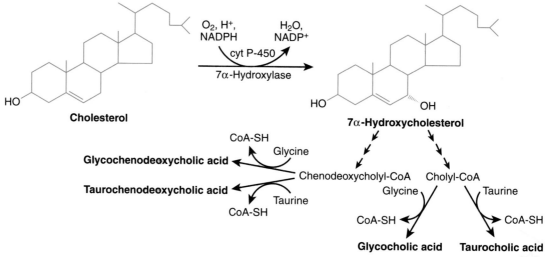

7-7: Synthesis of the primary bile acids by endoplasmic reticulum–associated enzymes in hepatocytes. cyt, cytochrome. *(From Meisenberg G, Simmons W: Principles of Medical Biochemistry, 2nd ed. Philadelphia, Mosby, 2006).*

3. Intestinal bacteria alter bile acids in the small intestine to produce secondary bile acids.
   a. Bile acids are converted into deoxycholic and lithocholic acid (glycine and taurine are removed).
   b. The enterohepatic circulation in the terminal ileum recycles about 95% of bile acids back to the liver.
   c. Secretion of reabsorbed bile acids is preceded by conjugation with taurine and glycine.
4. Bile salt deficiency leads to malabsorption of fat and fat-soluble vitamins (see Box 4-2 in Chapter 4).

D. **Steroid hormones in the adrenal cortex (Fig. 7-8)**
   1. Synthesis of steroid hormones begins with cleavage of the cholesterol side chain to yield pregnenolone, the C_{21} precursor of all the steroid hormones.
      a. ACTH stimulates conversion of cholesterol to pregnenolone in the adrenal cortex.
      b. Cytochrome P450 hydroxylases (mixed-function oxidases) catalyze the addition of hydroxyl groups in reactions that use O_2 and NADPH.
         (1) Other cytochrome P450 hydroxylases also function in detoxification of many drugs in the liver.
   2. Steroid hormones in the adrenal cortex contain 21 (C_{21}), 19 (C_{19}), or 18 (C_{18}) carbon atoms (see Fig. 7-8).
      a. Progesterone (C_{21}) is synthesized from pregnenolone.
         (1) Progesterone stimulates breast development, helps maintain pregnancy, and helps to regulate the menstrual cycle.
      b. Glucocorticoids (C_{21}) are synthesized in the zona fasciculata.
         (1) Cortisol promotes glycogenolysis and gluconeogenesis in the fasting state and has a negative feedback relationship with ACTH.
      c. Mineralocorticoids (C_{21}) are synthesized in the zona glomerulosa.
         (1) Aldosterone acts on the distal and the collecting tubules of the kidneys to promote sodium reabsorption and potassium and proton excretion.
         (2) Angiotensin II stimulates conversion of corticosterone into aldosterone.
         (3) 11-Deoxycorticosterone and corticosterone are weak mineralocorticoids.
      d. Androgens (C_{19}) are synthesized in the zona reticularis.
         (1) The 17-ketosteroids, dehydroepiandrosterone (DHEA) and androstenedione, are weak androgens.
         (2) Testosterone is responsible for the development of secondary sex characteristics in males.
         (3) Testosterone is converted to dihydrotestosterone by 5α-reductase and to estradiol by aromatase in peripheral tissues (e.g., prostate).
      e. Estrogens (C_{18}) are synthesized in the zona reticularis.
         (1) Estradiol is responsible for development of female secondary sex characteristics and the proliferative phase of the menstrual cycle.
         (2) Derived from conversion of testosterone to estradiol by aromatase in the granulosa cells of the developing follicle

All steroids derived from pregnenolone; pregnenolone derived from cholesterol

Zona fasciculata: synthesis of glucocorticoids (e.g., cortisol)

Zona glomerulosa: synthesis of mineralocorticoids (e.g., aldosterone)

Angiotensin II: stimulates conversion of corticosterone to aldosterone

Zona reticularis: synthesis of sex hormones (e.g., androstenedione, testosterone, estrogen)

Estradiol: conversion of testosterone to estradiol by aromatase in granulosa cells of the developing follicle

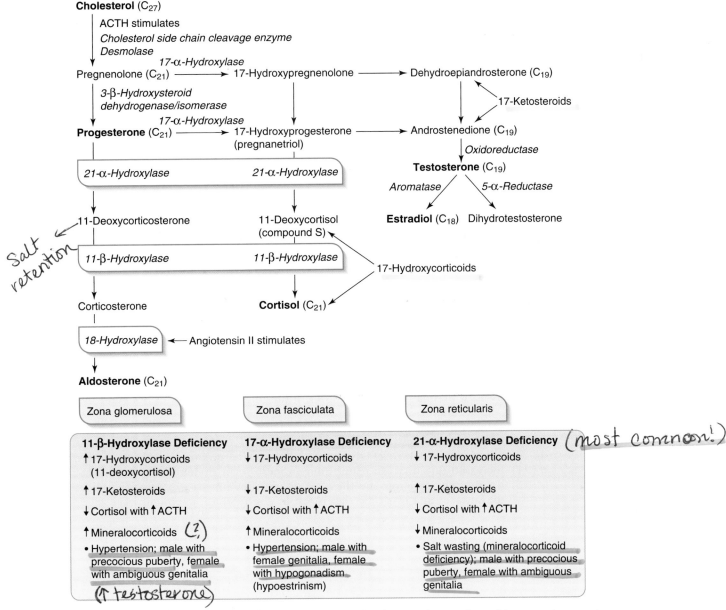

7-8: Overview of steroid hormone synthesis from cholesterol in the adrenal cortex. The outer layer of the cortex, the zona glomerulosa, synthesizes mineralocorticoids (e.g., aldosterone); the middle zona fasciculata synthesizes glucocorticoids (e.g., cortisol); and the inner zona reticularis synthesizes sex hormones (e.g., androstenedione, testosterone).

   f. The ovaries and testes contain only the 17α-hydroxylase enzyme, which favors conversion of progesterone to 17-ketosteroids, testosterone, 17-hydroxyprogesterone, and estrogen (by means of aromatization).
 E. **Adrenogenital syndrome (i.e., congenital adrenal hyperplasia)**
   1. The adrenogenital syndrome is a group of autosomal recessive disorders associated with deficiencies of enzymes involved in the synthesis of adrenal steroid hormones from cholesterol (see Fig. 7-8).
   2. Decreased cortisol production in all types of adrenogenital syndromes causes a compensatory increase in secretion of ACTH and subsequent bilateral adrenal hyperplasia.
      a. Enzyme deficiencies result in an increase in compounds proximal to the enzyme block; compounds distal to the block are decreased.
   3. 21α-Hydroxylase deficiency, the most common type of adrenogenital syndrome, exhibits variable clinical features depending on the extent of the enzyme deficiency.
      a. Less severe cases are marked only by masculinization due to increased androgen production (i.e., 17-ketosteroids and testosterone)

Adrenogenital syndrome: build up of steroid intermediates before block; deficiency of intermediates after block

b. More severe cases are also associated with deficiency of the mineralocorticoids, leading to sodium wasting and, if untreated, life-threatening volume depletion and shock.

   (1) The 17-hydroxysteroids are also decreased.

4. 11β-Hydroxylase deficiency is marked by salt retention (increase in 11-deoxycorticosterone), leading to hypertension, masculinization (increase in 17-ketosteroids and testosterone), and an increase in 11-deoxycortisol, a 17-hydroxycorticoid.

5. 17α-Hydroxylase deficiency is marked by increased production of the mineralocorticoids (hypertension) and decreased production of 17-ketosteroids and 17-hydroxycorticoids.

V. Plasma Lipoproteins

A. Overview

1. Plasma lipoproteins transport the low-solubility lipids, cholesterol, and triglycerides to and from the tissues.

2. Plasma lipoproteins are composed of apoproteins, phospholipids, and cholesterol.

3. Chylomicrons transport triacylglycerol from the diet while VLDL transport triacylglycerol synthesized in the liver.

4. Low-density lipoprotein delivers cholesterol to the tissues for use in membrane synthesis and repair.

5. High-density lipoprotein delivers cholesterol released during membrane repair to the liver (i.e., reverse cholesterol transport).

6. Hyperlipoproteinemias are produced from deficiencies in lipid transport components.

B. Structure and composition of lipoproteins

1. Spherical lipoprotein particles have a hydrophobic core of triacylglycerols and cholesteryl esters surrounded by a phospholipid layer associated with cholesterol and protein.

2. Four classes of plasma lipoproteins differ in the relative amounts of lipid and the protein they contain (Table 7-2).

   a. As the lipid-to-protein ratio decreases, particles become smaller and more dense in the following order: chylomicron > VLDL > LDL > HDL.

   b. A marked increase in triacylglycerol (>1000 mg/dL) produces turbidity in plasma.

      (1) Increased turbidity can result from an increase in chylomicrons or VLDL.

      (2) Because chylomicrons are the least dense, they form a supranate (i.e., float on the surface) in a test tube that is left in a refrigerator overnight.

21α-Hydroxylase deficiency: salt wasting; most common cause of adrenogenital syndrome

11β-Hydroxylase deficiency: salt retention, leads to hypertension

17-Hydroxylase deficiency: salt retention, leads to hypertension, decreased 17-hydroxycorticoids and ketosteroids

Increased level of chylomicrons produces a turbid supranate.

Increased levels of VLDL produce a turbid infranate.

TABLE 7-2. Plasma Lipoproteins

TYPE	COMPONENTS	FUNCTION AND METABOLISM
Chylomicron	Triacylglycerols: highest Cholesterol: lowest Protein: lowest Apolipoproteins: B-48, C-II, E	Transports dietary triacylglycerol to peripheral tissues (e.g., muscle, adipose tissue) and dietary cholesterol to liver
		Formed and secreted by intestinal mucosa; triacylglycerol-depleted remnants endocytosed by liver
VLDL	Triacylglycerols: moderate Cholesterol: moderate Protein: low Apolipoproteins: B-100, C-II, E	Transports liver-derived triacylglycerol to extrahepatic tissues (e.g., adipose tissue, muscle) Formed and secreted by liver; converted to LDL by hydrolysis of fatty acids by capillary lipoprotein lipase
LDL	Triacylglycerols: low	Delivers cholesterol from liver to extrahepatic tissues
	Cholesterol: highest Protein: moderate Apolipoprotein: B-100	Derived from VLDL; endocytosed by target cells with LDL receptors and degraded, releasing cholesterol, which decreases further uptake of cholesterol
HDL	Triacylglycerols: low Cholesterol: moderate Protein: high Apolipoproteins: A-I, C-II, E	Takes up cholesterol from cell membranes in periphery and returns it to liver (i.e., reverse cholesterol transport)
		Secreted by liver and intestine; activates LCAT to form cholesteryl esters; transfers apoC-II and apoE to nascent chylomicrons and VLDL
		"Good cholesterol"; the higher the concentration, the lower the risk for coronary artery disease

HDL, high-density lipoprotein; LCAT, lecithin cholesterol acyltransferase; LDL, low-density lipoprotein; VLDL, very-low-density lipoprotein.

    (3) An increase in VLDL produces an infranate in a test tube that is left in a refrigerator overnight because it is denser.

  3. Functions of apolipoproteins

    a. Apolipoprotein A-I (apoA-I):

      (1) Activates lecithin cholesterol acyltransferase (LCAT), which esterifies tissue cholesterol picked up by HDL

      (2) Major structural protein for HDL1

    b. Apolipoprotein C-II (apoC-II):

      (1) Activates capillary lipoprotein lipase, which releases fatty acids and glycerol from chylomicrons, VLDL, and IDL

    c. Apolipoprotein B-48 (apoB-48) is a component of chylomicrons.

    d. Apolipoprotein B-100 (apoB-100):

      (1) Contains the B-48 domain plus the LDL receptor recognition domain permitting binding to LDL receptors

      (2) Only structural protein in LDL

    e. Apolipoprotein E (apoE):

      (1) Mediates uptake of chylomicron remnants and intermediate-density lipoproteins (IDLs) by the liver

C. Functions and metabolism of lipoproteins

  1. Chylomicrons transport dietary lipids (e.g., long-chain fatty acids, fat-soluble vitamins) from the intestine to the peripheral tissues (Fig. 7-9).

    a. Step 1

      (1) Nascent chylomicrons formed in the intestinal mucosa are secreted into the lymph and eventually enter the subclavian vein through the thoracic duct.

      (2) Nascent chylomicrons are rich in dietary triacylglycerols (85%) and contain apoB-48, which is necessary for assembly and secretion of the chylomicron.

      (3) They contain only a minimal amount (<3%) of dietary cholesterol.

    b. Step 2

      (1) Addition of apoC-II and apoE from HDL leads to formation of mature chylomicrons.

    c. Step 3

      (1) Capillary lipoprotein lipase is activated by apoC-II and hydrolyzes triacylglycerols in chylomicrons, releasing glycerol and free fatty acids into the blood.

      (2) Glycerol is phosphorylated in the liver by glycerol kinase into glycerol 3-phosphate, which is used to synthesize more VLDL.

Margin notes:

Chylomicrons least dense; HDL most dense

ApoA-1: activates LCAT; structural protein for HDL

ApoC-II: activates capillary lipoprotein lipase

ApoB-48: component of chylomicrons

ApoB-100: structural protein of LDL

ApoE: mediates uptake of chylomicrons remnants and IDL remnants

Chylomicrons: contain diet-derived triacylglycerols

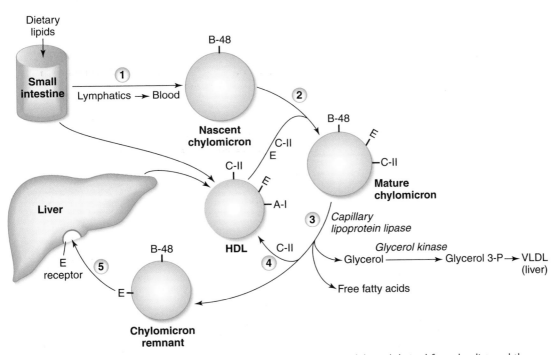

7-9: Transport of dietary lipids by chylomicrons. Chylomicrons represent triacylglycerol derived from the diet, and they are a source of fatty acids and glycerol for the synthesis of triacylglycerol in the liver.

(3) Free fatty acids enter the adipose tissue to produce triacylglycerols for storage.

(4) In muscle, the fatty acids are oxidized to provide energy.

d. Step 4

(1) ApoC-II returns to HDL.

e. Step 5

(1) Chylomicron remnants that remain after the removal of free fatty acids attach to apoE receptors in the liver and are endocytosed.

(2) Dietary cholesterol delivered to the liver by chylomicron remnants is used for bile acid synthesis and also depresses de novo cholesterol synthesis.

(3) Excess cholesterol is excreted in bile.

2. VLDL lipoproteins carry triacylglycerols synthesized in the liver to peripheral tissues (Fig. 7-10).

a. Step 1

(1) In addition to triacylglycerols, nascent VLDL particles formed in the liver also contain some cholesterol ($\approx 17\%$) and apoB-100; these VLDL particles obtain apoC-II and apoE from HDL.

b. Steps 2 and 3

(1) Conversion of circulating nascent VLDL particles into LDL particles proceeds by intermediate-density lipoprotein (IDL) particles.

(2) Degradation of triacylglycerols by apoC-II–activated capillary lipoprotein lipase converts nascent VLDL particles into IDL remnants, which are then converted into LDL particles.

(3) Fatty acids and glycerol are released into the bloodstream.

c. Steps 4 and 5

(1) LDL particles remaining after metabolism of VLDL and IDL are enriched in cholesterol (45%), which they deliver to peripheral tissues or to the liver (see Fig. 7-10).

(2) ApoB-100, the only apolipoprotein on LDL, binds to LDL receptors on the cell membrane of target cells in the liver and other tissues.

(3) After receptor-mediated endocytosis, LDL is degraded in lysosomes, releasing free cholesterol for use in membrane synthesis, bile salt synthesis (liver), or steroid hormone synthesis (endocrine tissues, ovaries, and testes).

(4) Excess cholesterol *not* needed by cells is esterified by acyl CoA:cholesterol acyltransferase (ACAT) and stored as cholesteryl esters.

VLDL: contain liver-derived triacylglycerols and cholesterol

Capillary lipoprotein lipase: hydrolyzes triacylglycerols and VLDL into fatty acids and glycerol

LDL: major carrier of cholesterol

7-10: Metabolism of very-low-density lipoprotein (VLDL), low-density lipoprotein (LDL), and high-density lipoprotein (HDL). VLDL is degraded by hydrolysis into LDL. HDL is a reservoir for apolipoproteins and transports cholesterol from tissue to the liver. CPL, capillary lipoprotein lipase; LCAT, lecithin cholesterol acyltransferase.

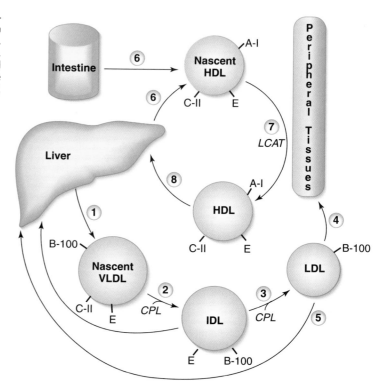

d. Free cholesterol in the cytosol has the following regulatory functions:
  (1) Activates ACAT
  (2) Suppresses HMG CoA reductase; decreases de novo synthesis of cholesterol
  (3) Suppresses further LDL receptor synthesis; decreases further uptake of LDL
e. Step 6
  (1) HDL, the "good cholesterol," is synthesized in the liver and small intestine and carries out reverse transport of cholesterol from extrahepatic tissues to the liver (see Fig. 7-10).
  (2) HDL also acts as a repository of apolipoproteins (e.g., apoC-II and apoE), which can be donated back to VLDL and chylomicrons.
f. Step 7: LCAT (lecithin-cholesterol acyltransferase) mediates esterification of free cholesterol removed from peripheral tissues by HDL.
  (1) HDL is converted from a discoid shape to a spherical shape when esterified cholesterol is transferred into the center of the molecule.
g. HDL transfers cholesteryl esters to VLDL in exchange for triacylglycerols, and VLDL transfers triacylglycerol to HDL.
  (1) The transfer is mediated by cholesteryl ester transfer protein (CETP).
  (2) This transfer explains why an increase in VLDL leads to a decrease in HDL cholesterol levels.
h. Step 8
  (1) Cholesteryl esters are returned to the liver by receptor-mediated endocytosis of HDL.
  (2) HDL is increased by estrogen (women therefore have higher HDL levels), exercise, weight loss, smoking cessation, *trans* fat elimination, monounsaturated fats, and soluble dietary fiber.

D. Hereditary disorders related to defective lipoprotein metabolism
  1. Abetalipoproteinemia is a rare autosomal recessive lipid disorder characterized by a lack of apoB lipoproteins.
    a. Chylomicrons, VLDL, and LDL are absent and levels of triacylglycerol and cholesterol are extremely low.
    b. Clinical findings include an accumulation of triacylglycerols in intestinal mucosal cells leading to malabsorption of fat and fat-soluble vitamins.
    c. Spinocerebellar ataxia, retinitis pigmentosa, and hemolytic anemia respond to megadoses of vitamin E.
  2. Genetic and acquired hyperlipoproteinemias (Table 7-3)

VI. Sphingolipid Degradation
  A. Overview
    1. Sphingolipids are essential components of membranes throughout the body and are particularly abundant in nervous tissue.
    2. Sphingolipids are named for the sphingosine backbone that is the counterpart of the glycerol backbone in phospholipids.
    3. Sphingolipidoses are hereditary lysosomal enzyme deficiency diseases involving lysosomal hydrolases; accumulation of sphingolipid substrate occurs.
  B. Ceramide
    1. Ceramide, a derivative of sphingosine (sphingosine + fatty acids = ceramide), is the immediate precursor of all the sphingolipids.
    2. Sphingomyelin contains phosphatidylcholine linked to ceramide.
    3. Cerebrosides, globosides, gangliosides, and sulfatides, the other classes of sphingolipids, all contain different types and numbers of sugars or sugar derivatives linked to ceramide.
  C. Sphingolipid degradation
    1. Lysosomal enzymes degrade sphingolipids to sphingosine by a series of irreversible hydrolytic reactions (Fig. 7-11).
  D. Sphingolipidoses
    1. Sphingolipidoses are a group of hereditary lysosomal enzyme deficiency diseases caused by a deficiency of one of the hydrolases in the degradative pathway (Table 7-4 and see Box 6-2 in Chapter 6).
    2. A block in the degradation of sphingolipids leads to accumulation of the substrate for the defective enzyme within lysosomes.
    3. Neurologic deterioration occurs in most of these diseases, leading to early death.
    4. Autosomal recessive inheritance is shown by most of the sphingolipidoses: Gaucher's disease, Krabbe's disease, metachromatic leukodystrophy, Niemann-Pick disease, and Tay-Sachs disease.
    5. Fabry's disease is the only X-linked recessive sphingolipidosis.

Margin notes:

HDL: reverse cholesterol transport, reservoir for apolipoproteins

CETP: transfers cholesterol from HDL to VLDL and triacylglycerols from VLDL to HDL

Increased level of VLDL always causes a decrease in HDL cholesterol.

Abetalipoproteinemia: rare hereditary lipid disorder; lack of apoB

Sphingolipidoses: lysosomal enzyme deficiencies caused by deficiency of a hydrolase in degradative pathway

Sphingolipidoses: Gaucher's disease, Krabbe's disease, metachromatic leukodystrophy, Niemann-Pick disease, and Tay-Sachs disease

TABLE 7-3. **Acquired and Genetic Hyperlipoproteinemias**

LIPID DISORDER AND PATHOGENESIS	CLINICAL ASSOCIATIONS	LABORATORY FINDINGS
Type I		
Familial lipoprotein lipase deficiency; ApoC-II deficiency Pathogenesis: inability to hydrolyze chylomicrons	Rare childhood disease	Increased chylomicron and triacylglycerol, normal cholesterol and LDL Standing chylomicron test: supranate but no infranate
Type II		
Familial hypercholesterolemia Pathogenesis: absent or defective LDL receptors	Autosomal dominant disorder with premature coronary artery disease Achilles tendon xanthomas are pathognomonic Acquired causes: diabetes, hypothyroidism, obstructive jaundice, nephrotic syndrome	Type IIa: increased LDL (often > 260 mg/dL) and cholesterol, normal triacylglycerol Type IIb: increased LDL, cholesterol, and triacylglycerol
Type III		
Familial dysbetalipoproteinemia "remnant disease" Pathogenesis: deficiency of apoE; chylomicron and IDL remnants are not metabolized in liver	Autosomal dominant Increased risk for coronary artery disease Hyperuricemia, obesity, diabetes	Cholesterol and triacylglycerol equally increased Increased chylomicron and IDL remnants
Type IV		
Familial hypertriglyceridemia	Autosomal dominant disorder	Increased triacylglycerol, slightly increased cholesterol
Pathogenesis: decreased catabolism or increased synthesis of VLDL	Most common hyperlipoproteinemia Increased triacylglycerol begins at puberty Increased incidence of coronary artery disease and peripheral vascular disease Acquired causes: alcoholism, diuretics, β-blockers, renal failure, oral contraceptive pills (estrogen effect)	Standing chylomicron test: turbid infranate Decreased HDL (inverse relationship with VLDL)
Type V		
Most commonly a familial hypertriglyceridemia with exacerbating factors Pathogenesis: combination of type I and type IV mechanisms	Particularly common in alcoholics and individuals with diabetic ketoacidosis Hyperchylomicronemia syndrome: abdominal pain, pancreatitis, dyspnea (impaired oxygen exchange), hepatosplenomegaly (fatty change), papules on skin	Much increased triacylglycerol, normal LDL Standing chylomicron test: supranate and infranate

HDL, high-density lipoprotein; IDL, intermediate-density lipoprotein; LDL, low-density lipoprotein; VLDL, very-low-density lipoprotein.

7-11: Overview of sphingolipid degradation.

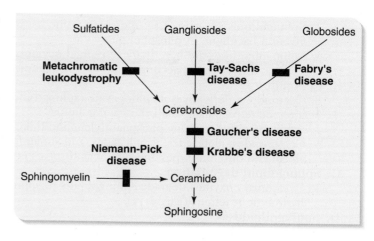

TABLE 7-4. **Sphingolipidoses: Lysosomal Storage Diseases**

DISEASE	ACCUMULATED MATERIAL (DEFICIENT ENZYME)	CLINICAL ASSOCIATIONS
Fabry's disease (X-linked recessive)	Ceramide trihexosides (α-galactosidase)	Paresthesia in extremities; reddish purple rash; cataracts; death due to kidney or heart failure
Gaucher's disease, adult (AR)	Glucocerebrosides (β-glucosidase)	Hepatosplenomegaly; macrophage accumulation in liver, spleen, bone marrow; crinkled paper–appearing macrophages; compatible with life
Krabbe's disease (AR)	Galactocerebrosides (β-galactosidase)	Progressive psychomotor retardation; abnormal myelin; large globoid bodies in brain white matter; fatal early in life
Metachromatic leukodystrophy (AR)	Sulfatides (arylsulfatase A)	Mental retardation; developmental delay; abnormal myelin; peripheral neuropathy; urine arylsulfatase decreased; death within first decade
Niemann-Pick disease (AR)	Sphingomyelin (sphingomyelinase)	Hepatosplenomegaly; mental retardation; "bubbly" appearance of macrophages; fatal early in life (foamy histiocytes in liver)
Tay-Sachs disease (AR)	GM₂ gangliosides (hexosaminidase A)	Muscle weakness and flaccidity; blindness, cherry-red macular spot; no hepatosplenomegaly; occurs primarily in eastern European Ashkenazi Jews; fatal at an early age

AR, autosomal recessive.

CHAPTER 8

NITROGEN METABOLISM

I. Biosynthesis of Nonessential Amino Acids

A. Overview

1. Eleven of the 20 amino acids are synthesized in the body (nonessential amino acids), and the remaining nine amino acids are required in the diet (essential amino acids).
2. Many of the nonessential amino acids are synthesized by transamination reactions, in which an amino group is added to an α-ketoacid to produce an amino acid.
   a. Ten of the nonessential amino acids are derived from glucose through intermediates derived from glycolysis and the citric acid cycle.
   b. For example, addition of an amino group from glutamate to the α-ketoacids pyruvate, oxaloacetate, and α-ketoglutarate produces alanine, aspartate, and glutamate, respectively.
3. Tyrosine is an exception in that it is derived from phenylalanine, which is an essential amino acid.
4. Cysteine receives its carbon skeleton from serine (product of 3-phosphoglycerate in glycolysis); however, its sulfur comes from the essential amino acid methionine.

B. Sources of the nonessential amino acids (Table 8-1)

II. Removal and Disposal of Amino Acid Nitrogen

A. Overview

1. Removal of the α-amino group from amino acids is the initial step in the catabolism of amino acids.
2. Nitrogen from the amino group is excreted as urea or incorporated into other compounds.
3. Nitrogen is removed by transamination or oxidative deamination.
4. Urea is formed in the liver in the urea cycle.
5. Ammonia that is not converted to urea is carried to the kidneys for secretion into the urine (i.e., acidifies urine).
6. Hyperammonemia is associated with encephalopathy producing feeding difficulties, vomiting, ataxia, lethargy, irritability, poor intellectual development, and coma.

B. Transamination and oxidative deamination (Fig. 8-1)

1. Step 1
   a. Transamination entails the transfer of the α-amino group of an α-amino acid to α-ketoglutarate, producing an α-keto acid from the amino acid and glutamate from α-ketoglutarate (see Fig. 8-1, left).
   b. Aminotransferases (transaminases) catalyze reversible transamination reactions that occur in the synthesis and the degradation of amino acids.
   c. The two most common aminotransferases transfer nitrogen from aspartate and alanine to α-ketoglutarate, providing α-ketoacids that are used as substrates for gluconeogenesis.
      (1) Aspartate aminotransferase (AST) reversibly transaminates aspartate to oxaloacetate.
      (2) Alanine aminotransferase (ALT) reversibly transaminates alanine to pyruvate.
   d. Pyridoxal phosphate (PLP), derived from vitamin B_6 (pyridoxine), is a required cofactor for all aminotransferases (see Chapter 4).
2. Step 2
   a. Oxidative deamination of glutamate (the product of transamination) is the major mechanism for the release of amino acid nitrogen as charged ammonia (NH_4^+), and it occurs primarily in the liver and kidneys (see Fig. 8-1, *right*).

Nonessential amino acids: most synthesized from intermediates of glycolysis and the citric acid cycle

Synthesized from essential amino acids: tyrosine (from phenylalanine) and cysteine (from methionine)

Transamination: reversible conversion of amino acids to their corresponding ketoacids

AST: reversible conversion of aspartate to oxaloacetate

ALT: reversible conversion of alanine to pyruvate

Plasma AST and ALT levels elevated in inflammatory liver diseases, such as viral hepatitis (ALT > AST) and alcoholic hepatitis (AST > ALT)

Oxidative deamination: primarily in liver and kidney; releases NH_4^+ from glutamate for conversion to urea

TABLE 8-1. **Synthesis of Nonessential Amino Acids**

AMINO ACID	SOURCE OF CARBON SKELETON	COMMENTS
Alanine	Pyruvate	Transamination of precursor
Arginine	Ornithine	Reversal of arginase reaction in urea cycle
Asparagine	Oxaloacetate	Amide group from glutamine
Aspartate	Oxaloacetate	Transamination of precursor
Cysteine*	Serine	Sulfur group from methionine
Glutamate	α-Ketoglutarate	Transamination of precursor
Glutamine	α-Ketoglutarate	Amide group from free NH_4^+
Glycine	3-Phosphoglycerate	From serine through transfer of methylene group to tetrahydrofolate (THF)
Proline	Glutamate	Cyclization of glutamate semialdehyde
Serine	3-Phosphoglycerate	Oxidation to keto acid, transamination, hydrolysis of phosphate
Tyrosine*	Phenylalanine	Hydroxylation by phenylalanine hydroxylase (tetrahydrobiopterin cofactor)

*Can be synthesized only if methionine and phenylalanine are available from the diet.

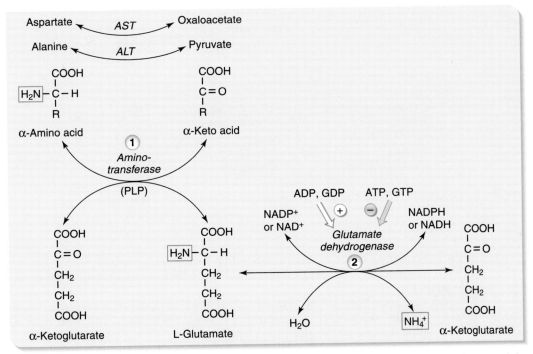

8-1: Transamination and oxidative deamination reactions. Transamination reactions *(left)* are used to synthesize and degrade amino acids. Oxidative deamination of glutamate *(right)*, the product of transamination, releases ammonia, which is disposed of in the urea cycle. ALT, alanine aminotransferase; AST, aspartate aminotransferase; PLP, pyridoxal phosphate.

    b. Glutamate dehydrogenase catalyzes this reversible reaction using NAD^+ or $NADP^+$.

    c. In amino acid catabolism, the enzyme reaction results in the conversion of glutamate to α-ketoglutarate and NH_4^+.

  3. Allosteric regulation of glutamate dehydrogenase favors release of NH_4^+ when the energy supply is inadequate.

    a. Adenosine triphosphate (ATP and) guanosine triphosphate (GTP) are signals of high-energy charge and inhibit the enzyme.

    b. Adenosine diphosphate (ADP) and guanosine diphosphate (GDP) are signals of low-energy charge and stimulate the release of nitrogen from amino acids, freeing their carbon skeletons for use as fuel.

C. Urea cycle (Fig. 8-2)

  1. The urea cycle functions mainly in the liver to convert highly toxic NH_4^+ to nontoxic urea.

  2. Glutamate is the primary source of NH_4^+ that is used in the urea cycle; however, ammonia is produced from other sources that are metabolized by the cycle.

Glutamate: primary source of NH_4^+

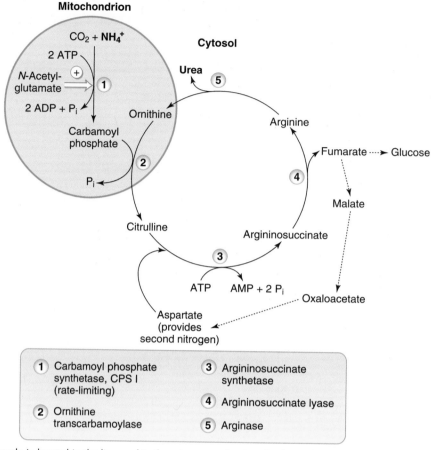

8-2: The urea cycle is located in the liver and is the primary mechanism for disposal of toxic ammonia.

Figure enzyme legend:
1. Carbamoyl phosphate synthetase, CPS I (rate-limiting)
2. Ornithine transcarbamoylase
3. Argininosuccinate synthetase
4. Argininosuccinate lyase
5. Arginase

Urea cycle: in liver, toxic NH$_4^+$ converted to nontoxic urea; CPS I is rate-limiting mitochondrial enzyme

N-Acetylglutamate: required activator of CPS I

Carbamoyl phosphate synthetase I: mitochondrial; urea cycle

Carbamoyl phosphate synthetase II: cytosol; nucleotide synthesis

Arginine is synthesized in the urea cycle.

In the laboratory, urea is measured as blood urea nitrogen (BUN).

Bacterial ureases release NH$_4^+$ from amino acids derived from dietary protein.

3. Urea cycle reactions occur in the mitochondrial matrix and cytosol.
   a. Two mitochondrial reactions generate citrulline, which is transported to the cytosol.
      (1) Step 1
          (a) Carbamoyl phosphate synthetase I (CPS I) catalyzes the first, rate-limiting step in which NH$_4^+$ (contains the first nitrogen), CO$_2$, and ATP react to produce carbamoyl phosphate (see Fig. 8-2).
          (b) N-Acetylglutamate is a required activator of CPS I and is in ample supply after eating a high-protein meal.
      (2) Step 2
          (a) Carbamoyl phosphate, with the addition of ornithine, is converted to citrulline by ornithine transcarbamoylase.
   b. Three cytosolic reactions incorporate nitrogen from aspartate to form ornithine, which reenters mitochondria, and urea, which leaves the cell.
      (1) Step 3
          (a) Citrulline reacts with aspartate (provides a nitrogen) and is converted to argininosuccinate by argininosuccinate synthetase.
      (2) Step 4
          (a) Argininosuccinate is converted to arginine by argininosuccinate lyase and releases fumarate, which enters the citric acid cycle to produce glucose or aspartate by transamination.
      (3) Step 5
          (a) Arginine is converted to urea and ornithine by arginase, which is an enzyme located only in the liver.
   c. Urea enters the blood and most of it is filtered and excreted in the urine.
      (1) A small amount, however, diffuses into the intestine, where it is converted by bacterial ureases into ammonia for elimination in the feces as charged ammonia (NH$_4^+$).

4. Regulation of the urea cycle involves short-term and long-term mechanisms.
   a. *N*-Acetylglutamate is a required allosteric activator of CPS I, providing short-term control.
   b. Elevated NH_4^+ causes increased expression of the urea cycle enzymes, providing long-term control (e.g., during prolonged starvation)

D. Ammonia metabolism
1. Ammonia is primarily converted to urea, with the exception of ammonia derived from glutamine, which is used to acidify urine.
2. Sources of ammonia
   a. Glutamate
      (1) Ammonia is derived from oxidative deamination of glutamate by glutamate dehydrogenase (see Fig. 8-1).
      (2) Glutamate receives amino groups from amino acids through transamination.
   b. Glutamine
      (1) In the proximal tubules of the kidneys, glutamine is converted by glutaminase into ammonia and glutamate.
   c. Monoamines
      (1) Amine oxidases release ammonia from epinephrine, serotonin, and histamine.
   d. Dietary protein
      (1) Bacterial ureases release ammonia from amino acids in dietary protein and from urea diffusing into the gut.
      (2) Depending on the pH, ammonia released by ureases is charged (NH_4^+) and nondiffusible through tissue or uncharged (NH_3) and diffusible through tissue.
      (3) At physiologic pH, NH_4^+ is produced, which is eliminated in the stool.
      (4) In alkalotic conditions (respiratory and metabolic alkalosis), NH_3 is produced (fewer protons available), which is reabsorbed into the portal vein for delivery to the liver urea cycle.
   e. Purines and pyrimidines
      (1) Ammonia is released from amino acids in the catabolism of these nucleotides.
3. Ammonia produced in extrahepatic tissues is toxic and is transported in the circulation primarily as urea and glutamine.
   a. Glutamine is synthesized from glutamate using the enzyme glutamine synthetase, which combines ammonia, ATP, and glutamate to form glutamine.
4. Ammonia carried by glutamine is important in acidifying urine.
   a. In the proximal renal tubules, glutamine is converted by glutaminase into glutamate and NH_4^+.
   b. The ammonia diffuses into the lumen of the collecting tubules as uncharged ammonia (NH_3), which combines with protons to produce ammonium chloride (i.e., acidifies the urine).
5. Hyperammonemia results primarily from the inability to detoxify NH_4^+ in the urea cycle, leading to elevated blood levels of ammonia.
   a. Hereditary hyperammonemia results from defects in urea cycle enzymes.
      (1) Deficiencies of enzymes that are used earlier in the cycle (i.e., CPS I and ornithine transcarbamoylase) are associated with higher blood ammonia levels and more severe clinical manifestations than deficiencies of enzymes that are used later in the cycle (e.g., arginase).
   b. Acquired hyperammonemia most commonly occurs in alcoholic cirrhosis and Reye's syndrome due to disruption of the urea cycle.
      (1) In cirrhosis, the architecture of the liver is distorted, leading to shunting of portal blood into the hepatic vein or backup of blood in the portal vein (i.e., portal hypertension).
      (2) Reye's syndrome occurs primarily in children with influenza or chickenpox who are given salicylates.
      (3) In Reye's syndrome, function of the urea cycle is disrupted by diffuse fatty change in hepatocytes and damage to the mitochondria by salicylates.
   c. Signs and symptoms of hyperammonemia include feeding difficulties, vomiting, ataxia, lethargy, irritability, poor intellectual development, and coma.
      (1) Death may result if signs and symptoms are not treated.
   d. Nonpharmacologic treatment for hyperammonemia is a low-protein diet.
      (1) This decreases the release of ammonia from amino acids by bacterial ureases.

Ammonia is converted to urea in the urea cycle.

Proximal tubules: glutamine converted to ammonia and glutamate by glutaminase

Sources of ammonia: glutamate, glutamine, amine oxidase action, bacterial ureases, and nucleotide catabolism

Bacterial ureases release ammonia from dietary protein.

NH_4^+ is nondiffusible; NH_3 is diffusible.

Glutamine carries ammonia in a nontoxic state; ammonia is released in the kidneys for urine acidification.

Hyperammonemia: inability to detoxify NH_4^+ in the urea cycle; produces encephalopathy

In cirrhosis, dysfunctional urea cycle leads to hyperammonemia and decreased BUN level.

In cirrhosis, serum ammonia is increased, and the serum BUN level is decreased.

Liver damage is measured by serum transaminase concentration.

Reye's syndrome: primarily in children; fatty liver; salicylates compromise urea cycle; high levels of serum transaminase

Low-protein diet deceases the serum ammonia level.

e. Pharmacologic treatment includes the following:
   (1) Oral intake of lactulose provides H^+ ions to combine with NH_3 to form NH_4^+, which is excreted.
   (2) Oral neomycin kills bacteria that release ammonia from amino acids.
   (3) Sodium benzoate forms an adduct with glycine to produce hippuric acid and pulls glycine out of the amino acid pool.
   (4) Phenylacetate forms an adduct with glutamine and pulls glutamine (glutamate plus ammonia) out of the amino acid pool.

III. Catabolic Pathways of Amino Acids

A. Overview

1. Transamination of amino acid nitrogen produces carbon skeletons of amino acids as α-keto acids that enter intermediary metabolism at various points.
2. Amino acids are classified as glucogenic (degraded to pyruvate or intermediates in citric acid cycle), ketogenic (degraded to acetyl CoA or acetoacetyl CoA), or both glucogenic and ketogenic.
3. Carbon skeletons remaining after removal of the α-amino group from amino acids are degraded to intermediates that can be used to produce energy in the citric acid cycle or to synthesize glucose, amino acids, fatty acids, or ketone bodies.
4. Tyrosine is converted to catecholamines, thyroid hormones, melanin, and dopamine, and it is degraded to homogentisate.
5. Branched-chain amino acids—leucine, isoleucine, and valine—are degraded to branched-chain α-ketoacids that can enter the citric acid cycle.
6. Methionine accepts a methyl group from methyl-folate to become *S*-adenosylmethionine, a common donor of a single carbon in metabolism.

B. Carbon skeletons of amino acids (Fig. 8-3)

1. Step 1
   a. Pyruvate is formed from six amino acids that are exclusively glucogenic (except for tryptophan, which is glucogenic and ketogenic): alanine, cysteine, glycine, serine, threonine, and tryptophan (see Fig. 8-3).
2. Step 2
   a. Acetyl CoA is formed from two amino acids: isoleucine (ketogenic and glucogenic) and leucine (exclusively ketogenic).

> Glucogenic amino acids are degraded to pyruvate or intermediates in the citric acid cycle.

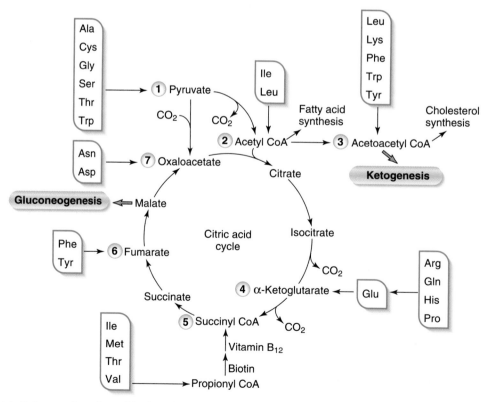

8-3: Metabolic intermediates formed by degradation of amino acids. Acetyl CoA and acetoacetyl CoA are ketogenic; all other products are glucogenic.

3. Step 3
    a. Acetoacetyl CoA, which is interconvertible with acetyl CoA, is formed from five amino acids: leucine and lysine (both exclusively ketogenic) and phenylalanine, tryptophan, and tyrosine (all are ketogenic and glucogenic).
4. Step 4
    a. α-Ketoglutarate is formed from five amino acids that are exclusively glucogenic: glutamate, glutamine, histidine, arginine, and proline.
5. Step 5
    a. Succinyl CoA is formed from four amino acids by means of propionyl CoA, which is a substrate for gluconeogenesis: isoleucine, valine, methionine, and threonine.
6. Step 6
    a. Fumarate is formed from two amino acids that are glucogenic and ketogenic: phenylalanine and tyrosine.
7. Step 7
    a. Oxaloacetate is formed from two amino acids that are exclusively glucogenic: aspartate and asparagine.

C. **Metabolism of phenylalanine and tyrosine (Fig. 8-4)**
1. Step 1
    a. Phenylalanine is converted to tyrosine by phenylalanine hydroxylase (see Fig. 8-4).
    b. The reaction requires tetrahydrobiopterin (BH_4) and oxygen.
2. Step 2
    a. Dihydrobiopterin (BH_2) is converted back into BH_4 by dihydrobiopterin reductase using NADPH as a cofactor.
    b. Phenylpyruvate, phenylacetate, and phenyllactate normally are not produced in large quantities unless there is a deficiency of phenylalanine hydroxylase.
    c. Deficiency of phenylalanine hydroxylase produces classic phenylketonuria (PKU) (Table 8-2).
    d. Deficiency of dihydrobiopterin reductase produces a variant of PKU called *malignant PKU* (see Table 8-2).
3. Step 3
    a. Tyrosine is converted after an intermediate reaction into homogentisate.
4. Step 4
    a. Tyrosine is converted by tyrosine hydroxylase into dopa, which is used to synthesize the catecholamines through a series of intermediate reactions.
    b. The reaction requires BH_4 and oxygen.

Ketogenic amino acids are degraded to acetyl CoA or acetoacetyl CoA; Leu and Lys are ketogenic.

Tetrahydrobiopterin (BH_4) is a cofactor in conversion of phenylalanine to tyrosine, tyrosine to dopa, and tryptophan to serotonin.

PKU is caused by a deficiency of phenylalanine hydroxylase; malignant PKU is caused by a deficiency of dihydrobiopterin reductase.

Tyrosine is used to synthesize catecholamines.

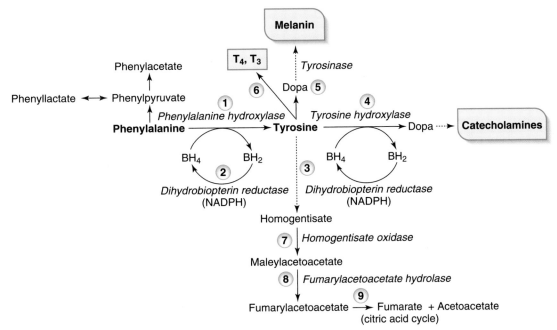

8-4: Metabolism of phenylalanine and tyrosine. Notice the role of tyrosine in the synthesis of thyroid hormones (triiodothyronine [T_3] and thyroxine [T_4]), melanin, and catecholamines. BH_2, dihydrobiopterin; BH_4, tetrahydrobiopterin.

TABLE 8-2. Genetic Disorders Associated with Degradation of Amino Acids

GENETIC DISORDER	ASSOCIATED ENZYME	CLINICAL ASSOCIATIONS
Classic PKU (AR)	Phenylalanine hydroxylase: catalyzes conversion of phenylalanine to tyrosine Deficiency leads to increased phenylalanine and neurotoxic phenylketones and acids and decreased tyrosine levels	Mental retardation; fair skin (decreased melanin synthesis from tyrosine) Mousy odor of affected individual Vomiting simulating congenital pyloric stenosis Must screen for phenylalanine after child is exposed to phenylalanine in breast milk Treatment: restrict phenylalanine, add tyrosine, and restrict aspartame (contains phenylalanine) from diet Pregnant women with PKU must restrict phenylalanine from diet to prevent neurotoxic damage to the fetus in utero
Malignant PKU (AR)	BH_2 reductase: cofactor for phenylalanine hydroxylase, which converts phenylalanine to tyrosine Deficiency leads to increased phenylalanine and neurotoxic byproducts and decreased tyrosine and BH_4 levels	Similar to classic PKU Neurologic problems occur regardless of restricting phenylalanine intake Inability to metabolize tryptophan or tyrosine (require BH_4), which causes decreased synthesis of neurotransmitters (serotonin and dopamine, respectively) Treatment: restrict phenylalanine in diet; administer L-dopa and 5-hydroxytryptophan to replace neurotransmitters and BH_4 replacement
Albinism (AR)	Tyrosinase: catalyzes a reaction converting tyrosine to dopa and dopa to melanin; melanocytes present but do not contain melanin pigment	Absence of melanin in hair (white hair), eyes (photophobia, nystagmus), and skin (pink skin with increased risk of UV light–related skin cancer)
Alkaptonuria (AR)	Homogentisate oxidase: catalyzes conversion of homogentisate to maleylacetoacetate Deficiency leads to increased homogentisate in urine (turns black when oxidized by light) Articular cartilage and sclera darken (ochronosis) due to homogentisate deposition	Degenerative arthritis in spine, hip, and knee
Tyrosinosis (AR)	Fumarylacetoacetate hydrolase: catalyzes conversion of maleylacetoacetate to fumarylacetoacetate Deficiency leads to increased tyrosine levels	Liver damage (hepatitis progressing to cirrhosis and hepatocellular carcinoma) and kidneys (aminoaciduria and renal tubular acidosis)
Maple syrup urine disease (AR)	Branched-chain α-ketoacid dehydrogenase: enzyme normally present in muscle and catalyzes the second step in degradation of isoleucine, leucine, and valine Deficiency leads to increased levels of branched-chain amino acids and their corresponding ketoacids in blood and urine	Feeding difficulties, vomiting, seizures, hypoglycemia, fatal without treatment Urine has odor of maple syrup Treatment: restrict intake of branched-chain amino acids to the amount required for protein synthesis
Homocystinuria (AR)	Cystathionine synthase: catalyzes conversion of homocysteine plus serine into cystathionine Deficiency leads to increased levels of homocysteine and methionine Homocysteine damages endothelial cells, causing thrombosis and thromboembolic disease	Similar to Marfan syndrome: dislocated lens, arachnodactyly (spider fingers), eunuchoid features (arm span > height) Distinctive features include mental retardation, vessel thrombosis (e.g., cerebral vessels), osteoporosis Treatment: high doses of vitamin B_6, restriction of methionine, addition of cysteine
Propionic acidemia (AR)	Propionyl carboxylase: catalyzes conversion of propionyl CoA to methylmalonyl CoA Deficiency leads to increased levels of propionic acid and odd-chain fatty acids in the liver	Neurologic and developmental complications Treatment: low-protein diet; L-carnitine (improves β-oxidation of fatty acids); increased intake of methionine, valine, isoleucine, and odd-chain fatty acids
Methylmalonic acidemia (AR)	Methylmalonyl CoA mutase: catalyzes conversion of methylmalonic acid to succinyl CoA, using vitamin B_{12} as a cofactor Deficiency leads to increased levels of methylmalonic and propionic acids	Neurologic and developmental complications Rule out vitamin B_{12} deficiency as a cause Treatment: same as for propionic acidemia

AR, autosomal recessive; BH_2, dihydrobiopterin; BH_4, tetrahydrobiopterin; PKU, phenylketonuria.

c. Dihydrobiopterin BH_2 is converted back into BH_4 by dihydrobiopterin reductase using NADPH as a cofactor.

5. Step 5
   a. Tyrosine can also be converted by tyrosinase into dopa and other intermediates to produce melanin.
   b. Deficiency of tyrosinase produces albinism, an autosomal recessive (AR) disorder (see Table 8-2).

6. Step 6
   a. Triiodothyronine (T_3) and thyroxine (T_4) synthesis in the thyroid gland begins with iodination of tyrosine residues.
   b. Condensation of iodinated tyrosine residues forms T_3 and T_4.

7. Step 7
   a. Homogentisate is converted to maleylacetoacetate by homogentisate oxidase.
   b. Deficiency of homogentisate oxidase produces alkaptonuria (see Table 8-2).

8. Step 8
   a. Maleylacetoacetate is converted to fumarylacetoacetate by fumarylacetoacetate hydrolase.
   b. Deficiency of fumarylacetoacetate hydrolase produces tyrosinosis (see Table 8-2).

9. Step 9
   a. Fumarylacetoacetate is converted to fumarate, which is a substrate in the citric acid cycle, and acetoacetate.

D. Metabolism of leucine, isoleucine, and valine: branched-chain amino acids
   1. Branched-chain amino acids are metabolized primarily in muscle and to a lesser extent in other extrahepatic tissues.
   2. Branched-chain amino acid metabolism involves a series of reactions resulting in the conversion of leucine (ketogenic) into acetyl CoA and acetoacetate; isoleucine (ketogenic and glucogenic) into acetyl CoA and succinyl CoA; and valine (glucogenic) into succinyl CoA.
   3. One of the enzymes used in the degradative process is branched-chain α-ketoacid dehydrogenase, which is deficient in maple syrup urine disease (see Table 8-2).
      a. Branched-chain ketoacids cause urine to have the odor of maple syrup.

E. Metabolism of methionine (Fig. 8-5)
   1. The essential amino acid methionine is the precursor of *S*-adenosylmethionine (SAM), which is the most important methyl group (CH_3) donor in biologic methylation

Melanin is derived from dopa.

Thyroid hormones are derived from tyrosine.

Albinism (AR): deficiency of tyrosinase

Alkaptonuria (AR): deficiency of homogentisate oxidase; homogentisate turns urine black

Fumarylacetoacetate hydrolase deficiency: tyrosinosis responsible for lethargy, drowsiness, irritability, and anorexia

Branched-chain amino acids: metabolized primarily in muscle, not liver

Maple syrup urine disease (AR): deficiency of branched-chain α-ketoacid dehydrogenase in muscle

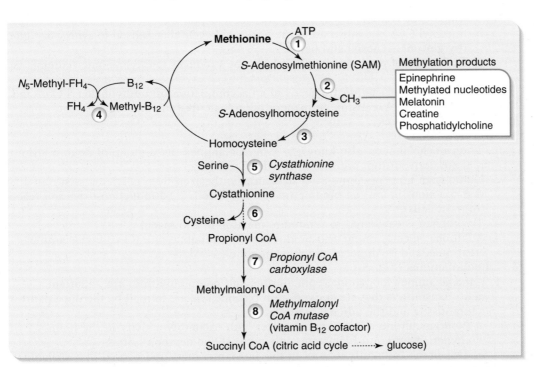

8-5: Metabolism of methionine. Notice the role of methionine in the donation of methyl groups, resynthesis by homocysteine with the aid of vitamin B_{12} and folate, synthesis of cysteine, and production of succinyl CoA in the citric acid cycle. FH_4, tetrahydrofolate; methyl-B_{12}, methylated vitamin B_{12}.

reactions (e.g., norepinephrine receives a methyl group from SAM to produce epinephrine).

2. Step 1
   a. SAM is formed by the transfer of the adenosyl group from ATP to methionine (see Fig. 8-5).
3. Step 2
   a. After donation of its methyl group, SAM becomes *S*-adenosylhomocysteine.
   b. The methyl group is transferred to a variety of acceptors (e.g., norepinephrine), resulting in a methylation product (e.g., epinephrine).
4. Step 3
   a. *S*-Adenosylhomocysteine is converted to homocysteine.
5. Step 4
   a. Homocysteine can resynthesize methionine with the aid of vitamin B$_{12}$ and folate (see Chapter 4).

   b. Vitamin B$_{12}$ removes the methyl group from N^5-methyltetrahydrofolate (N^5-methyl-FH$_4$) and produces tetrahydrofolate (FH$_4$).
   c. Methylated vitamin B$_{12}$ (methyl-B$_{12}$) transfers the methyl group to homocysteine, which produces methionine.
6. Step 5
   a. Homocysteine combined with serine is converted to cystathionine by cystathionine synthase (see Fig. 8-5).
   b. Deficiency of cystathionine synthase produces homocystinuria (see Table 8-2).
7. Step 6
   a. Cystathionine, after an intermediate reaction, is converted to propionyl CoA and cysteine.
   b. Propionyl CoA is also produced by the metabolism of odd-chain fatty acids (see Chapter 7) and is an intermediary product in the metabolism of the branched-chain amino acids valine and isoleucine.

8. Step 7
   a. Propionyl CoA is converted to methylmalonyl CoA by propionyl CoA carboxylase, which uses biotin as a cofactor (see Chapter 4).
   b. Deficiency of propionyl CoA carboxylase produces propionic acidemia (see Table 8-2).
9. Step 8
   a. Methylmalonyl CoA is converted to succinyl CoA by methylmalonyl CoA mutase, which uses vitamin B$_{12}$ as a cofactor (see Chapter 4).
   b. Succinyl CoA is a substrate in the citric acid cycle that is used to synthesize glucose (by gluconeogenesis) and heme.
   c. Deficiency of methylmalonyl CoA mutase produces methylmalonic acidemia (see Table 8-2).
   d. Deficiency of vitamin B$_{12}$ leads to an accumulation of methylmalonyl CoA and propionyl CoA, causing permanent neurologic dysfunction.

IV. **Amino Acid Derivatives**
   A. **Overview**
      1. The catecholamines, also known as biogenic amines, are a group of neurotransmitters that are derived from tyrosine in the dopa pathway.
      2. Heme is an iron-containing ring structure synthesized from glycine and succinyl CoA that functions in oxygen binding (in red blood cells) and reduction-oxidation reactions (in most cells).
      3. Tryptophan is converted to serotonin, melatonin, and niacin.
      4. Additional special amino acid products are γ-aminobutyrate (GABA), histamine, creatinine, and asymmetric dimethylarginine (ADMA).
   B. **Catecholamines (Fig. 8-6)**
      1. Catecholamines (dopamine, epinephrine, and norepinephrine) are important neurotransmitters that are derived from tyrosine and are formed by the dopa pathway in neural tissue and the adrenal medulla.

      a. Norepinephrine is a neurotransmitter with excitatory activity in the brain (hypothalamus and brainstem) and sympathetic nervous system.
      b. Dopamine (primarily located in the substantia nigra and ventral hypothalamus) is a neurotransmitter with multiple functions that affect behavior, especially reward responses.

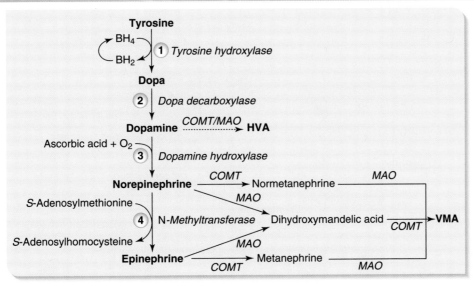

8-6: Catecholamine synthesis and degradation. Tyrosine plays a major role in the synthesis of catecholamines, which are important neurotransmitters. BH_2, dihydrobiopterin; BH_4, tetrahydrobiopterin; COMT, catechol-*O*-methyltransferase; HVA, homovanillic acid; MAO, monoamine oxidase; VMA, vanillylmandelic acid.

    c. Stimulation of the sympathetic nerves to the adrenal medulla causes the release of epinephrine and norepinephrine, which affect blood vessels (vasoconstriction is greater with norepinephrine than epinephrine); the heart (contraction is greater with epinephrine than norepinephrine); and the gastrointestinal tract (both inhibit peristalsis).

2. Step 1
    a. The reaction sequence for catecholamine synthesis begins with tyrosine, which is converted to dopa by tyrosine hydroxylase (copper-containing rate-limiting enzyme) in the cytoplasm (see Fig. 8-6).
    b. The reaction requires tetrahydrobiopterin (BH_4).
    c. Dihydrobiopterin (BH_2) is converted back into BH_4 by dihydrobiopterin reductase (see Fig. 8-4), using NADPH as a cofactor.

3. Step 2
    a. Dopa is converted to dopamine by dopa decarboxylase.
    b. This reaction occurs in storage vesicles in the adrenal medulla and synaptic vesicles in neurons.
    c. Catechol-*O*-methyltransferase (COMT) and monoamine oxidase (MAO) are involved in reactions that metabolize dopamine into homovanillic acid (HVA).
    d. Dopamine synthesis is deficient in idiopathic Parkinson's disease.

4. Step 3
    a. Dopamine is converted to norepinephrine by dopamine hydroxylase, a copper-containing enzyme, which uses ascorbic acid as a cofactor.
    b. COMT and MAO metabolize norepinephrine into vanillylmandelic acid (VMA).

5. Step 4
    a. Norepinephrine is converted to epinephrine by *N*-methyltransferase, using a methyl group donated by SAM.
    b. *N*-Methyltransferase is located only in the adrenal medulla; hence, epinephrine is synthesized only in the adrenal medulla.
    c. COMT metabolizes epinephrine into metanephrine, which is converted to VMA by MAO.
    d. HVA, VMA, and metanephrines are excreted in the urine, and levels are commonly measured to screen for tumors of the adrenal medulla or sympathetic nervous system.
      (1) Tumors of the adrenal medulla secrete excess catecholamines, leading to hypertension.
      (2) Pheochromocytomas are benign, unilateral tumors of the adrenal medulla that occur primarily in adults.
      (3) Neuroblastomas are malignant unilateral tumors of the adrenal medulla that occur primarily in children.

Dopa and dopamine: intermediates in conversion of tyrosine to norepinephrine (norepinephrine converted to epinephrine)

Dopamine synthesis: deficient in idiopathic Parkinson's disease

HVA: degradation product of dopamine

VMA: degradation product of norepinephrine and epinephrine

Metanephrine: degradation product of epinephrine

Pheochromocytoma (benign) and neuroblastoma (malignant): adrenal medulla tumors that secrete excess catecholamines

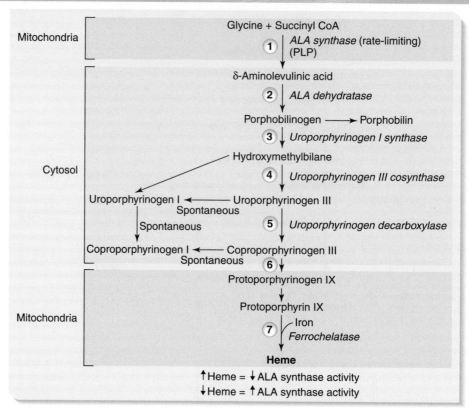

8-7: Porphyrin synthesis and metabolism. Heme is the most important porphyrin and has a major role in oxygen transfer reactions. Enzyme deficiencies in porphyrin synthesis result in various types of porphyria. ALA, δ-aminolevulinic acid; PLP, pyridoxal phosphate.

C. Heme synthesis and metabolism (Fig. 8-7)

1. Overview
   a. Heme is a component of hemoglobin, myoglobin, and cytochromes that is synthesized in most tissues of the body.
   b. Heme is a cyclic planar molecule (like a wheel) with an iron atom at the center (hub) and an asymmetric arrangement of side chains around the rim. The part of heme that serves as the rim is a porphyrin.
   c. Porphyrins are formed by the linking of pyrrole rings with methylene bridges to create ring compounds that bind iron in coordination bonds at their center.
   d. Porphyrinogens (i.e., porphyrin precursors) are colorless and nonfluorescent in the reduced state.
      (1) When porphyrinogen compounds in voided urine are oxidized and exposed to light, they become porphyrins, which have a red wine color and fluoresce under ultraviolet (UV) light.
      (2) Porphyrins in the peripheral circulation absorb ultraviolet light near the skin surface, becoming photosensitizing agents that damage skin and produce vesicles and bullae.

2. Heme synthesis begins in the mitochondria, moves into the cytosol, and then reenters the mitochondria.
   a. Step 1
      (1) Glycine and succinyl CoA are combined by the mitochondrial enzyme δ-aminolevulinic acid synthase (ALA synthase), which is a rate-limiting enzyme, to form δ-aminolevulinic acid (see Fig. 8-7).
      (2) The reaction requires PLP, derived from vitamin B_6, as a cofactor.
         (a) Deficiency of pyridoxine produces anemia related to a decrease in heme synthesis leading to a decrease in hemoglobin
      (3) An increase in heme suppresses ALA synthase; a decrease in heme (e.g., after metabolism of a drug in the liver) increases activity of the enzyme.

Heme synthesis begins in mitochondria, moves into the cytosol, and finishes in mitochondria.

Heme synthesis: ALA synthase is rate-limiting enzyme; feedback inhibition by heme

Drugs (e.g., alcohol, barbiturates) metabolized by the cytochrome P450 system decrease heme concentration and activate ALA synthase.

  b. Step 2
   (1) δ-Aminolevulinic acid is converted to porphobilinogen by the cytosolic enzyme δ-aminolevulinic acid dehydratase (ALA dehydratase).
   (2) Lead denatures ALA dehydratase, leading to an increase in δ-aminolevulinic acid.
  c. Step 3
   (1) Porphobilinogen is converted to hydroxymethylbilane by the cytosolic enzyme uroporphyrinogen I synthase (Table 8-3).
   (2) Uroporphyrinogen I synthase is deficient in acute intermittent porphyria.
  d. Step 4
   (1) Hydroxymethylbilane is converted to uroporphyrinogen III by the cytosolic enzyme uroporphyrinogen III cosynthase (see Table 8-3).
   (2) Some hydroxymethylbilane is nonenzymatically converted to uroporphyrinogen I, which is further converted to coproporphyrinogen I.

Lead denatures ALA dehydratase.

Acute intermittent porphyria (AD): deficiency of uroporphyrinogen I synthase; increase in porphobilinogen and δ-ALA in urine; neurologic problems

TABLE 8-3. Genetic Disorders Involving Porphyrin Synthesis

GENETIC DISORDER	ASSOCIATED ENZYME	CLINICAL ASSOCIATIONS
Acute intermittent porphyria (AD)	Uroporphyrinogen I synthase: catalyzes conversion of porphobilinogen to hydroxymethylbilane Deficiency leads to increased levels of PBG and δ-ALA in urine	Recurrent attacks of neurologically induced abdominal pain (mimics a surgical abdomen) Abdominal pain often leads to surgical exploration (bellyful of scars) without finding any cause Urine exposed to light develops a red wine color due to porphobilin (i.e., window sill test) Enzyme assay for RBCs is the confirming test when the patient is asymptomatic Attacks precipitated by drugs that induce the liver cytochrome P450 system (e.g., alcohol); drugs that induce ALA synthase (e.g., progesterone); and dietary restriction Treatment: carbohydrate loading and infusion of heme, both of which inhibit ALA synthase activity
Congenital erythropoietic porphyria (AR)	Uroporphyrinogen III cosynthase: catalyzes conversion of hydroxymethylbilane to uroporphyrinogen III Deficiency leads to increased levels of uroporphyrinogen I and its oxidation product uroporphyrin I	Hemolytic anemia and photosensitive skin lesions with vesicles and bullae Uroporphyrin I produces a red wine color in urine and teeth and induces a photosensitivity reaction in skin Treatment: protection of skin from light; bone marrow transplantation
Porphyria cutanea tarda (AD or acquired)	Uroporphyrinogen decarboxylase: catalyzes conversion of uroporphyrinogen III to coproporphyrinogen III Deficiency leads to accumulation of uroporphyrinogen III, which spontaneously converts into uroporphyrinogen I and coproporphyrinogen I and their respective oxidized porphyrins	Most common porphyria in United States Predominantly associated with photosensitive skin lesions consisting of vesicles and bullae and liver disease (e.g., cirrhosis) Exacerbating factors include iron therapy, alcohol, estrogens, and hepatitis C (most common acquired cause of PCT) Uroporphyrin I produces a red wine color in urine and predisposes to photosensitive skin lesions; PBG levels are normal Treatment: phlebotomy (reduce iron levels in the liver) and chloroquine
Lead poisoning (acquired)	Inhibits ALA dehydratase and ferrochelatase: ALA dehydratase catalyzes conversion of δ-ALA to PBG; inhibition leads to increased δ-ALA levels in urine Ferrochelatase combines iron with protoporphyrin IX to form heme Inhibition causes increased levels of RBC protoporphyrin IX and decreased heme Iron accumulates in mitochondria, causing a microcytic anemia with ringed sideroblasts (mitochondria around the RBC nucleus filled with iron) in the bone marrow Lead inhibits ribonuclease, causing persistence of ribosomes in peripheral blood RBCs (coarse basophilic stippling)	Causes include exposure to lead-based paint (e.g., pottery) and working in battery factories Children develop encephalopathy with convulsions, microcytic anemia, colicky abdominal pain Adults develop abdominal pain and diarrhea, peripheral neuropathies, and renal disease (e.g., aminoaciduria, renal tubular acidosis) Screen for blood lead levels Treatment: British antilewisite, calcium disodium edetate, and ᴅ-penicillamine

AD, autosomal dominant; ALA, aminolevulinic acid; AR, autosomal recessive; PBG, porphobilinogen; PCT, porphyria cutanea tarda; RBC, red blood cell.

(3) Uroporphyrinogen I and coproporphyrinogen I are spontaneously oxidized into uroporphyrin I and coproporphyrin I, respectively.

(4) Deficiency of uroporphyrinogen III cosynthase produces congenital erythropoietic porphyria.

e. Step 5

(1) Uroporphyrinogen III is converted to coproporphyrinogen III by the cytosolic enzyme uroporphyrinogen decarboxylase (see Table 8-3).

(2) Uroporphyrinogen III and coproporphyrinogen III can be spontaneously oxidized into uroporphyrinogen I and coproporphyrinogen I, respectively.

(3) Uroporphyrinogen decarboxylase is deficient in porphyria cutanea tarda, the most common porphyria in the United States.

f. Step 6

(1) Coproporphyrinogen III is converted to protoporphyrinogen IX and the latter into protoporphyrin IX by oxidase reactions that occur in the mitochondria.

g. Step 7

(1) Ferrochelatase combines iron with protoporphyrin IX to form heme (see Table 8-3).

(2) Lead inhibits ferrochelatase, leading to a decrease in heme and an increase in protoporphyrin IX (see Table 8-3).

3. Heme degradation (Fig. 8-8)

a. Most heme that is degraded comes from the hemoglobin of old erythrocytes, which are phagocytosed by macrophages (primarily in the spleen).

b. Step 1

(1) Oxidases convert free heme to bilirubin in macrophages located in the spleen.

c. Step 2

(1) Unconjugated bilirubin (indirect bilirubin) combines with albumin in the blood and is taken up into hepatocytes by binding proteins.

(2) Unconjugated bilirubin is not filtered in urine, because it is lipid soluble and bound to albumin.

d. Step 3

(1) In the hepatocytes, unconjugated bilirubin is conjugated by reacting with two molecules of glucuronic acid, a reaction that is catalyzed by uridine diphosphate glucuronyltransferase (UGT).

Deficiency of uroporphyrinogen III cosynthase: congenital erythropoietic porphyria

Porphyria cutanea tarda (AD): deficiency of uroporphyrinogen decarboxylase; skin photosensitivity

Lead poisoning: inhibition of ferrochelatase and ALA dehydratase; microcytic anemia with coarse basophilic stippling

Bilirubin: end product of heme degradation by macrophages

Unconjugated bilirubin: transported on albumin; hydrophobic

Conjugated bilirubin: transported in solution; found only in hepatitis and obstruction of bile duct (not normal serum component)

UGT: converts unconjugated bilirubin to conjugated bilirubin

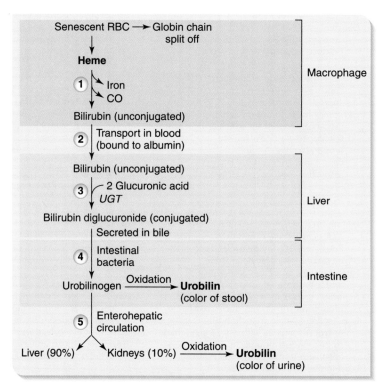

8-8: Heme degradation. The end products of heme degradation are bilirubin and its degradative product, urobilinogen. Oxidation of urobilinogen into urobilin provides the color in stool and urine. CO, carbon monoxide; UGT, uridine diphosphate glucuronyltransferase.

(2) Bilirubin diglucuronide, or conjugated (direct) bilirubin, is water soluble.

(3) Conjugated bilirubin does not have access to the blood unless there is inflammation in the liver (e.g., hepatitis) or obstruction to bile flow (e.g., gallstone in the common bile duct).

(4) Conjugated bilirubin is actively secreted into the bile ducts and stored in the gallbladder for eventual release into the duodenum.

e. Step 4

(1) Intestinal bacteria hydrolyze conjugated bilirubin and reduce free bilirubin to colorless urobilinogen.

(2) Oxidation of urobilinogen yields urobilin, which gives feces its characteristic brown color.

f. Step 5

(1) Approximately 20% of urobilinogen is reabsorbed back into the blood (i.e., enterohepatic circulation) in the terminal ileum and is recycled to the liver and kidneys.

(2) In urine, urobilinogen is oxidized into urobilin, which gives urine its yellow color.

(3) The color of stool and urine is caused by urobilin.

4. Hyperbilirubinemia results from overproduction or defective disposal of bilirubin and may lead to jaundice.

a. Measuring the serum concentration of conjugated bilirubin and unconjugated bilirubin provides clues to the cause of jaundice.

(1) Expressing the percent conjugated bilirubin of the total bilirubin (conjugated bilirubin divided by total bilirubin) is most often used in classifying the types of jaundice.

b. Predominantly unconjugated bilirubin (percent conjugated bilirubin is less than 20% of the total) is present in hemolytic anemias associated with macrophage destruction of RBCs (e.g., congenital spherocytosis) and problems with uptake and conjugation of bilirubin (e.g., Gilbert's disease, Crigler-Najjar syndrome).

(1) Hereditary spherocytosis is an autosomal dominant (AD) disorder with a defect in ankyrin, the contractile protein attached to the inner surface of an RBC that helps maintain its characteristic shape.

(2) Gilbert's disease is a benign autosomal dominant disorder with a defect in the uptake and conjugation of bilirubin and is second only to hepatitis as the most common cause of jaundice in the United States.

(3) Crigler-Najjar syndrome is a genetic disease associated with a partial (AD) or a total (AR) deficiency of UGT, the latter being incompatible with life.

c. Viral hepatitis is associated with a mixed hyperbilirubinemia (increase in unconjugated and conjugated bilirubin) due to problems with uptake, conjugation, and secretion of bilirubin into bile ducts.

(1) The percent conjugated bilirubin is between 20% and 50% of the total bilirubin.

d. Obstructive jaundice is primarily a conjugated type of hyperbilirubinemia (percent conjugated bilirubin > 50% of the total) and is caused by obstruction of bile ducts (e.g., gallstone in the common bile duct).

(1) Stools are light colored and urobilinogen is not present in urine, because bile containing bilirubin is prevented from reaching the small intestine.

D. Serotonin, melatonin, and niacin synthesis from tryptophan (Fig. 8-9)

1. The first reaction in the metabolism of tryptophan is catalyzed by tryptophan hydroxylase, which converts tryptophan to 5-hydroxytryptophan.

a. The reaction requires BH_4 as a cofactor.

2. 5-Hydroxytryptophan is converted to serotonin using pyridoxine (vitamin B_6) as a cofactor.

Margin notes:

Urobilinogen: colorless; converted to urobilin (brown color) in feces and urine

Jaundice: overproduction of bilirubin; decreased conjugation; disruption of bile ducts (hepatitis); bile duct obstruction

Congenital spherocytosis (AD): defect in spectrin; produces hemolysis; predominantly unconjugated bilirubin

Gilbert's disease (AD): defect in uptake and conjugation of bilirubin; predominantly unconjugated bilirubin

Crigler-Najjar syndrome: deficiency of UGT; partial deficiency is AD, total deficiency (AR) is lethal; predominantly unconjugated bilirubin

Mixed hyperbilirubinemia: between 20% and 50% conjugated bilirubin; viral hepatitis

Conjugated hyperbilirubinemia: conjugated bilirubin > 50% of the total; obstructive jaundice

Tryptophan: precursor of serotonin (neurotransmitter), melatonin (sleep-wake cycle), and niacin (deficiency causes pellagra)

8-9: Conversion of tryptophan to serotonin and melatonin. *(From Pelley JW: Elsevier's Integrated Biochemistry. Philadelphia, Mosby, 2007.)*

a. Serotonin (5-hydroxytryptamine) is synthesized primarily in the median raphe of the brainstem, pineal gland, and chromaffin cells of the gut.
b. Serotonin is a neurotransmitter that suppresses pain and helps control mood.
   (1) Deficiency of serotonin is associated with depression.
c. Serotonin stimulates contraction of smooth muscle in the gastrointestinal tract, increasing peristalsis, and it increases the formation of blood clots when released from platelets as a vasoconstrictor of arterioles.
3. Serotonin is converted to melatonin in the pineal gland using SAM as a methyl donor.
   a. Melatonin is involved in regulating the sleep-wake cycle.
4. Serotonin is excreted as 5-hydroxyindoleacetic acid (5-HIAA) in urine.
5. The carcinoid syndrome, involving an oversecretion of serotonin, typically occurs when a carcinoid tumor of the small intestine metastasizes to the liver.
   a. Serotonin produced by the metastatic nodules gains access to the systemic circulation through hepatic vein tributaries and causes flushing of the skin, sudden drops in blood pressure, watery diarrhea (i.e., hyperperistalsis), and an increase of 5-HIAA in urine.
6. Tryptophan is a precursor for the synthesis of niacin (not shown in Fig. 8-9).
   a. Deficiency of tryptophan or niacin produces pellagra (see Chapter 4).

E. Synthesis of γ-aminobutyrate (GABA) from glutamate
1. Glutamate is decarboxylated to GABA, an inhibitory neurotransmitter in the basal ganglia system.
2. GABA is increased in hepatic encephalopathy.

F. Synthesis of histamine from histidine
1. Histidine is decarboxylated to produce histamine, a potent vasodilator that is released by mast cells during type I hypersensitivity reactions.

G. Synthesis of creatine from arginine, glycine, and SAM
1. Creatine is combined with ATP in a reaction catalyzed by creatine kinase to produce creatine phosphate, which is a high-energy storage compound present in tissue, particularly muscle and brain.
2. Creatine phosphate provides a ready source of phosphate groups to regenerate ATP by the reverse reaction also catalyzed by creatine kinase.
3. Creatine is spontaneously converted to creatinine, which is excreted at a constant rate in urine, hence its usefulness in measuring the.

H. Asymmetric dimethylarginine (ADMA)
1. ADMA is formed as a metabolic byproduct of continuous protein turnover in all cells of the body and is a normal component of human blood plasma.
2. ADMA inhibits nitric oxide (NO) synthesis in the vascular endothelium and inhibits vasodilation produced by NO.
3. Biosynthesis of ADMA occurs during methylation of protein residues, which release unbound ADMA during their proteolytic degradation.
4. ADMA is implicated in hypertension and formation of atherosclerotic plaque.

CHAPTER 9

INTEGRATION OF METABOLISM

I. Hormonal Regulation of Metabolism

A. Overview

1. Hormones act by triggering intracellular signaling pathways leading to
   a. Coordinated activation or deactivation of key enzymes (usually by phosphorylation or dephosphorylation)
   b. Induction or repression of enzyme synthesis
2. Three hormones—insulin, glucagon, and epinephrine—play a critical role in integrating metabolism, especially energy metabolism, in different tissues (Table 9-1).
   a. Allosteric effectors, molecules that bind at a site other than the active site and activate or inhibit particular enzymes, are important in regulation of metabolic pathways (see Table 9-1).
3. Insulin and glucagon are the key hormones in the short-term regulation of blood glucose concentration under normal physiologic conditions.
   a. Insulin acts to reduce blood glucose levels (i.e., hypoglycemic effect).
   b. Glucagon acts to increase blood glucose levels (i.e., hyperglycemic effect).

B. Insulin action

1. Insulin is synthesized by pancreatic β cells as an inactive precursor, proinsulin.
2. Proteolytic cleavage of proinsulin yields C-peptide and active insulin, consisting of disulfide-linked A and B chains.
3. Secretion of insulin is regulated by circulating substrates and hormones.
   a. Stimulated by increased blood glucose (most important), increased individual amino acids (e.g., arginine, leucine), and gastrointestinal hormones (e.g., secretin), which are released after ingestion of food
   b. Inhibited by somatostatin, low glucose levels, and hypokalemia
4. Metabolic actions of insulin are most pronounced in liver, muscle, and adipose tissue.
   a. Overall effect is to promote storage of excess glucose as glycogen in liver and muscle and as triacylglycerols in adipose tissue.
5. The insulin receptor is a tetramer whose cytosolic domain has tyrosine kinase activity for generating second messengers (see Chapter 3).
   a. Insulin binding triggers signaling pathways that produce several cellular responses (i.e., post-receptor functions).
   b. Increased adipose tissue mass downregulates insulin receptor synthesis, and adipose weight loss upregulates receptor synthesis.
   c. Increased glucose uptake by muscle and adipose tissue is prompted by translocation of insulin-sensitive glucose transporter 4 (GLUT4) receptors to the cell surface.
   d. Dephosphorylation from insulin action activates energy-storage enzymes (e.g., glycogen synthase) and inactivates energy-mobilizing enzymes (e.g., glycogen phosphorylase).
   e. Increased enzyme synthesis from insulin action (e.g., glucokinase, phosphofructokinase) is caused by activation of gene transcription.

C. Glucagon and epinephrine action

1. Glucagon and epinephrine function to prevent fasting hypoglycemia.
2. Secretion of glucagon from pancreatic α cells is regulated by circulating substrates and hormones.
   a. Stimulated by increased amino acids, low glucose levels
   b. Inhibited by high glucose levels

Energy metabolism is regulated by insulin, glucagon, and epinephrine.

Insulin causes enzyme dephosphorylation; glucagon causes enzyme phosphorylation.

Proinsulin: active insulin + C-peptide

One C-peptide molecule is released with each active insulin molecule.

Insulin secretion: stimulated by high glucose levels; inhibited by somatostatin, low glucose levels, and hypokalemia

Insulin receptor: tyrosine kinase; autophosphorylation

GLUT4 receptors: insulin sensitive

Insulin action: dephosphorylation and increased synthesis of enzymes

Glucagon secretion: stimulated by amino acids; inhibited by high glucose levels

Epinephrine secretion is stimulated by central nervous system during stress or in hypoglycemia.

TABLE 9-1. Allosteric and Hormonal Regulation of Metabolic Pathways

METABOLIC PATHWAY	MAJOR REGULATORY ENZYMES	ALLOSTERIC EFFECTORS*	HORMONAL EFFECTS†
Glycolysis and pyruvate oxidation	Hexokinase	Glucose 6-P (−)	—
	Glucokinase (liver)	—	Induced by insulin
	Phosphofructokinase 1	Fructose 2,6-bisphosphate, adenosine monophosphate (AMP) (+); citrate (−)	Glucagon (↓) by decrease in fructose 2,6-bisphosphate
	Pyruvate kinase		
	Pyruvate dehydrogenase	Fructose 1,6-bisphosphate (+); adenosine triphosphate (ATP), alanine (−) Adenosine diphosphate (ADP) (+); acetyl CoA, NADH, ATP (−)	Glucagon (↓) Insulin (↑)
Citric acid cycle	Isocitrate dehydrogenase	ADP (+); ATP, NADH (−)	—
Glycogenesis	Glycogen synthase	Glucose 6-phosphate (+)	Insulin (↑); glucagon in liver, epinephrine in muscle (↓) Induced by insulin
Glycogenolysis	Glycogen phosphorylase	Ca²⁺ (+) in muscle	Glucagon in liver, epinephrine in muscle (↑)
Gluconeogenesis	Fructose 1,6-bisphosphatase	Citrate (+); fructose 2,6-bisphosphate BP, AMP (−)	Glucagon (↑) by decrease in fructose 2,6-BP All three enzymes induced by glucagon and cortisol; repressed by insulin
	Phosphoenolpyruvate (PEP) carboxykinase	—	
	Pyruvate carboxylase	Acetyl CoA (+)	
Pentose phosphate pathway	Glucose 6-phosphate dehydrogenase (G6PD)	NADPH (−)	—
Fatty acid synthesis	Acetyl CoA carboxylase	Citrate (+); palmitate (−)	Insulin (↑); glucagon (↓) Induced by insulin
Lipolysis	Hormone-sensitive lipase	—	Epinephrine (↑); insulin (↓)
β-Oxidation of fatty acids	Carnitine acyltransferase	Malonyl CoA (−)	—
Cholesterol synthesis	HMG CoA reductase	Cholesterol (−)	Insulin (↑); glucagon (↓)
Urea cycle	Carbamoyl phosphate synthetase I (CPS I)	N-Acetylglutamate (+)	—
Pyrimidine synthesis	Carbamoyl phosphate synthetase II (CPS II)	PRPP, ATP (+); uridine triphosphate (UTP)- (−)	—
Purine synthesis	Phosphoribosyl-1-pyrophosphate (PRPP) amidotransferase	PRPP (+); inosine monophosphate (IMP), AMP, guanosine monophosphate (GMP) (−)	—
Heme synthesis	Aminolevulinic acid (ALA) synthase	Enzyme synthesis repressed by heme	—

*Stimulates (+) or inhibits (−) enzyme activity.
†Promotes formation of active form (↑) or inactive form (↓) of enzyme by phosphorylation or dephosphorylation.

Epinephrine supplements the hyperglycemic effect of glucagon.

3. Secretion of epinephrine from the adrenal medulla is triggered by release of acetylcholine from preganglionic sympathetic nerves in response to stress, prolonged exercise, or trauma.
4. Metabolic actions of glucagon and epinephrine reinforce each other and counteract insulin action.
   a. Glucagon acts primarily on the liver to promote glycogenolysis and gluconeogenesis.
   b. Epinephrine stimulates glycogenolysis in muscle and the liver and the release of free fatty acids (lipolysis) in adipose tissue.

Glucagon and epinephrine receptors: stimulatory G proteins; increased phosphorylation

5. Glucagon and epinephrine receptors are coupled to stimulatory G proteins (see Chapter 3).
   a. Hormone binding activates adenylate cyclase, leading to an increase in cAMP, which activates protein kinase A.

b. Subsequent phosphorylation by protein kinase A results in activation of energy-mobilizing enzymes (e.g., glycogen phosphorylase, hormone-sensitive lipase) and inactivation of energy-storage enzymes (e.g., glycogen synthase, acetyl CoA carboxylase in fatty acid synthesis).

II. The Well-Fed State

A. Overview

1. The metabolic activity of various tissues interacts to store energy when ingested fuel is plentiful (i.e., well-fed state) and to draw on energy stores to maintain blood glucose during fasting or starvation.

2. The period from about 1 to 3 hours after ingestion of a normal meal is marked by a high insulin-to-glucagon ratio and elevated blood glucose levels due to circulating absorbed dietary glucose (Table 9-2).

Well-fed state: high insulin-to-glucagon ratio; elevated blood glucose levels

B. Liver metabolism: well-fed state (Fig. 9-1)

1. After a meal, the hepatic portal vein delivers venous blood containing absorbed nutrients (with the exception of long-chain fatty acids) and elevated levels of insulin directly to the liver.

TABLE 9-2. **Comparison of the Well-Fed, Fasting, and Starvation States**

PROCESS	WELL-FED STATE	FASTING STATE	STARVATION STATE
Glycogenesis	Increased	None	None
Glycogenolysis	Decreased; none in the liver, some in muscle	Increased; early supply of glucose derived from liver, not muscle	None; glycogen depleted
Gluconeogenesis	None	Increased; primary source of glucose after glycogenolysis	Decreased; just enough to supply red blood cells (RBCs)
Triacylglycerol synthesis in liver, adipose tissue	Increased	None	None
Lipolysis	None	Increased	Increased
Fate of glycerol	Synthesize more triacylglycerol in liver	Substrate for gluconeogenesis	Substrate for gluconeogenesis
β-Oxidation of fatty acids	None	Increased	Markedly increased; primary fuel for muscle
Muscle catabolism	None; increased protein synthesis and uptake of amino acids	Increased; supply amino acids for gluconeogenesis	Decreased; conserve muscle for important body functions
Urea synthesis, excretion	Remains constant; handles NH_4^+ load from protein degradation in gut by bacteria	Increased; deamination of amino acids used for gluconeogenesis increases urea synthesis	Decreased; less muscle breakdown of protein with fewer amino acids to degrade
Ketone body synthesis	None	Increased	Markedly increased; byproduct of acetyl CoA from increased β-oxidation of fatty acids
Muscle use of glucose for fuel	Primary fuel	Decreased	None; mainly uses fatty acids
Muscle use of fatty acids for fuel	None	Increased; primary fuel	Markedly increased; primary fuel
Muscle use of ketones for fuel	None	Some; alternative fuel	None; allows the brain to use ketones for fuel
Brain use of glucose for fuel	Remains constant	Remains constant	Decreased; allows RBCs to primarily use glucose for fuel
Brain use of ketones for fuel	None	None	Increased; primary fuel
RBC use of glucose for fuel	Remains constant	Remains constant	Remains constant

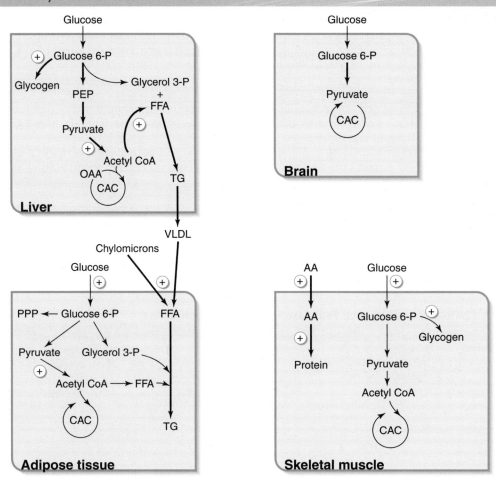

9-1: Overview of metabolism in the well-fed state. *Thick arrows* indicate pathways that are prominent; the plus sign (+) indicates steps that insulin directly or indirectly promotes. AA, amino acid; CAC, citric acid cycle; FFA, free fatty acid; PEP, phosphoenolpyruvate; TG, triacylglycerol; VLDL, very-low-density lipoprotein.

2. Glucokinase traps most of the large glucose influx from the portal vein as glucose 6-phosphate.
   a. In contrast to hexokinase, which is present in most tissues, liver glucokinase is active only at high glucose concentrations and is not inhibited by glucose 6-phosphate (see Table 6-1).
   b. Elevated glucose 6-phosphate immediately stimulates the less active phosphorylated form of glycogen synthase, which increases glycogen synthesis.
3. Active (dephosphorylated) forms of glycogen synthase and pyruvate dehydrogenase are favored by a high insulin-to-glucagon ratio.
   a. Increased pyruvate dehydrogenase activity provides abundant acetyl CoA for synthesis of free fatty acids, which are esterified as triacylglycerols in hepatocytes and transported to adipose tissue (as very-low-density lipoproteins [VLDLs]) for synthesis of triacylglycerol for storage.
4. The oxidative branch of the pentose phosphate pathway provides NADPH, which is required for fatty acid synthesis.
5. Dihydroxyacetone phosphate produced from glucose 6-phosphate is converted into glycerol 3-phosphate, which is the carbohydrate backbone for triacylglycerol synthesis.

C. **Adipose tissue metabolism: well-fed state**
   1. High insulin levels stimulate triacylglycerol synthesis through several reactions.
   2. Increased glucose uptake by insulin-sensitive GLUT4 provides glycerol 3-phosphate for esterification of free fatty acids and storage of triacylglycerol.
   3. Increased lipoprotein lipase activity promotes release and uptake of free fatty acids from chylomicrons and VLDL.
   4. Inhibition of hormone-sensitive lipase by insulin prevents fat mobilization.

Glucokinase in liver traps influx of glucose after meals.

Active fatty acid synthesis from glucose after meals

High insulin levels (fed state): ingested fuels stored as glycogen (liver, muscle), triacylglycerols (adipose, liver), and protein (muscle)

Insulin increases lipoprotein lipase activity and inhibits hormone-sensitive lipase.

D. Muscle metabolism: well-fed state

1. High insulin increases glucose uptake by insulin-sensitive GLUT4 and activation of glycogen synthase leading to the formation of glycogen.
   a. Glucose is the primary fuel for muscle in the fed state.
2. High insulin levels increase amino acid uptake and protein synthesis.
   a. Designed to store carbon skeletons for use as an energy source when needed

Muscle stores glycogen and protein after a meal.

E. Brain metabolism: well-fed state

1. Glucose is the exclusive fuel for brain tissue, except during extreme starvation, when it can use ketone bodies.
2. The brain normally relies on the aerobic metabolism of glucose, so hypoxia and severe hypoglycemia produce similar symptoms (e.g., confusion, motor weakness, visual disturbances).

Glucose is the primary fuel for the brain and can use ketone bodies in starvation.

III. The Fasting State

A. Overview

1. The period extending from 3 to 36 hours after a meal is marked by decreasing levels of absorbed nutrients in the bloodstream and a declining insulin-to-glucagon ratio.
2. Metabolism initially shifts to increasing reliance on glycogenolysis and then to gluconeogenesis to maintain blood glucose in the absence of nutrient absorption from the gut (see Table 9-2).

Fasting state: low insulin-to-glucagon ratio; normal blood glucose level (produced by liver); mobilization of free fatty acids

B. Liver metabolism: fasting state (Fig. 9-2)

1. Glucose 6-phosphatase, a gluconeogenic enzyme that is present in liver but not in muscle, converts glucose 6-phosphate from glycogenolysis and gluconeogenesis to glucose, which is released into blood.
2. Glycogen degradation (glycogenolysis) is stimulated by glucagon-induced activation of glycogen phosphorylase and inhibition of glycogen synthase, which prevents futile recycling of glucose 1-phosphate.
3. Gluconeogenesis is stimulated by glucagon through several mechanisms.
   a. Reduction in fructose 2,6-bisphosphate concentration relieves inhibition of fructose 1,6-bisphosphatase (rate-limiting enzyme) and reduces activation of phosphofructokinase 1.
      (1) The net result is increased gluconeogenesis and decreased glycolysis.
   b. Inactivation of pyruvate kinase (by protein kinase A) reduces futile recycling of phosphoenolpyruvate (PEP).
      (1) Pyruvate kinase is also allosterically inhibited by high adenosine triphosphate (ATP) and alanine levels.
   c. Increased liver uptake of amino acids (derived from protein catabolism in skeletal muscle) provides carbon skeletons for gluconeogenesis (alanine is transaminated into pyruvate).
   d. Increased synthesis of urea cycle enzymes disposes of nitrogen from amino acids and increases the excretion of urea in the urine.

Glucose 6-phosphatase: located in liver, a blood sugar–regulating organ, not in muscle, a glucose-consuming tissue

Gluconeogenesis: stimulated in fasting; activation of fructose 1,6-bisphosphatase; inactivation of pyruvate kinase

Increased amino acid mobilization and urea cycle during fasting

Low insulin levels (fasting): stored fuels mobilized from glycogen (liver, muscle), fat (adipose), and protein (muscle); adequate blood glucose maintained for brain

4. Hepatic oxidation of free fatty acids (derived from lipolysis in adipose tissue) elevates the concentration of ATP, acetyl CoA, and citrate.
   a. Citrate allosterically stimulates fructose 1,6-bisphosphatase (increases gluconeogenesis) and inhibits phosphofructokinase 1 (decreases glycolysis).
   b. Acetyl CoA activates pyruvate carboxylase, which converts pyruvate to oxaloacetate (OAA) for use in the gluconeogenic pathway.
   c. Inhibition of pyruvate dehydrogenase by acetyl CoA also increases shunting of pyruvate toward oxaloacetate.
   d. Increased ATP concentration resulting from oxidation of fatty acid–derived acetyl CoA in the citric acid cycle inhibits glycolysis and supplies energy for gluconeogenesis.
5. Glycerol derived from lipolysis in adipose tissue is phosphorylated in the liver by glycerol kinase and contributes carbon skeletons for hepatic gluconeogenesis.
6. Some ketogenesis occurs in the liver, with ketone bodies primarily transported to muscle as an alternative fuel, sparing blood glucose.

Fat oxidation maintains cellular energy levels during fasting.

C. Adipose tissue metabolism: fasting state

1. Low insulin levels and elevated epinephrine levels promote the active form of hormone-sensitive lipase, which splits triacylglycerols into glycerol and free fatty acids.
2. Free fatty acids are transported in blood bound to serum albumin.
   a. Liver and muscle use the free fatty acids released by β-oxidation in the mitochondria as the primary energy source during fasting.
3. Glycerol is converted into glycerol 3-phosphate in the liver and is used as a substrate for gluconeogenesis.

Glycerol from mobilized triacylglycerols is used by the liver to make glucose.

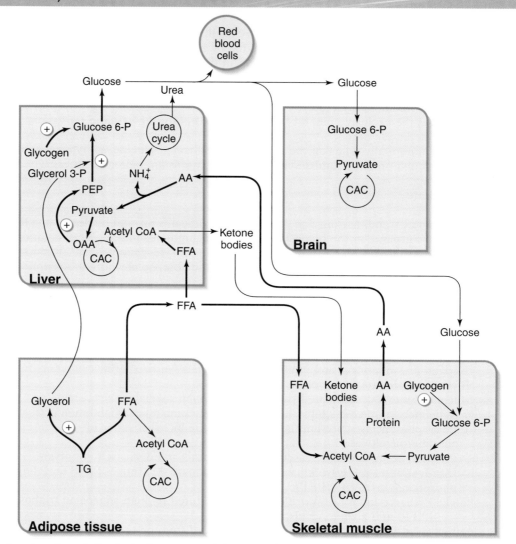

9-2: Overview of metabolism in the fasting state. *Thick arrows* indicate pathways that are prominent; the plus sign (+) indicates steps that are promoted directly or indirectly by glucagon in the liver and epinephrine in adipose tissue and muscle. Some glucogenic amino acids (AA) are converted to citric acid cycle (CAC) intermediates in the liver. FFA, free fatty acid; PEP, phosphoenolpyruvate; TG, triacylglycerol.

D. Muscle metabolism: fasting state

1. Degradation of muscle protein provides carbon skeletons for hepatic gluconeogenesis.
   a. Most amino acids released from muscle protein are transported directly to the liver, where they are transaminated and converted to glucose.
   b. Branched-chain amino acids (i.e., isoleucine, leucine, and valine) are converted to their α-keto acids in muscle by transamination of pyruvate, yielding alanine, which is transported to the liver.
      (1) The alanine cycle, which disposes of nitrogen from branched-chain amino acids, results in no net production of glucose for use by other tissues (Fig. 9-3).

Fasting muscle: glycogen, branched-chain amino acids, and fatty acids for energy

9-3: Alanine cycle for disposing of nitrogen from branched-chain amino acids (BCAA). In contrast to other amino acids, the three BCAA (i.e., isoleucine, leucine, and valine) are metabolized to alanine and branched-chain keto acids (BCKA) in skeletal muscle (not the liver, which lacks the necessary enzymes). Glucose produced in the liver is returned to muscle to regenerate the pyruvate supply for transamination of more BCAA, resulting in no net production of glucose for use by other tissues.

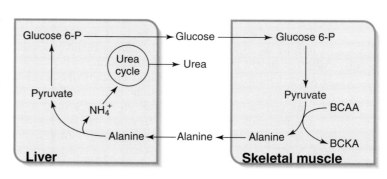

2. Free fatty acids are the primary fuel source for muscle during fasting.
3. Glycogen degradation can provide glucose as fuel for muscle for short periods of exertion.
   a. Skeletal muscle lacks glucose 6-phosphatase; therefore, degradation of muscle glycogen cannot contribute to blood glucose.

E. **Brain metabolism: fasting state**
   1. Brain tissue continues to use glucose as an energy source during periods of fasting.

IV. **The Starvation State**
   A. **Overview**
      1. After 3 to 5 days of fasting, increasing reliance on fatty acids and ketone bodies for fuel enables the body to maintain the blood glucose level at 60 to 65 mg/dL and to spare muscle protein for prolonged periods without food (see Table 9-2).
      2. Less NH_4^+ is produced; therefore, less urea is excreted in the urine.
   B. **Liver metabolism: starvation state (Fig. 9-4)**
      1. The rate of gluconeogenesis decreases as the supply of amino acid carbon skeletons from muscle protein catabolism decreases.
         a. Glycerol released by lipolysis in adipose tissue supports a low level of gluconeogenesis in the liver (kidneys), which is the only tissue that contains glycerol kinase:

$$\text{Glycerol} \rightarrow \text{Glycerol 3-P} \rightarrow \text{DHAP} \rightarrow \rightarrow \rightarrow \text{Glucose}$$

      2. Fatty acid oxidation continues at a high level.
      3. Acetyl CoA accumulates as the citric acid cycle slows down.
         a. Elevated acetyl CoA is shunted to produce ketone bodies, which consist of acetoacetate and β-hydroxybutyrate (acetone is not a ketone body).

> Fatty acids are the primary source of fuel for muscle in the fasting state.

> Starvation state: sustained very low insulin-to-glucagon ratio; very low blood glucose levels; ketosis

> Starvation: gluconeogenesis from glycerol, not amino acids; high fatty acid oxidation and ketone production

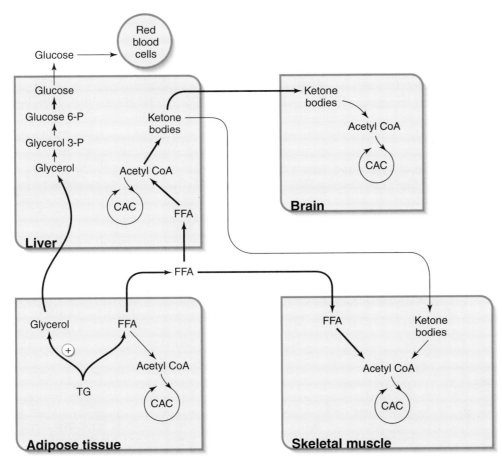

9-4: Overview of metabolism in the starvation state. As starvation persists, the use of ketone bodies by skeletal muscle decreases, sparing this fuel source for brain tissue.

b. Ketoacidosis resulting from increased hepatic production of ketone bodies is the hallmark of starvation.
  (1) Acetoacetate and β-hydroxybutyrate are metabolized to acetyl CoA and used for energy production by many tissues (e.g., muscle, brain, kidney) but not by the red blood cells (RBCs) or the liver.
  (2) Acetone, which is not metabolized, gives a fruity odor to the breath.

C. **Adipose tissue metabolism: starvation state**
  1. The elevated epinephrine level caused by the stress of starvation coupled with very reduced levels of insulin increases the activity of hormone-sensitive lipase, which further stimulates the mobilization of fatty acids from stored fat.

D. **Muscle metabolism: starvation state**
  1. Degradation of muscle protein decreases as the demand for blood glucose is reduced due to a reduction in gluconeogenesis.
  2. Free fatty acids and ketone bodies are used as energy sources in early starvation.
  3. As starvation persists, muscle relies increasingly on free fatty acids, sparing ketone bodies for use by the brain.

E. **Brain metabolism: starvation state**
  1. Increasing ketone body use by the brain spares blood glucose for use by RBCs, which rely solely on glucose for energy production.
  2. Decreasing glucose use by the brain reduces the need for hepatic gluconeogenesis and indirectly spares muscle protein.

V. **Diabetes Mellitus**
  A. **Overview**
    1. Hyperglycemia leads to a similar pathology in type 1 and type 2 diabetes mellitus (DM), but the two conditions have different underlying causes, acute manifestations, and treatments (Table 9-3).
  B. **Type 1 DM**
    1. Type 1 DM is caused by autoimmune destruction of pancreatic β cells.
      a. Cell-mediated immunity and antibodies directed against insulin and islet cells
      b. Total absence of endogenous insulin production eventually results, accompanied by onset of clinical symptoms.
      c. Human leukocyte antigen (HLA) genes are involved with the autoimmune destruction of β cells.
    2. Polydipsia, polyuria, and polyphagia, the classic triad of presenting symptoms, are usually accompanied by weight loss, fatigue, and weakness.
    3. Metabolic changes in untreated type 1 DM resemble, but are distinct from, those in starvation and lead to four characteristic metabolic abnormalities (Fig. 9-5).

TABLE 9-3. **Comparison of Type 1 and Type 2 Diabetes Mellitus**

CHARACTERISTIC	TYPE 1 DM	TYPE 2 DM
Proportion of diagnosed diabetics	5-10%	90-95%
Usual time of onset (exceptions are common)	Childhood, adolescence, early adulthood	Patients older than 40 years; frequently associated with obesity
Cause	Gradual elimination of insulin production due to autoimmune destruction of β cells; human leukocyte antigen (HLA) relationship	Relative insulin deficiency; insulin resistance of target tissues due to decreased insulin receptors; post-receptor defects (e.g., tyrosine kinase defects); no HLA relationship
Plasma insulin (basal)	Absent	Normal to high
Metabolic disorders	Hyperglycemia, ketoacidosis, lactic acidosis from shock, hypertriglyceridemia, muscle wasting	Hyperglycemia, hyperosmolarity; no ketosis; lactic acidosis from shock
Symptoms	Rapid onset of polydipsia, polyuria, polyphagia	Insidious onset
Insulin therapy	Always necessary	May be necessary; diet, exercise, and oral glucose-lowering agents used primarily
Long-term complications	Atherosclerosis, microvascular disease, peripheral neuropathy, retinopathy, nephropathy	Similar to type 1, but slower onset

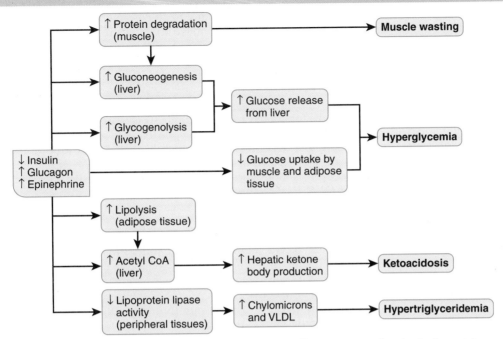

9-5: Mechanisms of metabolic changes in untreated type 1 diabetes mellitus. VLDL, Very-low-density lipoprotein.

a. Muscle wasting results from exaggerated degradation of muscle protein.
  (1) Lack of insulin reduces uptake of amino acids and protein synthesis by muscle but promotes protein degradation.
  (2) Released amino acids are used for muscle energy production (along with free fatty acids from adipose tissue) and also are transported to the liver for gluconeogenesis.
  (3) By comparison, during starvation muscle protein is spared (less gluconeogenesis).
b. Hyperglycemia is caused by increased hepatic glucose production and reduced glucose uptake by insulin-sensitive GLUT4 in adipose tissue and muscle.
  (1) Very low insulin-to-glucagon ratio causes increased gluconeogenesis (most important) and increased glycogenolysis in the liver despite high blood glucose.
  (2) Osmotic diuresis from glucosuria results in hypovolemic shock due to significant losses of sodium in the urine.
  (3) For comparison, during starvation blood glucose is usually maintained near the lower end of the normal range due to decreased gluconeogenesis.
c. Ketoacidosis results from excessive mobilization of fatty acids (lipolysis) from adipose tissue.
  (1) Elevated acetyl CoA resulting from β-oxidation of fatty acids in the liver leads to accelerated ketone body production, which is much greater than in starvation.
  (2) Because blood glucose is high, brain tissue uses glucose instead of ketone bodies (as in starvation), which contributes to ketoacidosis.
d. Lactic acidosis from shock related to osmotic diuresis related to glucosuria
e. Hypertriglyceridemia is caused by reduced lipoprotein lipase activity in adipose tissue and excessive fatty acid esterification in the liver.
  (1) Reduced lipoprotein lipase activity in adipose tissue, due to lack of insulin, leads to elevated plasma levels of chylomicrons from ingested fats and VLDL from hepatic triacylglycerol production, which often results in a type V hyperlipoproteinemia (see Table 7-3).
4. Insulin therapy reduces clinical symptoms of type 1 DM and alleviates life-threatening ketoacidosis.
  a. Successful treatment of type 1 DM also reduces the risk for long-term complications, which are thought to result primarily from prolonged hyperglycemia.
  b. Glucose-related damage is due primarily to two mechanisms.
  (1) Nonenzymatic glycosylation (glucose covalently linked to protein) of the basement membranes of arterioles and capillaries

Type 1 DM: hyperglycemia from increased hepatic gluconeogenesis

Type 1 DM: ketoacidosis, lactic acidosis, and hyperglycemia

Starvation: ketosis and hypoglycemia

Type 1 DM: hypertriglyceridemia (reduced lipoprotein lipase) as elevated chylomicrons and VLDL

Nonenzymatic glycosylation increases basement membrane permeability to proteins in the plasma.

(2) Osmotic damage to the lens in the eye due to glucose conversion into sorbitol by aldose reductase; produces cataracts, damage to Schwann cells (i.e., peripheral neuropathy), weakening of retinal vessels (i.e., microaneurysms with potential for blindness)

c. The most common complication of insulin therapy is hypoglycemia and subsequent insulin coma due to excess insulin.

C. Type 2 DM

1. Type 2 DM results from dysfunctional pancreatic β cells (i.e., relative insulin deficiency) and insulin resistance (i.e., reduced responsiveness of target tissues to insulin at the receptor and post-receptor level), which is associated with obesity.
   a. Obesity leads to downregulation of insulin receptor synthesis.
   b. Examples of post-receptor abnormalities include defects in tyrosine kinase on the insulin receptor and GLUT4 dysfunction.
2. Basal insulin level is normal to high, but insulin release in response to glucose is insufficient to prevent hyperglycemia.
   a. Insulin levels are sufficient to prevent ketoacidosis.
3. Metabolic changes in untreated type 2 DM usually are milder than in type 1 DM, partly because the imbalance in insulin and glucagon levels is not as extreme.
   a. Ketoacidosis does not occur, but hyperglycemia may be greater than in type 1 DM (e.g., hyperosmolar nonketotic coma).
   b. Lactic acidosis does occur, because patients are frequently in shock from osmotic diuresis related to glucosuria.
4. Weight loss (upregulates insulin receptor synthesis), exercise, and dietary modifications can adequately control blood glucose in some type 2 diabetics.
5. In more difficult cases, insulin therapy or an oral hypoglycemic agent is necessary to reduce hyperglycemia.

VI. Alcohol Metabolism

A. Overview

1. Metabolism of alcohol occurs primarily in the liver by two pathways, depending on the ethanol concentration.
2. Ethanol oxidation is similar to fat oxidation in the production of high levels of NADH.

B. Low concentrations of ethanol

1. At low concentrations of ethanol, oxidation by two dehydrogenases results in an increased $NADH/NAD^+$ ratio (Fig. 9-6).
2. Alcohol dehydrogenase in the cytosol and aldehyde dehydrogenase in mitochondria transfer electrons to NAD^+ to form NADH.
   a. Disulfiram, which inhibits aldehyde dehydrogenase and is used in the treatment of chronic alcoholism, causes severe perspiration, abdominal cramping, and nausea after ingestion of alcohol.

Margin notes:

Type 1 DM: hypoglycemia most common complication

Type 2 DM: relative insulin deficiency, insulin resistance, post-receptor defects, hyperglycemia, no ketosis, and obesity

Type 1 DM: managed with exogenous insulin

Type 2 DM: managed with weight loss, exercise, dietary modifications, oral hypoglycemic agents, and insulin

Ethanol oxidation: primarily in liver

Disulfiram: inhibits aldehyde dehydrogenase

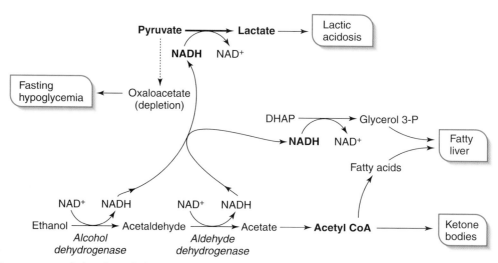

9-6: Consequences of ethanol metabolism in the liver. Reduction in the pyruvate level as it is converted to lactate and depletion of oxaloacetate eventually lead to fasting hypoglycemia. The higher $NADH/NAD^+$ ratio also increases the conversion of dihydroxyacetone phosphate (DHAP) to glycerol 3-phosphate, leading to increased synthesis of triacylglycerol.

3. Lactic acidosis results from shunting of pyruvate into lactate due to a high NADH/NAD$^+$ ratio.
4. Fasting hypoglycemia results from depletion of pyruvate, which leads to reduced gluconeogenesis.
   a. Alcohol-induced hypoglycemia most likely occurs in individuals who have not eaten recently.
   b. In this situation, liver glycogen stores are depleted, and maintenance of blood glucose depends entirely on gluconeogenesis.
5. Fatty liver in alcoholics is a consequence of prolonged imbalance in the NADH/NAD$^+$ ratio.
   a. High NADH promotes increased formation of glycerol 3-phosphate from dihydroxyacetone phosphate, which is esterified to yield triacylglycerols.
   b. Reduction in the citric acid cycle leads to shunting of acetyl CoA and to increased fatty acid synthesis.
   c. Impaired protein synthesis prevents assembly and secretion of VLDL, causing triacylglycerols to accumulate in the liver.
6. Ketone bodies are produced by increased shunting of acetyl CoA from the citric acid cycle.
   a. High NADH promotes increased conversion of acetoacetic acid to β-hydroxybutyric acid (β-OHB)
   b. β-OHB is not detected by the standard tests for ketone bodies performed on urine and blood.

C. Higher concentrations of ethanol
1. At higher concentrations of ethanol, cytochrome P450 enzymes, which have a relatively high K_m for ethanol, are active in the smooth endoplasmic reticulum (SER).

$$\text{Ethanol} + \text{NADPH} + O_2 \rightarrow \text{Acetaldehyde} + \text{NADP}^+ + H_2O$$

2. This microsomal-ethanol oxidizing system also detoxifies drugs such as barbiturates.
   a. Alcohol causes the SER in liver cells to increase, which results in an increased synthesis of γ-glutamyltransferase (GGT), an enzyme located in the SER that is an excellent marker of alcohol ingestion.
3. Because ethanol oxidation by the microsomal-ethanol oxidizing system does not affect the NADH/NAD$^+$ ratio substantially, it does not have the metabolic effects described for low concentrations of ethanol.

Side notes:

Ethanol metabolism in liver: high NADH levels leading to lactic acidosis and hypoglycemia

Increased NADH from ethanol pushes lactate and β-OHB formation.

Cytochrome P450 in liver SER oxidizes excessive ethanol and increases serum GGT levels.

I. Overview

A. Nucleotides are composed of ribose phosphate (donated by phosphoribosyl pyrophosphate [PRPP]) and a purine or pyrimidine base.

B. Several kinases use adenosine triphosphate (ATP) to generate nucleoside diphosphates and triphosphates.

C. Purines are synthesized by addition of groups to the PRPP.

D. Pyrimidines are synthesized by condensation of carbamoyl phosphate and aspartate; the ribose phosphate is donated later in the process.

E. Synthesis of purines and pyrimidines is allosterically regulated to produce balanced amounts of each nucleotide.

F. Degradation pathways are designed to salvage some nucleotides and to produce uric acid (purines).

G. Several genetic diseases are caused by blocks in the degradation pathways.

II. Nucleotide Structure

A. Nucleotides consist of an organic base bound by an *N*-glycosyl linkage to a phosphorylated pentose (Fig. 10-1A).

B. The nucleotide base has a purine (adenine, guanine) or pyrimidine (cytosine, thymine, uracil) ring structure (see Fig. 10-1B).

1. RNA and DNA contain adenine (A), guanine (G), and cytosine (C).

2. Uracil (U) occurs only in RNA, and thymine (T) occurs only in DNA.

C. The nucleotide pentose is ribose in RNA or deoxyribose in DNA.

1. Ribonucleotide reductase reduces nucleoside diphosphates (ADP, GDP, CDP, and UDP) to their deoxy forms (e.g., dADP) in a reaction that requires reduced thioredoxin as a cofactor.

2. Oxidized thioredoxin must be converted back into reduced thioredoxin for continued conversion of ribonucleotides to their deoxy forms.

   a. Thioredoxin reductase, using NADPH as a cofactor, is required for this conversion.

   b. Notice the similarity of this reaction with conversion of BH_2 back to BH_4 by dihydrobiopterin reductase (See Chapter 8).

III. Purine Synthesis (Fig. 10-2)

A. Step 1

1. Formation of 5-phosphoribosylamine from 5-phosphoribosyl 1-pyrophosphate (PRPP) and glutamine is a committed step in the *de novo* synthesis of purines (see Fig. 10-2).

2. End-product feedback inhibition of this reaction by guanosine monophosphate (GMP), adenosine monophosphate (AMP), and inosine monophosphate (IMP) prevents overproduction of purines; high PRPP concentrations overcome this inhibition.

3. Ribose phosphate pyrophosphokinase (PRPP synthetase), the rate-limiting enzyme, produces PRPP.

4. PRPP donates the ribose 5-phosphate on which the newly synthesized purine ring is assembled.

5. PRPP, an activated form of ribose 5-phosphate, is also used in the purine salvage pathway and in pyrimidine synthesis.

B. Step 2

1. Conversion of IMP to AMP or GMP is controlled by two mechanisms to achieve balanced production of AMP and GMP.

RNA: contains ribose phosphate and A, G, C, and U

Deoxynucleotide formation requires thioredoxin and NADPH.

DNA: contains deoxyribose and A, G, C, and T

Purine synthesis: PRPP synthetase is rate-limiting enzyme; purine ring assembled on ribose 5-phosphate supplied by PRPP

Total amount of purines regulated by feedback inhibition of 5-phosphoribosylamine formation

---

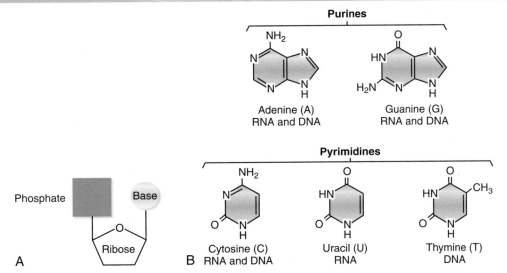

10-1: Nucleotide structure. **A,** All nucleotides have three components. **B,** Five bases are most commonly found in nucleotides. RNA and DNA contain adenine (A), guanine (G), and cytosine (C); uracil (U) is present only in RNA and thymine (T) only in DNA. Ribose in nucleoside diphosphates can also be reduced to deoxyribose.

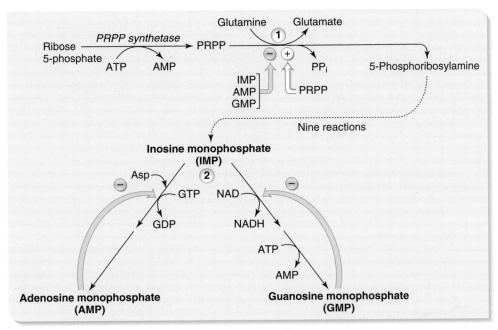

10-2: Overview of purine synthesis. The purine ring is assembled on a ribose 5-phosphate molecule supplied by 5-phosphoribosyl 1-pyrophosphate (PRPP).

    2. One mechanism is cross-regulation, in which the end product of one pathway is required for the other pathway.
       a. For example, the GMP pathway requires ATP, and the AMP pathway requires guanosine triphosphate (GTP).
    3. The other mechanism for balanced production is end-product inhibition of each pathway.
  C. Purine salvage pathways produce nucleotides from preformed bases by transfer of ribose 5-phosphate groups from PRPP to free bases.

IV. Pyrimidine Synthesis (Fig. 10-3)
  A. In contrast to the purine ring, which is assembled on a ribose 5-phosphate molecule, the pyrimidine ring is synthesized before the ribose molecule is supplied by PRPP.
  B. Chemical reaction sequence of pyrimidine synthesis
    1. Step 1
      a. Carbamoyl phosphate synthetase II (CPS II) catalyzes the formation of carbamoyl phosphate from glutamine, carbon dioxide (CO_2), and 2 ATP (see Fig. 10-3).

Pyrimidine synthesis: CPS II is rate-limiting cytosolic enzyme; ring synthesized before ribose 5-phosphate molecule supplied by PRPP

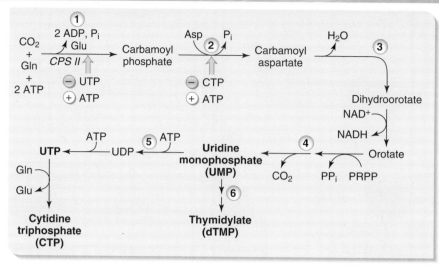

10-3: Overview of pyrimidine synthesis. The pyrimidine ring is synthesized before the ribose 5-phosphate molecule is supplied by 5-phosphoribosyl 1-pyrophosphate (PRPP). Asp, aspartate; CPS II, carbamoyl phosphate synthetase II; Gln, glutamine; Glu, glutamate.

In orotic acidemia, deficiency of carbamoyl phosphate synthetase in hyperammonemia elevates cytosolic carbamoyl phosphate and pushes pathway to elevate orotate, which spills into blood.

  b. Unlike CPS I, which is a mitochondrial enzyme in urea synthesis (See Chapter 8), CPS II is located in the cytosol and is not activated by *N*-acetylglutamate.
  c. This reaction, the committed step in pyrimidine synthesis, is inhibited by uridine triphosphate (UTP) and activated by ATP and PRPP.
  2. Step 2
    a. Aspartate transcarbamoylase adds aspartate to carbamoyl phosphate, producing carbamoyl aspartate.
  3. Step 3
    a. Dihydroorotase converts carbamoyl aspartate into dihydroorotate, which is oxidized to produce orotic acid.
  4. Step 4
    a. Orotate is converted to uridine monophosphate (UMP).
  5. Step 5
    a. UMP is the common precursor for synthesis of the remaining pyrimidines uridine triphosphate (UTP) and cytidine triphosphate (CTP).
  6. Step 6
    a. UMP is converted to deoxyuridine monophosphate (dUMP) before conversion to thymidylate (dTMP).
    b. Thymidylate synthase methylates dUMP to produce dTMP, using methylene tetrahydrofolate as the carbon donor (see Fig. 4-3 in Chapter 4).
  7. Salvage of uracil and thymine (but *not* cytosine) and their conversion into nucleotides is catalyzed by pyrimidine phosphoribosyl transferase, which uses PRPP as the source of ribose 5-phosphate.

Anticancer drugs inhibiting nucleotide synthesis: methotrexate, 5-fluorouracil, and hydroxyurea

V. Anticancer Drugs Inhibiting Nucleotide Synthesis (Table 10-1)

TABLE 10-1. **Anticancer Drugs Inhibiting Nucleotide Synthesis**

DRUG	MECHANISM OF ACTION	TREATMENT
Methotrexate (MTX) (folate analogue)	Competitively inhibits dihydrofolate reductase (DHR), which catalyzes reduction of dihydrofolate to tetrahydrofolate (FH$_4$), required for synthesis of thymidine and purine nucleotides (see Chapter 4)	Molar disease (hydatidiform moles and choriocarcinoma), leukemias and lymphomas, osteogenic sarcoma, and rheumatoid arthritis
5-Fluorouracil (5-FU)	Converted by tumor cell enzymes to 5-fluorodeoxyuridine monophosphate (FdUMP), which irreversibly inhibits thymidylate synthase and prevents the synthesis of deoxythymidine monophosphate (dTMP) from deoxyuridine monophosphate (dUMP)	Breast, stomach, and colon cancers
Hydroxyurea	Inhibits ribonucleotide reductase, which converts ribonucleotides to deoxyribonucleotides, decreasing deoxyribonucleotides and DNA synthesis	Chronic myelogenous leukemia and sickle cell anemia, in which it increases synthesis of hemoglobin F and decreases sickling

VI. Degradation of Nucleotides and Purine Salvage (Fig. 10-4)

A. Purine nucleotides are degraded to uric acid, which is excreted in urine.

B. Chemical reactions involved in the degradation of purines to uric acid

1. Step 1
   a. The first set of reactions in Figure 10-4 shows IMP synthesis for perspective (see Fig. 10-2).
2. Step 2
   a. Adenosine monophosphate (AMP) is converted to IMP by AMP deaminase; AMP is also converted to adenosine.
3. Step 3
   a. Adenosine is converted to inosine by adenosine deaminase.
   b. Adenosine deaminase also catalyzes a reaction converting 2-deoxyadenosine to 2-deoxyinosine.
   c. Deficiency of adenosine deaminase is associated with severe combined immunodeficiency (SCID) (Table 10-2).
4. Step 4
   a. IMP is converted into inosine, and GMP is converted into guanosine by 5'-nucleotidase.
5. Step 5
   a. Inosine is converted into hypoxanthine (purine base) and guanosine into guanine (purine base) by purine nucleoside phosphorylase.
6. Step 6
   a. Guanine and hypoxanthine are converted into xanthine, with the latter converted by means of xanthine oxidase.
7. Step 7
   a. Xanthine oxidase further converts xanthine into uric acid.
   b. Allopurinol inhibits xanthine oxidase, resulting in reduced synthesis of uric acid and an increase in xanthine and hypoxanthine levels.
   c. Underexcretion (most common) or overproduction of uric acid may produce gout (Table 10-2).

> Production of uric acid:
> AMP → IMP → inosine → hypoxanthine → uric acid

> Adenosine deaminase deficiency (AR): combined B- and T-cell deficiency (SCID); first gene therapy experiment

> Allopurinol inhibits xanthine oxidase.

> Gout: hyperuricemia due to overproduction or underexcretion of uric acid

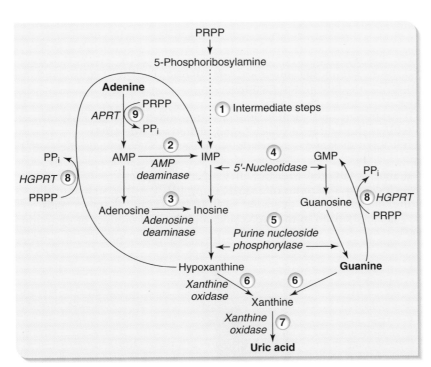

10-4: Overview of purine degradation and purine salvage. Hypoxanthine-guanine phosphoribosyl transferase (HGPRT) and adenine phosphoribosyl transferase (APRT) are involved in salvaging purines for their conversion into nucleotides. Uric acid is the end product of purine degradation. PP$_i$, pyrophosphate; PRPP, 5-phosphoribosyl 1-pyrophosphate.

TABLE 10-2. **Genetic Disorders Involving Nucleotides**

GENETIC DISORDER	PATHOGENESIS	CLINICAL ASSOCIATIONS
Severe combined immunodeficiency (SCID) (autosomal recessive)	Variant of SCID due to deficiency of adenosine deaminase (ADA), which catalyzes conversion of adenosine to inosine Deficiency leads to an accumulation of adenosine (toxic to B and T lymphocytes) and adenosine monophosphate (AMP), which is converted by AMP deaminase to inosine monophosphate (IMP) ADA also catalyzes conversion of 2-deoxyadenosine to 2-deoxyinosine Deficiency of ADA leads to an accumulation of 2-deoxyadenosine and deoxyadenosine monophosphate (dAMP) dAMP is converted to dADP and dATP, which accumulates in the cell; dATP inhibits ribonucleotide reductase, which reduces conversion of ribonucleotides to deoxynucleotides, resulting in decreased DNA synthesis in B and T cells	Combined B- and T-cell deficiency leads to recurrent infections involving bacteria, viruses, fungi, and protozoa due to loss of humoral and cellular immunity
Gout (acquired or genetic)	Associated with overproduction or underexcretion (most common) of uric acid, end product of purine degradation Overproduction may be caused by overactivity of ribose phosphate pyrophosphokinase (PRPP synthetase) or deficiency of hypoxanthine-guanine phosphoribosyl transferase (HGPRT), a purine salvage enzyme Underexcretion is caused by defects in renal excretion of uric acid	Acute gout most often occurs in the metatarsophalangeal joint of the large toe Recurrent attacks are common hallmarks of disease: hyperuricemia with deposition of monosodium urate (MSU) crystals in synovial fluid Treatment: reduce intake of red meats and alcohol Treat underexcretors with uricosuric agents (e.g., probenecid) and overproducers with allopurinol (blocks xanthine oxidase) Untreated, MSU accumulates in soft tissue (tophus)
Lesch-Nyhan syndrome (X-linked recessive)	Due to total deficiency of HGPRT, the salvage enzyme for hypoxanthine and guanine Loss of the enzyme results in conversion of hypoxanthine and guanine into xanthine, which is converted into uric acid, leading to hyperuricemia	Severe mental retardation, self-mutilating behavior, spasticity, gout, and urate deposition in the kidney, leading to renal failure

C. Hypoxanthine-guanine phosphoribosyl transferase (HGPRT) is involved in the salvage of the purines hypoxanthine and guanine.
   1. Step 8
      a. Hypoxanthine and guanine are converted into IMP and GMP, respectively, using PRPP as the source of ribose 5-phosphate (see Fig. 10-4).
      b. HGPRT deficiency is associated with Lesch-Nyhan syndrome (see Table 10-2).
   2. Step 9
      a. Adenine is salvaged with the enzyme adenine phosphoribosyl transferase (APRT), which converts adenine into AMP, using PRPP as the source of ribose 5-phosphate.
D. Pyrimidine nucleotides are degraded to CO_2, NH_4^+, and β-amino acids.
   1. NH_4^+ is metabolized in the urea cycle, and the end product of pyrimidine degradation is urea.

Lesch-Nyhan syndrome: deficiency of HGPRT; manifests as hyperuricemia with mental retardation

End product of pyrimidine degradation is urea.

CHAPTER 11
ORGANIZATION, SYNTHESIS, AND REPAIR OF DNA

I. DNA Organization

A. Overview

1. Two characteristics of DNA that affect its organization in the nucleus: extreme length and predominance of noncoding sequences
2. Supercoiling allows compaction of DNA while allowing selective expression of genetic information.
3. More than 98% of the genome is noncoding:
    a. Introns divide the coding regions (exons).
    b. Regulation of gene expression as promoters and enhancers (or silencers)
    c. Pseudogenes
    d. Repetitive DNA

B. Nucleosomes

1. Chromatin is a complex of double-strand DNA (Fig. 11-1) associated with histone proteins and other proteins that folds and packs to form the chromosomes.
    a. Histones are basic proteins (positively charged) that associate tightly with acidic DNA (negatively charged).
    b. The nucleosome, the basic structural unit of chromatin, is made up of a core containing a total of eight histones (two molecules each of histones H2A, H2B, H3, and H4), around which DNA is wrapped twice (Fig. 11-2).
    c. Linker DNA, which is associated with a fifth type of histone (H1), connects adjacent nucleosomes and produces the "beads on a string" form of chromatin.
    d. Supercoiling, mediated by histone H1, produces the more compact 30-nm chromatin fiber.
    e. Higher-order coiling and folding of chromatin permits compaction of DNA into chromosomes.
        (1) Fluoroquinolone antibiotics, such as ciprofloxacin, which block DNA compaction by inhibiting bacterial DNA gyrase (topoisomerase II), have little effect on the equivalent human enzyme.

C. Pseudogenes

1. A pseudogene is produced by a retrovirus that makes a DNA copy of a messenger RNA.
2. Pseudogenes cannot be expressed because they lack a promoter and introns.
3. Pseudogenes are replicated along with the genome at cell division, but they remain dormant and have no ability to be expressed.

D. Repetitive DNA and transposons

1. Repetitive DNA is constituted from repeated sequences (from two base pairs to several thousand) that are arranged in tandem; they constitute about 30% of the genome.
2. An alternate term for repetitive DNA is satellite DNA (i.e., appears as a satellite band in centrifugation).
3. Unequal crossover events produce repetitive DNA.
4. Transposons, also referred to as jumping genes, are DNA sequences that can jump to different areas of the genome within the same cell.
5. Retrotransposons are a class of jumping genes that use reverse transcription by a cellular reverse transcriptase (see Section II, E later) encoded in the genome of normal, uninfected cells.
    a. Insertion of a jumping gene into an active gene causes an insertion mutation, which prevents expression of normal protein.

The DNA helix is composed of two strands that are antiparallel and complementary.

DNA: predominance of noncoding sequences; 98% of the genome

Basic histones bind to acidic DNA

Antibodies against histones are increased in drug-induced lupus erythematosus.

Nucleosome: DNA wrapped twice around histone core; connected by linker DNA

Chromatin fibers: nucleosomes arranged in supercoils

Chromosomes: compaction of chromatin fibers

Fluoroquinolone antibiotics inhibit DNA compaction and inhibit gyrase.

Pseudogenes: DNA copy of a messenger RNA; lack promoter and introns

Repetitive DNA: repeated sequences arranged in tandem; 30% of the genome

Transposons: jumping genes; insert themselves into the genome

Retrotransposons: cellular reverse transcriptase; DNA inserts created from mRNA; inactivates genes

11-1: DNA double helix. Strands are complementary (i.e., A is always paired with T, and G is paired with C) and antiparallel (i.e., one strand runs in the $3' \rightarrow 5'$ direction; the other runs in the $5' \rightarrow 3'$). Adenine-thymine (A-T) pairs have two hydrogen bonds; guanine-cytosine (G-C) pairs have three hydrogen bonds.

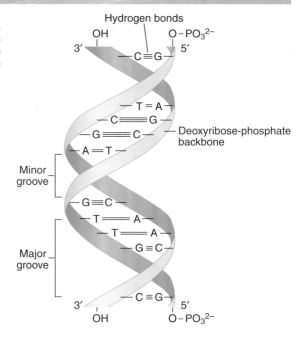

11-2: Schematic model depicting the organization of human DNA. **A** and **B,** Double-strand DNA wraps around histone cores to form nucleosomes, which are connected by intervening linker DNA to which another histone is bound. **C,** Supercoiling produces a more compact 30-nm chromatin fiber. **D,** In interphase chromosomes, long stretches of 30-nm chromatin loop out from a protein scaffold. **E** and **F,** Further folding of the scaffold yields the highly condensed structure of a metaphase chromosome.

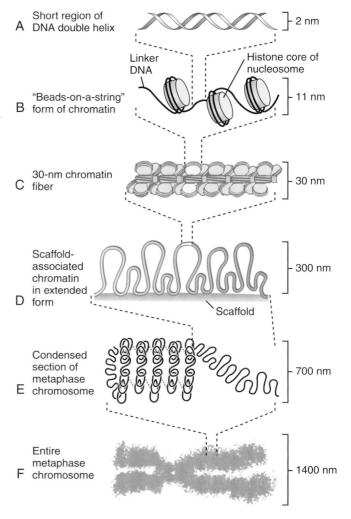

b. Jumping gene insertion can cause genetic diseases such as hemophilia and Duchenne muscular dystrophy.

c. The most abundant type of retrotransposon in the human genome retrotransposons is the L1 family.

II. DNA Synthesis

A. Overview

1. Two signals initiate DNA synthesis: cell division and DNA repair.
2. DNA must be physically exposed to the enzymes and substrates for replication or repair.
3. DNA repair and DNA replication occur at distinct times during the cell cycle.

B. The cell cycle (Fig. 11-3)

1. The cell cycle is an ordered sequence of events during which proliferating cells replicate their chromosomes and separate into daughter cells.
2. The cell cycle consists of interphase, followed by mitosis (nuclear replication) and cytokinesis (cell division).
   a. Interphase consists of a G_1 (first gap or growth) phase, an S (synthesis) phase, and a G_2 phase.
      (1) The G_1 and G_2 phases are periods of RNA and protein synthesis for general growth of organelles and mitotic structures; no DNA synthesis occurs.
      (2) Resting cells in the G_0 phase reenter the cell cycle early in the G_1 phase, the most variable phase of the cell cycle.
      (3) Replication of DNA occurs during the S phase of the cycle. During mitosis and cytokinesis of the M phase, replicated chromosomes align on the mitotic spindle and then segregate evenly to the daughter cells.
3. Cell cycle control proteins regulate progression through the cell cycle.
   a. Cyclin-dependent protein kinases (CDKs) phosphorylate various target proteins, thereby modulating their activity at different stages of the cell cycle.
   b. Cyclins activate CDKs by binding to them. Concentrations of various cyclins rise and fall throughout the phases of the cell cycle, leading to formation of specific cyclin-CDK complexes that control entry into the S phase and entry into and exit from mitosis.

G_1 to S phase is the most important phase to control.

G_1 and G_2 phases: gap (G) periods of RNA and protein synthesis

Drugs that inhibit the G_2 phase, in which tubulin is synthesized, include etoposide and bleomycin.

S phase: DNA synthesis

Cell cycle: interphase, consisting of G_1, S, and G_2 phases, followed by mitosis and cytokinesis

CDKs control entry into stages of the cell cycle.

Cyclins activate CDKs, and concentrations change during the cell cycle.

Arrests at checkpoints prevent abnormalities and permit repair of damaged DNA.

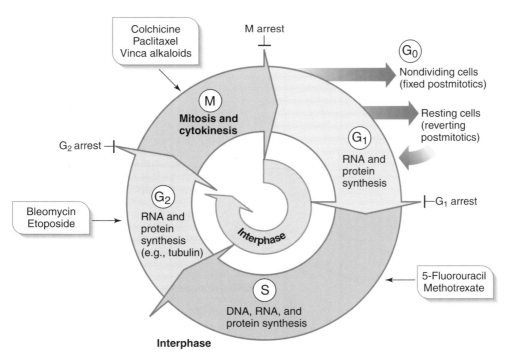

Interphase

11-3: Cell cycle in eukaryotic cells. DNA synthesis occurs only during the synthesis (S) phase. The gap or growth (G) phases are periods of growth of organelles and mitotic structures. Anticancer drugs exert their effect at different stages, depending on their mode of action. Damage to DNA or improper spindle formation leads to arrest, preventing entry to the next stage.

4. Checkpoints are positions in the cell cycle where progress can be arrested to prevent formation of abnormal daughter cells (see M, G_1, and G_2 arrests) (see Fig. 11-3).

a. When DNA is damaged, the G_1 checkpoint prevents entry into the S phase, and the G_2 checkpoint prevents entry into mitosis.

(1) Expression of *TP53*, a tumor-suppressor gene (on chromosome 17) that is required for operation of the G_1 and G_2 checkpoints, codes for a protein product that inhibits activated CDKs.

(2) Arrest of the cell cycle allows time for DNA repair and reduces the chances for perpetuation of mutations.

(3) Inactivation of the *TP53* suppressor gene is seen in most human cancers.

b. The M checkpoint prevents exit from mitosis when the spindle is abnormal, thereby preventing dysfunctional segregation of chromatids and formation of daughter cells with an abnormal number of chromosomes.

C. The replication fork

1. DNA synthesis begins at specific nucleotide sequences, called replication origins, and proceeds in both directions (i.e., bidirectional action).

a. Replication is semiconservative; that is, each new DNA double helix contains one parental (old copy) strand and one daughter (new copy) strand.

b. DNA is exposed to replication enzymes by action of DNA gyrase (topoisomerase II); DNA is relaxed with negative supercoils and unwound.

2. Multiple proteins participate in initiation and elongation of daughter DNA strands at the replication fork (Fig. 11-4).

a. Helicase separates the two strands of the relaxed DNA helix, and single strand–binding proteins attach to each strand; the point of separation is the replication fork.

b. Helicase unwinding causes a physical strain (positive supercoiling) immediately ahead of the moving replication fork; topoisomerase I relieves this strain by breaking one of the DNA strands and allowing it to rotate around the unbroken strand before rejoining the broken ends.

c. DNA polymerases catalyze the addition of 5′-deoxyribonucleoside triphosphates to the free 3′-OH end of the growing strands with a loss of pyrophosphate (PP_i); this always leaves a free 3′-hydroxy group to accept the next addition.

(1) The parental (template) strand is always read in the $3' \rightarrow 5'$ direction.

(2) The daughter (new) strand is always synthesized in the $5' \rightarrow 3'$ direction.

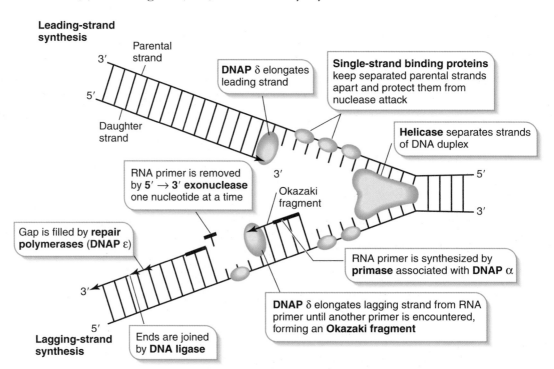

11-4: Schematic depiction of the replication fork and synthesis of leading and lagging strands in eukaryotic cells. Major proteins and their functions in DNA synthesis are indicated; various accessory proteins that are also required are not shown. DNAP, DNA polymerase.

(3) Because of the directionality of DNA synthesis, one daughter strand is synthesized continuously (i.e., leading strand), and the other is synthesized discontinuously (i.e., lagging strand) (see Fig. 11-4).

3. Different processes are involved in the synthesis of leading and lagging strands.
   a. Leading-strand synthesis, catalyzed primarily by DNA polymerase δ, proceeds continuously from a single RNA primer synthesized by DNA polymerase α.
   b. Lagging-strand synthesis proceeds discontinuously from multiple RNA primers synthesized by DNA polymerase α, forming short, 1000 base pair Okazaki fragments.
      (1) The Okazaki fragments are extended by DNA polymerase δ, as in leading-strand elongation.
      (2) Primer is removed from the fragments by a 5′→ 3′ exonuclease, and gaps are filled by DNA polymerase ε, which also functions in repair.
      (3) DNA ligase connects the ends of the adjoining fragments.
4. The proofreading activity of DNA polymerases δ and ε usually detects and corrects base-pairing errors that occur during DNA synthesis (e.g., formation of an A-C pair rather than the correct A-T pair).
   a. The 3′ → 5′ exonuclease activity of the DNA polymerases removes a mismatched nucleotide, and synthesis then continues with insertion of the correct nucleotide.

D. Telomerase
1. Shortening of chromosomes during DNA replication occurs in cells that lack telomerase.
2. Dissociation of DNA polymerase from the 3′ end of the daughter strand during normal lagging strand synthesis prevents completion of the last Okazaki fragment.
   a. The result is a newly synthesized DNA strand that is shortened at one end.
3. Telomerase is an enzyme that adds repeat sequences beyond the coding region to constitute the telomeric regions at the ends of chromosomes.
   a. Telomerase is a large ribonucleoprotein that has reverse transcriptase activity and contains in its structure an RNA molecule that acts as a template for synthesis of the telomeric region.
   b. Telomerase is expressed in germline cells and neoplastic cells.
4. Senescence of somatic cells, which do not express telomerase, is related to progressive shortening of chromosomes resulting from successive rounds of replication.

E. Reverse transcriptase
1. Reverse transcriptase mediates the synthesis of DNA using an RNA template.
2. Human immunodeficiency virus (HIV) and other retroviruses use viral reverse transcriptase to produce a DNA copy of the viral RNA genome. Viral DNA is then integrated into the genome of the infected cell.
   a. High mutation rate in HIV and emergence of new strains is caused in part by a lack of proofreading activity in reverse transcriptase.
   b. Zidovudine (AZT) is a thymidine analogue that inhibits reverse transcriptase, thereby blocking infection of cells.
      (1) AZT has a minimal effect on nuclear DNA polymerases, but the drug inhibits mitochondrial DNA replication.
      (2) Aerobic metabolism is impaired in patients treated with AZT because mitochondria are produced by a similar process to cell division and must replicate their DNA before division.

III. DNA Repair
A. Overview
1. A mutation is a change in the base sequence of DNA that escapes detection and correction by proofreading or repairing enzymes and is permanently "locked in" to a cell's genome at the next cell division.
2. Inherited diseases originate from mutations in germ cells that are passed on to the offspring and incorporated into all of their cells (e.g., inactivation of the *RB1* suppressor gene on chromosome 13 causes retinoblastoma).

B. Mismatch repair (Fig. 11-5)
1. Mismatch repair enzymes detect distortions due to mismatched bases in the DNA helix; the normal helix is a perfect (but bendable) cylinder.
2. The new strand of DNA is identified by an enzyme that locates a GATC sequence that has not been methylated; this enzyme makes a cut in the new strand with the incorrect base.
3. After the damaged area is removed, the repair enzymes fill in the proper sequence, and ligase seals the gap.

Margin notes:

DNA polymerases always synthesize daughter strands in the 5′ → 3′ direction.

Creates fragmented, lagging strand; Okazaki fragments linked together with DNA ligase

Proofreading detects base pairing errors; repaired by excision and continuation of normal synthesis

Telomerase: preserves length of chromosomes by allowing completion of last lagging strand fragment

Upregulation of telomerase activity is a characteristic of malignant cells.

Absence of telomerase in differentiated somatic cells is associated with cell aging.

Reverse transcriptase produces a DNA copy from RNA.

Viral reverse transcriptase converts the viral RNA genome to DNA for integration into the host genome.

Zidovudine (AZT) inhibits reverse transcriptase in HIV and blocks replication of mitochondrial DNA.

DNA damage produces mutation when DNA is replicated; repair preserves the original sequence.

Mutations in germ cells are inherited.

Mismatched bases distort DNA; mismatch repair detects distortion.

11-5: Mismatch repair. A GATC sequence directs GATC endonuclease to cut the damaged new daughter strand so that it can be repaired. *(From Pelley JW: Elsevier's Integrated Biochemistry. Philadelphia, Mosby, 2007.)*

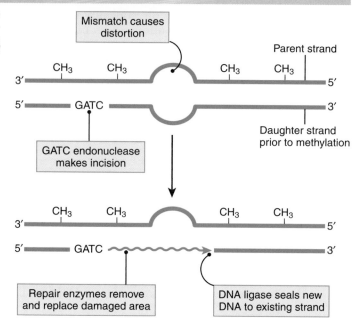

Cytosine deamination produces uracil (unmethylated thymine); uracil is excised and replaced with cytosine.

Pyrimidine dimers (thymine-thymine) are produced by excessive UVB light exposure and lead to skin cancer.

Nucleotide excision repair finds abnormal dimers (on same strand) and excises strand on both sides; damage is repaired with normal polymerase enzymes.

Sequence of DNA repair enzymes: endonuclease, exonuclease, DNA polymerase, and ligase

Methylated guanine (from alkylating agents) is repaired by MGMT, which becomes inactivated by covalent bonding to the methyl group.

Double-strand repair rejoins the helix when both strands are broken; breaks are produced by ionizing radiation or free radical oxidation.

Certain cancer drugs block the cell cycle or reduce the accuracy of DNA replication.

5-FU and methotrexate reduce dTMP synthesis to block DNA replication.

Bleomycin and etoposide prevent strand separation.

C. Base excision repair (Fig. 11-6)
1. Cytosine spontaneously deaminates to uracil (C → U) at a constant rate; this changes a GC pair to an AT pair.
2. The uracil is detected as unmethylated thymine by uracil-DNA glycosidase, which excises the uracil base.
3. After an apurinic or apyrimidinic (AP) endonuclease nicks the site of the mutation, the repair enzymes fill in the missing cytosine, and DNA ligase seals the gap.

D. Nucleotide excision repair (Fig. 11-7)
1. Pyrimidine dimers (e.g., thymine-thymine), caused by exposure to ultraviolet (UV) light, distort the DNA helix.
   a. Pyrimidine dimers are responsible for UV-induced cancers of the skin (e.g., basal cell carcinoma, squamous cell carcinoma).
2. The distortion in the DNA helix is detected by an excision endonuclease, which nicks the damaged strand on both sides of the lesion.
3. The repair enzymes fill in the missing sequence and DNA ligase seals the gap.

E. Direct repair (Fig. 11-8)
1. Guanine residues in DNA can become methylated when exposed to alkylating agents, such as antineoplastic drugs (see Section III, H).
2. The DNA repair enzyme, O^6-methylguanine-DNA methyltransferase (MGMT) repairs the altered guanine by removing the methyl group directly (Fig. 11-9).
3. MGMT becomes permanently inactivated, and new enzyme must be synthesized for continued effectiveness.
   a. MGMT inhibits the antineoplastic alkylating agent, temozolomide, but it can be depleted by continued dosing with the agent over a period of 7 weeks.

F. Double-strand repair
1. If both DNA strands are broken by ionizing radiation or free radical oxidation, they can be rejoined by unwinding both ends with helicase activity to create short, single-stranded ends.
2. The free ends are allowed to base pair with normal polymerase activity and ligase activity to seal the gaps.

G. Antineoplastic drug action (see Fig. 11-3)
1. 5-Fluorouracil (5-FU) and methotrexate are S phase–specific antimetabolites that cause reduced synthesis of dTMP and decreased DNA replication.
2. Bleomycin and etoposide inhibit human topoisomerase II, which normally acts in the S and G_2 phases to facilitate separation of intertwined DNA molecules during and after replication and during repair of double-strand breaks.

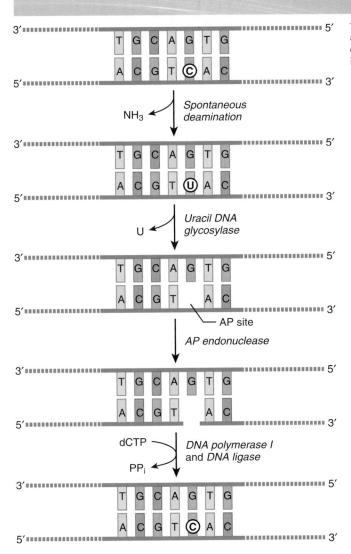

11-6: Base excision repair. Uracil is recognized as an incorrect base in DNA by uracil-DNA glycosidase and is removed so that cytosine can be inserted. *(From Pelley JW: Elsevier's Integrated Biochemistry. Philadelphia, Mosby, 2007.)*

3. Alkylating agents such as cyclophosphamide and nitrosoureas lead to copying errors during DNA replication and expression of abnormal proteins.
4. Doxorubicin and actinomycin D intercalate into the DNA structure and block DNA and RNA polymerases.
5. Vinca alkaloids (e.g., vincristine) and paclitaxel are M phase–specific agents that inhibit mitotic spindle formation and breakdown, respectively.

IV. DNA damage and cancer

A. Overview
1. Most cancers are noninherited disorders resulting from the accumulation over time of multiple mutations in somatic cells.
2. Mutations in protooncogenes, tumor-suppressor genes, and DNA-repair genes are primarily associated with the development of cancer.

B. Protooncogenes
1. Protooncogenes encode proteins that promote cell proliferation or inhibit apoptosis (i.e., programmed cell death) (Table 11-1).
2. Gain-of-function mutations occur in protooncogenes, converting them into oncogenes, which act dominantly.
   a. Oncogenes code for oncoproteins, and only one copy of an oncogene is needed to produce an effect.
3. Oncoproteins may be constitutively active forms of normal proteins or normal proteins that are expressed in excessive amounts, in the wrong tissue, or at the wrong time during development.

Cyclophosphamide and nitrosoureas lead to copying errors and expression of abnormal proteins.

Doxorubicin and actinomycin D block DNA and RNA polymerases.

Vincristine inhibits mitotic spindle formation, and paclitaxel inhibits mitotic spindle breakdown.

Cancers result from multiple mutations.

Genes associated with cancer: protooncogenes, tumor-suppressor genes, and DNA-repair genes

Gain-of-function mutations in protooncogenes (e.g., *RAS*) exhibit dominance.

Oncoproteins: normal proteins expressed in excessive amounts

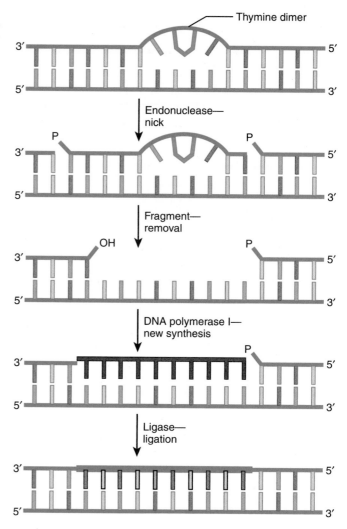

11-7: Nucleotide excision repair. The dimer is excised by an excision endonuclease and removed so that the region can be resynthesized. *(From Pelley JW: Elsevier's Integrated Biochemistry. Philadelphia, Mosby, 2007.)*

11-8: Methylation of guanine by alkylating agents. Methylated guanine can form a base pair with thymine to create a point mutation. *(From Pelley JW: Elsevier's Integrated Biochemistry. Philadelphia, Mosby, 2007.)*

11-9: Direct repair of guanine by O^6-methylguanine-DNA methyltransferase (MGMT). *(From Pelley JW: Elsevier's Integrated Biochemistry. Philadelphia, Mosby, 2007.)*

TABLE 11-1. **Selected Human Protooncogenes and Tumor-Suppressor Genes**

GENE NAME	FUNCTION OF GENE PRODUCT	TUMORS ASSOCIATED WITH MUTATION
Protooncogenes		
BCL2	Inhibits apoptosis (programmed cell death)	Follicular B-cell lymphoma
ERBB2	Binds epidermal growth factor (cell-surface receptor)	Breast, ovarian, stomach cancers
MYC	Activates transcription	Burkitt's lymphoma, neuroblastoma
RAS	Transduces signal from many growth factors; is a G protein	Bladder, lung, colon, and pancreatic cancers
Tumor-Suppressor Genes		
APC	Inhibits MYC expression	Familial adenomatous polyposis; colorectal carcinoma
BRCA1		
BRCA2	Involved in DNA repair	BRCA1: breast, ovary, colon, prostate cancers
		BRCA2: male and female breast cancer
NF1		
NF2	Helps control intracellular signaling	Neurofibromatosis: optic nerve glioma, meningioma, acoustic neuroma, pheochromocytoma
TP53	Induces proteins that mediate G_1 arrest or apoptosis of cells with damaged DNA; called the guardian of the cell	Most human cancers, Li-Fraumeni syndrome, breast and brain cancers, leukemia, sarcomas
RB1	Induces protein that mediates G_1 arrest	Retinoblastoma, osteosarcoma, breast cancer
WT1	Represses transcription in developing kidney	Wilms' tumor

C. Tumor-suppressor genes

1. Tumor-suppressor genes encode proteins that inhibit cell proliferation or promote apoptosis (see Table 11-1).
2. Loss-of-function mutations in tumor-suppressor genes act recessively.
   a. Both alleles of a tumor-suppressor gene must be mutated to permit development of a cancer.
3. Genetic predisposition to certain cancers results from inheritance of one mutant allele (from either parent) of various tumor-suppressor genes.
   a. In familial cancers associated with tumor-suppressor genes (e.g., retinoblastoma, breast carcinomas, Wilms' tumor), an individual inherits one mutant allele.
   b. Only one somatic mutation in the remaining normal allele is required for the cancer to develop.
   c. Sporadic forms of these cancers have a much lower frequency and later onset than familial forms.
   d. Two somatic mutations, one in each allele of the involved gene within a given cell lineage, must occur before the disease manifests.

D. Defects in DNA-repair enzymes

1. Defects in DNA-repair systems cause several inherited diseases that are marked by increased cancer susceptibility (Table 11-2).

Loss-of-function mutations in suppressor genes (e.g., TP53) act recessively.

Genetic predisposition to cancer: inheritance of one mutant tumor-suppressor allele

Genetic defects in DNA-repair enzymes: increased sensitivity to agents that damage DNA; associated with increased cancer risk

TABLE 11-2. **Inherited Diseases Associated with Defective DNA-Repair Enzymes**

DISEASE	TYPE OF CANCER SUSCEPTIBILITY	CLINICAL FEATURES
Ataxia telangiectasia	Lymphomas	Cerebellar ataxia; dilation of blood vessels in skin and eyes; immunodeficiency (B and T cells)
Bloom's syndrome	Carcinomas, leukemias, lymphomas	Facial telangiectasia, growth retardation, immunodeficiency
Fanconi's anemia	Leukemias	Progressive aplastic anemia, pancytopenia, numerous congenital anomalies
Hereditary nonpolyposis colorectal cancer (HNPCC)	Colon, ovary	Tumors usually develop before 40 years of age
Xeroderma pigmentosum	Basal cell carcinoma, squamous cell carcinoma, malignant melanomas	Severe skin lesions (absent DNA repair enzymes)

CHAPTER 12
GENE EXPRESSION

I. RNA transcription

A. Overview

1. RNA is a single-stranded molecule that often contains internal base-paired regions, forming stem loops.
2. Bacteria have one RNA polymerase, and eukaryotes have three.
3. Prokaryotic mRNA is synthesized as a final product, but eukaryotic mRNA is synthesized as a primary transcript that must be further processed to remove introns.
4. Prokaryotic and eukaryotic genes have promoter regions that point RNA polymerase in the right direction and position it to start transcribing at the correct nucleotide base.

B. Types of RNA

1. The three major types of RNA have specific functions in protein synthesis.
   a. Messenger RNAs (mRNAs) provide the blueprint for an amino acid sequence; mRNA is about 5% of total RNA.
   b. Transfer RNAs (tRNAs) match genetic information to an amino acid sequence; tRNA is about 15% of total RNA.
      (1) tRNAs have a cloverleaf shape with an acceptor arm, which receives the amino acid, and a three-base anticodon, which can form base pairs with the complementary codon in mRNA.
   c. Ribosomal RNAs (rRNAs) self-assemble with basic proteins to form ribosomes; rRNA is about 80% of total RNA.

C. RNA polymerase

1. RNA polymerase unwinds the DNA helix and transcribes the DNA template strand into RNA, beginning at the transcription start site.
   a. The DNA strand complementary to the template strand is the coding, or sense, strand.
2. The RNA chain is elongated in the $5' \rightarrow 3'$ direction.
   a. Ribonucleoside triphosphates are added to the free 3'-OH end of the RNA, producing a new 3'-OH terminal.
3. RNA polymerases are different in eukaryotes and prokaryotes.
   a. Eukaryotes contain three nuclear RNA polymerases that produce different types of RNA.
      (1) RNA polymerase I produces rRNA, the most abundant RNA.
      (2) RNA polymerase II produces mRNA, the largest RNA.
      (3) RNA polymerase III produces tRNA (and other small RNAs), the smallest RNA.
   b. Prokaryotes (bacteria) contain a single RNA polymerase that produces all the types of RNA.
      (1) Rifampin, an antibiotic used in the treatment of tuberculosis, inhibits RNA polymerase in bacteria but not in eukaryotes.

D. Prokaryotic transcription (Fig. 12-1)

1. Prokaryotic genes are organized as transcription units preceded by regulatory sequences called promoters.
2. The core enzyme, RNA polymerase, binds first to the promoter region in a precise orientation and then begins synthesis of RNA.
3. A transcription bubble (Fig. 12-2) is created as the helix opens for RNA polymerase transcription with closure of the bubble as the DNA reassociates.
4. RNA polymerase does not proofread, in contrast to DNA polymerase.
5. Either strand can serve as a template, but synthesis is always in the 5' to 3' direction.

Margin notes:

RNA: single-stranded polynucleotide synthesized by RNA polymerase

tRNA: high degree of secondary structure; three base anticodon, acceptor arm for amino acid attachment

mRNAs are the blueprints for protein synthesis; tRNAs match codons to amino acids; and rRNAs contribute to ribosome structure.

Template strand: read by RNA polymerase

Sense strand: complementary to template strand; matches new RNA sequence

RNA polymerase I: rRNA

RNA polymerase II: mRNA

RNA polymerase III: tRNA plus other small RNAs

α-Amanitin, produced by the mushroom *Amanita phalloides*, strongly inhibits RNA polymerase II.

Rifampin inhibits bacterial RNA polymerase but not eukaryotic RNA polymerases.

RNA polymerase binds to promoter to initiate transcription.

Transcription bubble: melted area of DNA helix containing RNA polymerase and growing RNA

No proofreading by RNA polymerase

Chain termination signal (hairpin loop) releases RNA from transcription bubble.

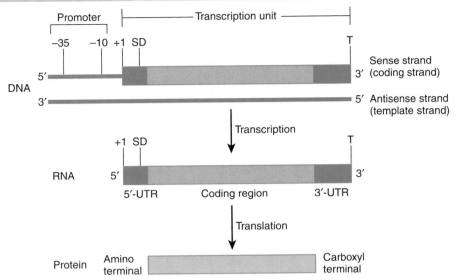

12-1: Components of a prokaryotic gene. SD, Shine-Dalgarno sequence; T, termination signal; UTR, untranslated region. *(From Pelley JW: Elsevier's Integrated Biochemistry. Philadelphia, Mosby, 2007.)*

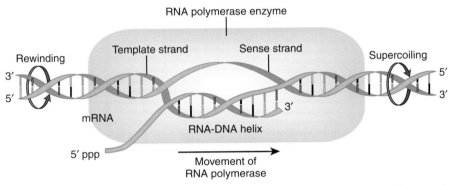

12-2: The transcription bubble is formed when RNA polymerase separates the helix to read the template strand. The helix reforms the original base pairing as RNA polymerase moves along the helix. *(From Pelley JW: Elsevier's Integrated Biochemistry. Philadelphia, Mosby, 2007.)*

   6. RNA synthesis stops when RNA polymerase reaches a sequence called the chain termination signal that causes the new RNA to form a hairpin loop; this causes the RNA polymerase to release from the DNA.
 E. Eukaryotic transcription of mRNA (Fig. 12-3)
   1. mRNA synthesis in eukaryotes begins with the binding of RNA polymerase II at specific DNA sequences known as promoters.
   2. The initiation complex, containing RNA polymerase II and several basal transcription factors that bind RNA polymerase II, is assembled near the transcription start site (Fig. 12-4).
      a. The TATA box, a conserved promoter sequence about 25 bases upstream of the start site, helps orient the RNA polymerase in many eukaryotic genes.
      b. The CAAT box, another conserved sequence within the promoter region, is often present about 70 bases upstream of the start site.
      c. A CG box that binds a specific transcription factor (SP-1).
   3. Elongation of the RNA strand by RNA polymerase II continues until the enzyme reaches a termination signal.
   4. The primary mRNA transcript, also called hnRNA (heterogeneous nuclear RNA), is initially produced; it contains coding and noncoding sequences.
      a. Exons are expressed (coding) sequences within a gene.
      b. Introns are intervening (noncoding) sequences within a gene.
 F. Processing of the primary mRNA transcript
   1. Processing of hnRNA occurs in the nucleus, yielding functional mRNA, which is exported to the cytosol for translation (see Fig. 12-3).

Initiation complex: RNA polymerase II plus transcription factors

Common sequences in eukaryotic promoter: TATA box, CAAT box, and CG box

Primary mRNA transcript contains introns that must be removed by splicing.

Exons specify amino acids in a protein; introns carry no genetic information.

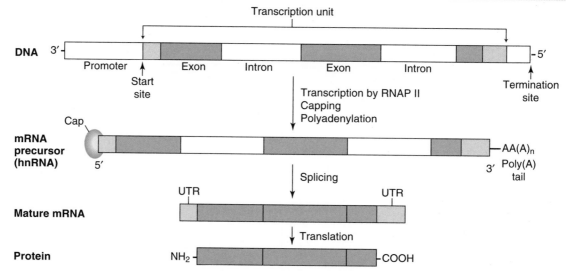

12-3: Eukaryotic protein-coding gene and its conversion into RNA products and encoded protein. Only the template (missense) strand of DNA is depicted. Untranslated regions (UTRs) at each end of a transcription unit and mRNA *(gray regions)* do not appear in the protein. A, adenylate; hnRNA, heterogeneous nuclear RNA; RNAP II, RNA polymerase.

12-4: Sequences found in the eukaryotic promoter. Each promoter binds a specific transcription factor to form the basal transcription apparatus. The TATA box binds TBP transcription factor; the CAAT box binds NF-1/CTF transcription factors; and the GC box binds SP-1 transcription factor. *(From Pelley JW: Elsevier's Integrated Biochemistry. Philadelphia, Mosby, 2007.)*

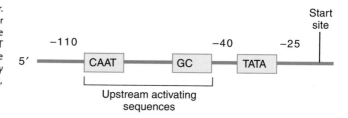

Primary transcript is capped and tailed during processing.

Capping orients mRNA on ribosome, and tailing provides a longer half-life.

snRNPs: spliced out introns

2. A 7-methylguanosine cap is added to the 5′ end of the growing mRNA before transcription is completed.
3. Polyadenylation at the poly(A) site leads to formation of the poly(A) tail consisting of 20 to 250 adenylate residues at the 3′ end of the transcript.
4. Splicing removes introns from capped, tailed transcripts and rejoins exons in a continuous coding sequence.
   a. Splice sites, the junctions between exons and introns, usually are marked by consensus sequences: GU at the 5′ end of an intron and AG at the 3′ end.
   b. Small nuclear ribonucleoproteins (snRNPs) assemble at splice sites in hnRNA, forming a spliceosome.
      (1) Two transesterification reactions catalyzed by snRNPs result in excision of the intron as a lariat structure and joining of exons.
      (2) Systemic lupus erythematosus is an autoimmune disease in which the patient produces antibodies to snRNPs that have been released into the blood due to tissue damage.

Systemic lupus erythematosus: autoimmune disease; antibodies produced to snRNPs, which function in mRNA splicing

II. Transcriptional Control of Gene Expression
A. Overview

Transcription of a gene is controlled by controlling access of RNA polymerase or by modification of the mRNA.

1. The prokaryotic *lac* operon demonstrates positive control (i.e., factors required to initiate RNA synthesis) and negative control (i.e., factors required to prevent initiation of RNA synthesis).
2. Eukaryotic gene expression is regulated by controlling physical access to the DNA by the RNA polymerase and by controlling the rate of its binding.
3. Unequal crossing over can produce extra copies of the same gene and amplify its expression.
4. Some exons can be spliced in or spliced out as a way of including or excluding domains; the final product is same gene product with different properties (e.g., membrane bound or soluble).
5. Editing of the final mRNA allows elimination of domains from carboxyl end of protein if the editing inserts a termination codon.
6. RNA interference takes advantage of an anti-RNA virus dicer enzyme that uses antisense RNA of a known sequence to digest target mRNA molecules.

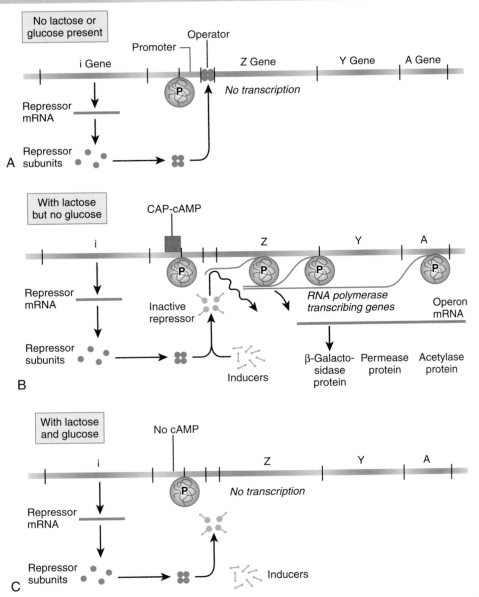

12-5: Expression of the *lac* operon with different energy sources. **A,** Expression is not needed; the repressor prevents expression. **B,** Expression is needed; the repressor is inactivated, and the activator protein (CAP) is active. **C,** Expression is not needed; the repressor is inactivated, but CAP is not active. P, RNA polymerase. *(From Pelley JW: Elsevier's Integrated Biochemistry. Philadelphia, Mosby, 2007.)*

B. Prokaryotic control of gene expression (Fig. 12-5)

1. Regulation of the *lac* operon illustrates several modes of regulation of RNA transcription in bacteria:

    a. Negative control occurs when only glucose is available as an energy source and expression of the *lac* operon is not needed.

    (1) The *lac* repressor protein is produced constitutively and binds to the operon to prevent transcription.

    b. Positive control occurs when lactose is present (with or without glucose present) and expression of the *lac* operon is needed.

    (1) If lactose is present, the inducer, allolactose, is formed followed by binding the repressor to inactivate it.

    (2) If no glucose is present, cAMP is elevated and combines with the catabolite activator protein (CAP), which binds to the promoter stimulating binding of RNA polymerase.

    (3) If glucose is also present, cAMP is not present and the CAP-cAMP complex cannot stimulate transcription; this allows glucose to be used until it is depleted.

Positive control: activator protein needed to start transcription

Negative control: repressor protein needed to inhibit transcription

C. Eukaryotic control of gene expression

1. Eukaryotic gene expression is regulated by controlling physical access to the DNA by the RNA polymerase and by controlling the rate of its binding.

2. Physical accessibility of DNA to transcription factors and RNA polymerase II depends on the degree of compaction of chromatin in the region of a gene.

   a. Heterochromatin is highly condensed, DNase resistant, and inactive in transcription; the DNA has CpG islands that are highly methylated.

      (1) CpG islands are associated with promoters of housekeeping genes, which encode proteins that are expressed at a low basal rate in all cells (e.g., metabolic enzymes, cytoskeleton proteins).

   b. Euchromatin is dispersed, DNase digestible, and active in transcription; the histones are highly acetylated (i.e., changes packing of nucleosomes to cause local unwinding).

3. The rate of binding RNA polymerase to the promoter is regulated by transcription factors that bind to special DNA sequences.

   a. Sequences in the promoter that bind transcription factors are those mentioned above (see Section I, E, 2): the CAAT box, the TATA box, and the CG box (the CG box is not the same as CpG islands).

   b. Other regulatory elements in the DNA sequence, called enhancers or silencers, bind DNA regulatory proteins (i.e., specific transcription factors) that may activate or repress transcription by RNA polymerase II.

      (1) Enhancers generally are located at a considerable distance upstream or downstream from the promoter region of a gene, but they also may be located within introns.

      (2) The complex made up of the enhancer or silencer and activator protein folds back to interact with the basal transcription apparatus (BTA) at the promoter region, greatly increasing/decreasing the rate of transcription of regulated genes (Fig. 12-6).

4. Steroid hormones, when complexed to their intracellular receptors, function as transcriptional activators (see Chapter 3).

   a. Each hormone-receptor complex binds to a particular regulatory element for the gene being regulated.

   b. Tamoxifen, used in the treatment of breast cancer, is an estrogen antagonist that blocks the binding of estrogen to its intracellular receptor in estrogen-sensitive tumor cells in breast tissue.

5. Many polypeptide hormones and growth factors also regulate gene expression by triggering intracellular signaling pathways leading to activation (or less often to repression) of transcription (see Chapter 3).

Heterochromatin: highly condensed, DNase resistant, highly methylated, and inactive

Euchromatin: dispersed, DNase digestible, highly acetylated, and active

Promoter sequences for binding transcription factors: CAAT box, TATA box, and CG box (not CpG islands)

Enhancers: sequences away from the promoter that bind transcription activation factors

Silencers: sequences away from the promoter that bind transcription inhibition factors

Steroid hormone–receptor complex functions as a transcription factor.

Tamoxifen: estrogen antagonist used to treat breast cancer

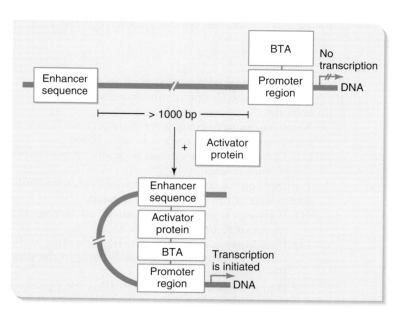

12-6: Regulation of transcription initiation. Transcription initiation depends on the interaction of an activator protein (bound to an enhancer sequence) with components of the basal transcription apparatus (BTA). Nonregulated housekeeping genes require only assembly of the BTA for transcription. bp, base pairs.

D. Gene amplification

1. Gene duplication from unequal crossing over can produce 100 to 1000 copies of a gene as a way of increasing the rate of its expression.
2. Drug resistance is developed in cancer cells because of amplification of a gene that is being targeted by the chemotherapy.
   a. The gene for dihydrofolate reductase, an enzyme inhibited by the drug methotrexate, can be increased to a level that requires 3000 times the lethal dose of methotrexate for normal cells.

E. Alternative splicing

1. Multiple protein species can be generated from an RNA transcript of a single gene by splicing out different exons during the processing of the primary transcript.
2. Fibronectin has three forms: extracellular matrix, cell surface, and soluble, all of which are transcribed from the same mRNA.
3. Alternative tailing is a special type of splicing that can splice out a carboxy-terminal membrane binding domain to convert a protein from a membrane protein to a soluble protein.
   a. The IgG heavy chain in B cells is an example of alternative splicing.

F. Editing of mRNA

1. The sequence in a processed mRNA can be edited to include a termination sequence before a domain.
2. Apoprotein B is edited differently in gut and liver to include or exclude the low-density lipoprotein (LDL) receptor binding domain.
   a. ApoB-48 is produced in the gut to assemble chylomicrons that do not bind to the LDL receptor; it is edited so that the LDL binding domain is not included during protein synthesis.
   b. ApoB-100 is produced in the liver to produce very-low-density lipoprotein (VLDL) (which becomes LDL) that must bind to the LDL receptor in peripheral tissues; it is not edited so that the LDL receptor binding domain is also synthesized.

G. RNA interference and gene silencing

1. RNA interference is carried out by components that are found in all cells to protect against RNA viruses; however, known sequences can be introduced to direct the digestion of unwanted mRNA.
2. A dicer protein binds small, 22-base fragments of antisense RNA with a sequence matching the target mRNA.
3. The dicer-RNA complex then recognizes and digests corresponding target mRNAs.

III. The genetic code and mutation

A. Overview

1. The genetic code is a nucleotide sequence that specifies a polypeptide sequence; it is nonoverlapping, unambiguous, degenerate (due to wobble), and universal.
2. Aminoacyl-tRNA synthetases represent the point at which the genetic code and amino acid structure are simultaneously recognized.
3. Mutations can lead to no change in function (silent), variable change in function (missense), or no function due to premature termination (nonsense).
4. Frameshift mutations usually lead to loss of function.
5. Trinucleotide repeat mutations within a gene can increase in size, causing symptoms to appear earlier in future generations, a process called anticipation.
6. Translocations can relocate genes to different promoters and cause inappropriate activation or inactivation; genes may also be inactivated due to loss of primary structure.
7. Other mutations affecting chromosomes are a change in chromosome number for individual chromosomes (due to nondisjunction) and microdeletion.

B. The genetic code

1. The genetic code is a triplet code in which sequences (codons) made up of three nucleotides specify amino acids and the start and stop commands. The code has several important features:
   a. Nonoverlapping and commaless: codons are contiguous, and no nucleotide sequences represent spacers (exceptions include some viruses).
   b. Unambiguous: each codon corresponds to only one amino acid.
   c. Degenerate: some amino acids are coded by more than one codon.
      (1) Wobble at the third base of the codon allows the same anticodon on aminoacyl-tRNA to pair with more than one codon (Fig. 12-7).
   d. Universal: the same code is used in all protein synthesis, with minor exceptions in the mitochondria and a few microorganisms and plants.

Unequal crossing over can amplify gene expression by increasing the number of genes being transcribed

Dihydrofolate reductase gene is amplified to produce methotrexate resistance.

Alternative splicing includes or excludes domains to change properties of protein.

When ApoB mRNA is read without editing, apoB-100 includes the LDL receptor binding domain.

When ApoB mRNA is read after editing for termination, apoB-48 lacks the LDL receptor binding domain.

Dicer-RNA complex digests target mRNA that matches introduced antisense RNA.

Mutation alters the genetic code, producing a change in the function of a protein or a loss of function.

Genetic code: nucleotide sequence that specifies a polypeptide sequence

The code can be nonoverlapping, unambiguous, degenerate (due to wobble), or universal.

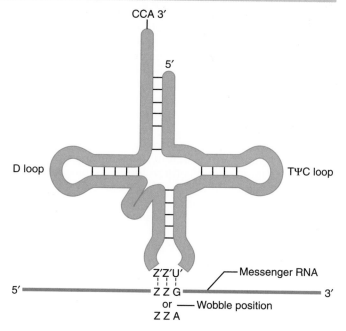

12-7: The wobble position in the anticodon. Notice that the anticodon can form base pairs with two different codons. *(From Pelley JW: Elsevier's Integrated Biochemistry. Philadelphia, Mosby, 2007.)*

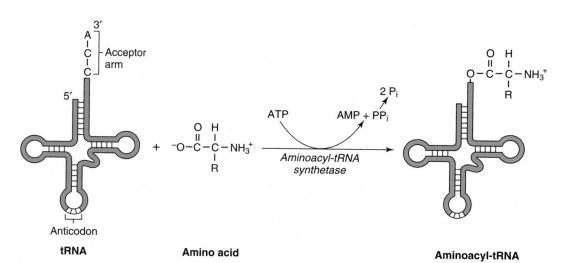

12-8: Charging of tRNA by aminoacyl-tRNA synthetases. These enzymes recognize the amino acid that corresponds to the anticodon in a particular tRNA and link it by a high-energy bond to the 3'-adenylate residue of the acceptor arm.

C. Aminoacyl-tRNA synthesis

Aminoacyl-tRNA synthetases: genetic code and amino acid structure simultaneously recognized

1. Transfer RNAs are charged with an amino acid by aminoacyl-tRNA synthetases (Fig. 12-8).
2. Each aminoacyl-tRNA synthetase attaches one type of amino acid to all of the tRNAs specific for that amino acid.
3. Proofreading by aminoacyl-tRNA synthetases permits detection and removal of amino acids that are linked to the incorrect tRNA (Fig. 12-9).

D. Effects of mutations

Mutation: inherited change in base sequence

Single-gene defects obey mendelian inheritance patterns.

Many diseases, such as type 2 diabetes, show polygenic inheritance.

1. A mutation is a change in the base sequence of DNA that escapes detection and correction by proofreading or repairing enzymes and is permanently locked into a cell's genome at the next cell division.
   a. Inherited diseases originate from mutations in germ cells that are passed on to the offspring and incorporated into all of their cells (e.g., inactivation of the *RB1* suppressor gene on chromosome 13 causes retinoblastoma).
   b. Mendelian inheritance is shown by inherited diseases resulting from single-gene defects in nuclear DNA (Box 12-1).
2. Alterations in a single nucleotide base, known as a point mutation, within a protein-coding gene can have three possible outcomes (Fig. 12-10).

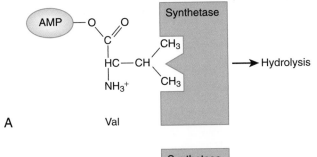

12-9: Proofreading by isoleucyl tRNA synthetase. Valine fits the hydrolytic site and is removed; isoleucine is not affected. *(From Pelley JW: Elsevier's Integrated Biochemistry. Philadelphia, Mosby, 2007.)*

BOX 12-1   MENDELIAN INHERITANCE PATTERNS

Autosomal Dominant Pattern

One defective copy of the gene on an autosomal (non-sex) chromosome produces the disease. A child has a 50% chance of having the disease if one of the parents is heterozygous. Males and females are affected. Clinical symptoms may not develop until adulthood. Some patients with the abnormal gene do not develop clinical symptoms (<100% penetrance), but all can transmit the disease to their children.

Examples include familial hypercholesterolemia, Huntington's disease, Marfan syndrome, osteogenesis imperfecta, von Willebrand's disease, congenital spherocytosis, adult polycystic kidney disease, and neurofibromatosis.

Autosomal Recessive Pattern

Two defective copies (alleles) of the gene, each on an autosomal chromosome, are needed to produce the disease. Each child of two heterozygous parents (i.e., asymptomatic carriers) has a 25% chance of having the disease (homozygous). Males and females are affected. Clinical symptoms usually develop in infancy or childhood and commonly are more severe than in dominant disorders.

Examples include most inborn errors of metabolism (e.g., glycogen storage diseases, maple syrup urine disease, phenylketonuria, Tay-Sachs disease), sickle cell disease, cystic fibrosis, adrenogenital syndrome, and Wilson's disease.

X-linked Recessive Pattern

Each son of a heterozygous mother, who is usually an asymptomatic carrier, has a 50% chance of being affected. Affected males transmit the abnormal X chromosome to all of their daughters (carriers) but to none of their sons. Heterozygous females may show minor effects, but males, who have only one X chromosome, manifest clinical symptoms.

Examples include Duchenne-type muscular dystrophy, Fabry's disease, hemophilia A and B, Hunter's syndrome, Lesch-Nyhan syndrome, and glucose 6-phosphate dehydrogenase (G6PD) deficiency.

---

a. Silent mutations occur when the altered codon specifies the same amino acid; therefore, there is no phenotypic effect.

b. Missense mutations occur when the altered codon specifies a different amino acid, causing variable phenotypic effects.

(1) For example, altered hemoglobins resulting from different missense mutations cause symptoms ranging from benign (HbC, which causes a mild, chronic anemia) to moderate (HbS, sickle cell hemoglobin) to severe (HbM, associated with methemoglobinemia) (see Chapter 2).

c. Nonsense mutations occur when the mutation creates a stop codon that causes premature termination during protein synthesis, producing a nonfunctional protein.

Point mutation: silent (no change in function), missense (variable change in function), or nonsense (loss of function)

Point mutations cause silent, missense, or nonsense mutations.

Frameshift mutation: altered reading frame that produces nonfunctional primary structure

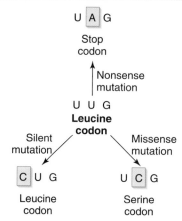

12-10: Possible outcomes resulting from a point mutation that alters a single nucleotide base in the coding region of mRNA. As seen here for the leucine codon, a point mutation can result in no change in the amino acid (i.e., silent mutation), a change in the amino acid (i.e., missense mutation), or a premature termination of synthesis of the polypeptide (i.e., nonsense mutation).

12-11: Effect of mutations resulting from a base addition or base substitution of a single nucleotide in the coding region of mRNA. With the base substitution of C for G, CCC codes for Pro rather than Ala (i.e., missense mutation). When U is added, UAG becomes a stop codon (i.e., nonsense mutation). Had the addition of U not produced a stop codon, a frameshift mutation would have occurred, resulting in changes in the amino acid sequence of the protein (e.g., UUA, leucine; UUU, phenylalanine). The normal mRNA sequence and corresponding amino acid sequence are shown in the middle.

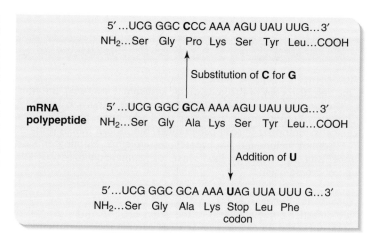

3. Frameshift mutations are caused by insertion or deletion of any number of nucleotides not divisible by three producing a shift in the reading frame during translation of mRNA (Fig. 12-11).
   a. A frameshift mutation results in a randomly incorrect amino acid sequence on the carboxyl side of the mutation and usually a truncated protein caused by the introduction of a premature stop codon.
   b. Example: Tay-Sachs disease is the result of a four-nucleotide insertion that causes a frameshift mutation and defective hexosaminidase.
4. Trinucleotide repeat mutations are caused by errors in DNA replication that lead to an abnormal number of repeated trinucleotides (e.g., CGG, CAG, CTG) within a gene.
   a. Huntington's disease, Friedreich's ataxia, fragile X syndrome, and myotonic dystrophy are caused by this type of mutation.
   b. These disorders worsen in future generations due to the addition of trinucleotides; increasing severity with each successive generation is called anticipation.
5. Translocation mutations are caused by movement of all or part of a gene from its normal position to another position in the genome.
   a. Translocation mutations usually are caused by mistakes during homologous recombination, a normal process that occurs in meiosis and in repair of double-strand breaks in DNA.
   b. These mutations cause various effects ranging from no protein production to reduced expression of normal protein.
   c. Examples include Burkitt's lymphoma (t8;14) and chronic myelogenous leukemia (t9;22).

Tay-Sachs disease: frameshift mutation caused by a four-nucleotide insertion

Trinucleotide repeat mutation: abnormal number of repeated trinucleotides

Earlier appearance in future generations is called anticipation.

Translocation mutation: all or part of gene moved by break and reattachment of chromosome

Down syndrome: caused by an extra copy of chromosome 21; nondisjunction

6. Mutations may involve a gain or a loss of an entire chromosome.
   a. For example, trisomy 21, with three copies of chromosome 21, causes Down syndrome.
   b. Most chromosomal number disorders result from nondisjunction, when homologous chromosomes or chromatids fail to separate properly in the first phase of meiosis, resulting in one or more extra chromosomes in some gametes and fewer chromosomes in other gametes.
   c. Loss from a paternal-origin chromosome can produce different disease than loss from a maternal-origin chromosome (i.e., Prader-Willi and Angelman syndromes, respectively)

7. Mutations may involve a loss of a small portion of a chromosome (i.e., microdeletion).
   a. Microdeletion on chromosome 15 may result in Prader-Willi syndrome if the abnormal chromosome is of paternal origin.
   b. Angelman syndrome may result if the particular abnormal chromosome is of maternal origin.
   c. The mechanism of Prader-Willi and Angelman syndromes is genomic imprinting, meaning that the phenotypic outcome depends on whether the chromosome is maternal or paternal in origin.

> Microdeletion: loss of small portion of a chromosome

IV. Protein Synthesis and Degradation

A. Overview

1. Proteins are synthesized as polypeptides on ribosome that are assembled from two subunits during the initiation stage.
2. Prokaryotic and eukaryotic polypeptide synthesis have an initiation stage that aligns the mRNA and the sites on the ribosome so that the next aminoacyl tRNA can bind.
3. Translation proceeds until a stop codon is reached and termination factors release the polypeptide followed by dissociation of the ribosome and mRNA.
4. Several antibiotics have their effect at various points in the polypeptide synthesis cycle.
5. Secreted proteins are pushed through a translocon in the membrane of the endoplasmic reticulum and are stored in the lumen after any posttranslational modifications.
6. Proteasomes digest proteins labeled with ubiquitin to control the amount of that protein at any one time.

> Protein synthesis and degradation are controlled to maintain needed amounts of specific proteins.

B. Ribosomes

1. Ribosomes are particle-like ribonucleoprotein complexes composed of small and large subunits, whose composition and function in protein synthesis are different.
2. Small (40S) subunit (equivalent to the 30S subunit in prokaryotes)
   a. Binds mRNA and aminoacyl-tRNAs
   b. Locates AUG start codon on mRNA
3. Large (60S) subunit (equivalent to the 50S subunit in prokaryotes)
   a. Binds to the small subunit after the start codon is located
   b. Has peptidyltransferase activity

> A mall subunit forms the initiation complex.
>
> A large subunit joins the initiation complex to form a whole ribosome.
>
> Peptidyl transferase: ribozyme composed of large-subunit ribosomal RNA

C. Polypeptide synthesis: prokaryotic example

1. Translation of the mRNA message into a protein involves three stages: initiation, chain elongation, and termination (Fig. 12-12).
2. Initiation of protein synthesis
   a. Formation of the 30S initiation complex
      (1) The start codon (AUG), at which synthesis of all proteins begins, specifies methionine (formylmethionine [f-Met] in bacteria).
      (2) Initiation factors (IFs) are required to bind the mRNA and beginning methionyl-tRNA (or fMet-tRNA) to the small (30S) ribosomal subunit.
   b. Formation of the 70S initiation complex occurs with binding of the large (50S) subunit to form a complete ribosome containing the first aminoacyl tRNA correctly aligned with the mRNA.
3. Elongation of the protein chain
   a. Synthesis of the protein chain proceeds from the amino end to the carboxyl end.
      (1) During translation, the ribosome shifts by one codon in the 5′ → 3′ direction along the mRNA, in a process known as translocation.
   b. Peptide bond formation is catalyzed by peptidyltransferase activity of an rRNA component with enzyme function (a ribozyme) in the large ribosomal subunit.
   c. Elongation factors (EFs) are needed for alignment of the incoming aminoacyl-tRNA and to translocate the mRNA and peptidyl tRNA along the mRNA.
4. Termination of protein synthesis
   a. Stop codons (UGA, UAG, or UAA) signal termination of chain elongation.

> 30S initiation complex: alignment of start codon (AUG) and fMet-tRNA
>
> 70S initiation complex: binding of aminoacyl tRNA and peptide bond formation, followed by translocation
>
> Elongation factors: proteins that bind aminoacyl-tRNA to next codon and translocate peptidyl-tRNA to an open aminoacyl site

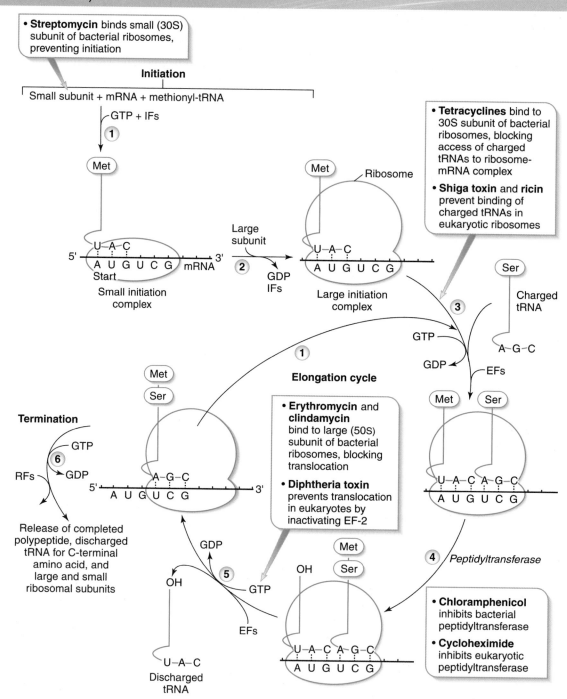

• **Streptomycin** binds small (30S) subunit of bacterial ribosomes, preventing initiation

Initiation

Small subunit + mRNA + methionyl-tRNA

GTP + IFs
1

• **Tetracyclines** bind to 30S subunit of bacterial ribosomes, blocking access of charged tRNAs to ribosome-mRNA complex
• **Shiga toxin** and **ricin** prevent binding of charged tRNAs in eukaryotic ribosomes

Met

Met   Ribosome

Large subunit

5'   U–A–C
A U G U C G   mRNA   3'
Start

2   GDP
IFs

Small initiation complex

U–A–C
A U G U C G

Large initiation complex

3

Ser

Charged tRNA

GTP

GDP

A–G–C

EFs

Met

Met   Ser

Elongation cycle

1

Ser

Termination

• **Erythromycin** and **clindamycin** bind to large (50S) subunit of bacterial ribosomes, blocking translocation
• **Diphtheria toxin** prevents translocation in eukaryotes by inactivating EF-2

GTP
6
RFs   GDP

5'   A–G–C
A U G U C G   3'

U–A–C–A–G–C
A U G U C G

Release of completed polypeptide, discharged tRNA for C-terminal amino acid, and large and small ribosomal subunits

GDP

5

OH

GTP

EFs

OH   Met

Ser

4   *Peptidyltransferase*

U–A–C   Discharged tRNA

U–A–C–A–G–C
A U G U C G

• **Chloramphenicol** inhibits bacterial peptidyltransferase
• **Cycloheximide** inhibits eukaryotic peptidyltransferase

12-12: Overview of translation, indicating steps inhibited by various antibiotics and toxins. Specific protein initiation factors (IFs), elongation factors (EFs), and release factors (RFs) are required in prokaryotes and eukaryotes. During step 5 (translocation), the ribosome moves toward the 3' end of the mRNA by one codon, freeing the site for another aminoacyl-tRNA. The polypeptide chain is elongated by repetition of steps 3, 4, and 5 with different charged tRNAs. Termination occurs when the ribosome encounters a stop codon.

Release factors: no tRNA exists for stop codons; recognized by protein factors

Polyribosomes: synthesis of multiple copies of the same protein concurrently

(1) No tRNAs exist that have anticodons complementary to stop codons.
   b. Release factors (RFs) bind to a stop codon and stimulate the release of the newly synthesized polypeptide chain, the final tRNA, and the dissociated ribosome.

D. Polyribosomes
  1. Polyribosomes (i.e., polysomes) are extended arrays of individual ribosomes all associated with a single strand of mRNA.
  2. By carrying out synthesis of multiple copies of the same protein concurrently, polyribosomes increase the rate of protein synthesis.
  3. Prokaryotes and eukaryotes form polyribosomes.

E. Bacterial antibiotic action

1. Several antibiotics act by selectively inhibiting bacterial protein synthesis (see Fig. 12-12).
2. Streptomycin binds to the small (30S) ribosomal subunits of bacteria and interferes with formation of the initiation complex.
   a. Tetracyclines bind to the small ribosomal subunits of bacteria and block the binding of incoming aminoacyl-tRNAs.
   b. Aminoglycosides and spectinomycin are inhibitors of small ribosomal subunits.
3. Chloramphenicol inhibits peptidyltransferase activity of the large (50S) ribosomal subunits of bacteria.
4. Erythromycin and clindamycin also bind to the large ribosomal subunits, preventing translocation of the ribosome along the bacterial mRNA.

F. Eukaryotic antibiotic action

1. Some toxins and antibiotics disrupt eukaryotic protein synthesis (see Fig. 12-12).
2. Shiga toxin (*Shigella dysenteriae*) and Shiga-like toxin (enterohemorrhagic *Escherichia coli*) cleave the 28S rRNA in the large (60S) ribosomal subunits of eukaryotes, preventing binding of incoming aminoacyl-tRNAs.
   a. Ricin, a protein found in castor beans, inactivates large ribosomal subunits in the same manner as Shiga toxin.
3. Diphtheria toxin (*Corynebacterium diphtheriae*) inactivates an elongation factor, EF-2, thereby preventing translocation in the eukaryotic ribosome.
4. Cycloheximide, which is used as a fungicide and rat repellent, inhibits eukaryotic peptidyltransferase.

G. Secreted proteins (Fig. 12-13)

1. Synthesis of secreted, lysosomal, and membrane proteins begins in the cytosol but is completed on the rough endoplasmic reticulum (RER).
2. A signal sequence at the amino terminus of the nascent polypeptide guides the nascent polypeptide to the RER membrane and ultimately into the RER lumen.
   a. Steps 1 and 2
      (1) After synthesis of the signal sequence on a free ribosome in the cytosol, the ribosome attaches to the RER membrane with the aid of a signal recognition particle (SRP) and SRP receptor.
   b. Step 3
      (1) The SRP dissociates, the ribosome–nascent protein associates with a translocon in the RER membrane, and the polypeptide chain enters the open channel in the translocon.

Many antibiotics target the differences between prokaryotic and eukaryotic ribosomes, selectively inhibiting bacterial protein synthesis.

30S subunit antibiotics: streptomycin, tetracyclines, aminoglycosides, and spectinomycin

50S subunit antibiotics: chloramphenicol, erythromycin, and clindamycin

60S subunit antibiotics: Shiga toxin, Shiga-like toxin, and ricin

Diphtheria toxin: elongation factor 2 inactivated

Peptidyl transferase antibiotic: cycloheximide

Eukaryotic protein synthesis: targeted by certain toxins and antibiotics

Signal sequence directs polypeptide to signal recognition particle.

Signal recognition particle directs polypeptide to translocon.

Translocon directs polypeptide to lumen of ER.

Signal sequence: digested in ER lumen

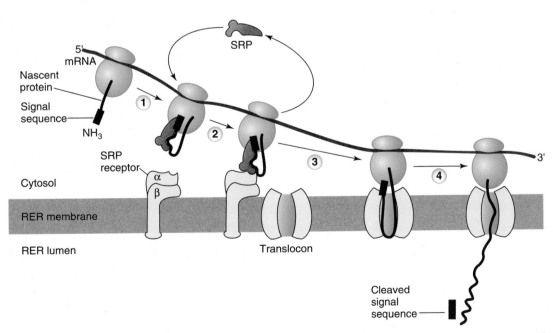

12-13: Synthesis of proteins on rough endoplasmic reticulum (RER). A signal sequence on a nascent protein interacts with a signal-recognition particle (SRP) and an SRP receptor on the RER. The growing polypeptide passes through a translocon into the RER lumen, where it may be glycosylated.

TABLE 12-1. **Posttranslational Modifications in Proteins**

TYPE OF MODIFICATION	EXAMPLE
Proteolytic cleavage	Digestive enzymes (e.g., trypsinogen → trypsin)
	Clotting factors (e.g., prothrombin → thrombin)
	Hormones (e.g., proinsulin → insulin + C-peptide)
Disulfide bond formation	Secreted proteins
Attachment of prosthetic group	Covalent linkage of biotin to carboxylases
Glycosylation (sugars attached to amino acids: serine, threonine, and asparagine residues)	Membrane and secreted proteins (e.g., glycosylated membrane proteins, ABO blood group determinants, immunoglobulins)
Phosphorylation (serine, threonine, and tyrosine residues)	Hormonally regulated enzymes (e.g., glycogen synthase, acetyl CoA carboxylase, pyruvate kinase)
Hydroxylation (lysine and proline residues)	Collagen α-chains
γ-Carboxylation (glutamine residues)	Clotting factors II, VII, IX, X, proteins C and S

c. Step 4
   (1) As translation continues, the elongating polypeptide is translocated into the RER lumen, amino end first, and the signal sequence is cleaved and digested.
   (2) After translation is completed, the new protein may be glycosylated in the RER lumen.
3. Transport vesicles that bud off from the RER carry newly formed proteins to the Golgi apparatus, where they undergo various posttranslational modifications.
   a. Posttranslational modifications of many proteins are required for their biologic activity (Table 12-1).
4. Hydrolytic enzymes destined for lysosomes are tagged with mannose 6-phosphate residues by enzymes in the lumen of the Golgi apparatus.
   a. The mannose 6-phosphate residues on the enzymes bind tightly to specific receptors in the Golgi membrane, and vesicles containing the bound enzymes bud off and eventually fuse with lysosomes.
   b. I-cell disease is a lysosomal storage disease that results from an inherited deficiency of the phosphotransferase needed to form the mannose 6-phosphate tag on lysosomal enzymes (see Box 6-3 in Chapter 6).
      (1) Lysosomal enzymes are produced normally, but because they lack the mannose 6-phosphate tag, they are secreted into the blood rather than being directed to lysosomes.
      (2) Other lysosomal storage diseases, such as mucopolysaccharidoses, Pompe's disease, and sphingolipidoses, are caused by inherited defects in the lysosomal enzymes themselves (see Chapters 6 and 7).
H. **Protein degradation**
   1. All proteins are degraded at a rate that allows control over the amount of protein at any point in time.
   2. The pathway for protein degradation is the ubiquitin-proteasome system.
      a. Proteins to be digested are labeled by the covalent attachment of the protein, ubiquitin.
      b. Ubiquinated proteins are digested by a barrel-shaped multisubunit complex called the proteasome.
      c. Small peptides from the proteasome are then digested to amino acids.

Posttranslational modifications: proteolytic cleavage, disulfide bonds, prosthetic groups, phosphorylation, hydroxylation, and γ-carboxylation

I-cell disease: cannot form mannose 6-phosphate on enzymes targeted for lysosomes; lysosomes have no enzymes

Ubiquitin: signal protein attached to proteins to be digested

Proteasome: organelle that digests ubiquinated proteins

CHAPTER 13
DNA TECHNOLOGY

I. Recombinant DNA and DNA Cloning
A. Overview
1. Cloning is a means of obtaining large quantities of specific DNA fragments (i.e., target DNA) for sequencing, genetic engineering, and expression of the encoded proteins.
2. Target DNA is obtained from genomic DNA by digestion with restriction enzymes or from complementary DNA (cDNA) that is produced from mRNA with reverse transcriptase.
3. Restriction endonucleases produce DNA fragments with complementary ("sticky") ends that can be rejoined.
4. Cloning vectors are able to recombine with restriction fragments and replicate within a host cell to amplify the target DNA.
5. Vectors may be plasmids that grow within living host cells, phage vectors that destroy the host bacterium after replicating, or larger artificial chromosomes from bacteria and yeast.
6. DNA cloning in plasmids involves growing a host culture containing a mixture of recombinant plasmids and then detecting the target DNA in individual host cells by using probes; the host cells containing the target are then grown to produce large quantities of the target DNA.
7. A bacterial host culture containing an assortment of recombinant plasmids can be used as a library for finding different target DNAs.

B. Target DNA
1. Target DNA to be cloned is inserted into a cloning vector, forming recombinant DNA, which is introduced (transfected) into rapidly growing cells (e.g., *Escherichia coli* cells) and replicated along with the host-cell DNA.

C. Genomic DNA and complementary DNA
1. Two methods are commonly used to produce target DNA pieces suitable for cloning: genomic DNA and complementary DNA (cDNA).
   a. Genomic DNA is digested with restriction endonucleases producing a mixture of all or part of a particular gene, including introns and other noncoding sequences.
   b. cDNA is produced from messenger RNA (mRNA) isolated from a particular tissue, and it does not contain introns.
      (1) Reverse transcriptase produces a DNA strand complementary to the mRNA.
      (2) After the resulting DNA-mRNA hybrid is denatured, DNA polymerase is used to produce double-strand cDNA from the single-strand DNA copy of the mRNA.

D. Restriction endonucleases
1. Restriction endonucleases are bacterial enzymes that are highly specific for short nucleotide sequences (i.e., restriction sites) and cleave both DNA strands within this region.
2. Staggered cuts are made by most restriction enzymes, producing DNA fragments with single-strand ends.
   a. Restriction sites are palindromes; both strands of DNA have the same sequence when read in a $5' \rightarrow 3'$ direction.
   b. Complementary single-strand sticky ends, which can form base pairs, result from a staggered cut in a palindromic sequence.
3. DNA fragments from different sources produced by cleavage with the same restriction enzyme and containing complementary sticky ends anneal (i.e., form base pairs) with each other.

Target DNA: specific DNA fragments obtained for sequencing, genetic engineering, and expression of the encoded proteins

Vectors: DNA molecules that are able to recombine with restriction fragments and replicate within a host cell to amplify the target DNA

Genomic library: a collection of host bacterial cells, each containing random restriction fragments of genomic DNA

Cloning target DNA involves insertion into a cloning vector to form recombinant DNA.

cDNA: double-strand DNA that contains all the coding sequences (exons) corresponding to a single mRNA

Genomic DNA: a restriction digest of DNA isolated from an organism

cDNA: a collection of DNA molecules synthesized on an mRNA template using reverse transcriptase

Restriction endonucleases produce DNA fragments with sticky ends that are used to produce recombinant DNA.

Restriction sites are palindromes, allowing complementary annealing with any other DNA cut at the same site.

Newly recombined restriction fragments must always be sealed with DNA ligase to make a continuous DNA molecule.

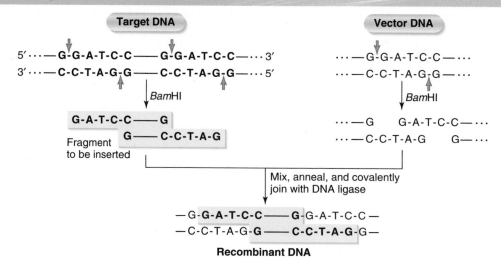

13-1: Restriction endonucleases and the production of recombinant DNA. Target DNA to be inserted and vector DNA are cleaved with the same restriction enzyme (*Bam*HI), producing complementary sticky ends.

Cloning vector: autonomously replicating DNA molecule used to carry a DNA fragment into a host cell

Cloning vectors: selectable features (e.g., antibiotic resistance, color change) that allow differentiating host cells with a vector from those without one

Plasmid vectors replicate within host without killing it; they are isolated after the host is grown in culture.

Plasmids are naturally occurring DNA molecules that are taken up by bacteria and often confer antibiotic resistance.

4. Joining of the free ends with DNA ligase yields stable recombinant DNA molecules made up of segments from two or more sources (Fig. 13-1).
E. **Cloning vectors**
 1. Cloning vectors must possess three essential properties to function in DNA cloning:
  a. Presence of a restriction site permitting insertion of target DNA
  b. Autonomous replication in a host cell permitting amplification of target DNA independently of host-cell DNA synthesis
  c. A selectable feature that permits transfected host cells containing the vector to be distinguished from cells that do not contain the vector
F. **Plasmid vectors (Fig. 13-2)**
 1. Plasmid vectors are constructed from small, circular, extrachromosomal DNAs (i.e., plasmids) that are spontaneously taken up by host bacterial cells.
 2. Target DNA up to 5 kilobases (kb) long can be inserted into plasmid vectors.
 3. Plasmids used as vectors naturally contain or are engineered to contain
  a. Known restriction sites
  b. An origin that permits rapid replication in host cells
  c. An antibiotic-resistance gene that confers antibiotic resistance on host cells
 4. Naturally occurring plasmids, which often carry antibiotic-resistance genes, are readily transferred from one bacterial cell to another, promoting the rapid spread of antibiotic resistance throughout a bacterial population.
G. **Other vectors**
 1. Other vectors can hold larger DNA fragments and are more efficient than plasmids in transfecting host cells.

13-2: Formation of a recombinant plasmid vector. Plasmids *(left)* used as cloning vectors contain a replication origin *(gray)*, at least one known restriction site, and a selectable gene *(dark blue)* that confers antibiotic resistance (e.g., *Amp*). Two restriction sites are shown, *Pst*I and *Eco*RI, at which target DNA could be inserted without disrupting the origin or selectable gene. When plasmids *(right)* digested with a restriction enzyme (e.g., *Pst*I) are mixed with an excess of target DNA digested with the same enzyme, a large proportion of the plasmids recombines with target DNA fragments. Host cells that take up a recombinant plasmid become resistant to ampicillin.

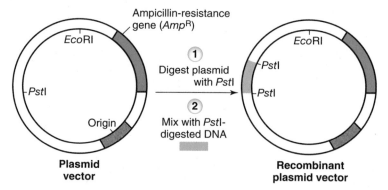

2. Lambda-phage vectors, produced from lambda (λ) bacteriophage, can hold DNA fragments up to 20 kb long.

3. Yeast artificial chromosomes (YACs) and bacterial artificial chromosomes (BACs) can hold DNA fragments as long as 1 to 2 Mb.

   a. YACs and BACs have been particularly useful in studying human and other eukaryotic genes, which usually are quite long because of the presence of introns.

H. Basic steps in DNA cloning

1. Summary of basic steps in DNA cloning, which are similar for all types of vectors, is depicted in Figure 13-3 for a plasmid vector.

2. Recombinant vectors are transfected into appropriate host cells under conditions that favor incorporation of only one vector per cell.

3. Each host cell that takes up a plasmid multiplies into a colony of genetically identical cells (a clone), with each cell containing the same recombinant vector.

I. DNA libraries

1. DNA libraries are collections of restriction fragments cloned within suitable host cells.

2. A genomic DNA library is obtained from restriction enzyme digestion of the entire genome of an organism and theoretically contains all of the nuclear DNA sequences.

3. A cDNA library is obtained from the mixture of mRNAs expressed by a particular tissue under given physiologic conditions.

   a. Each clone in a cDNA library carries an intact gene that lacks introns and a promoter region.

Plasmid vectors: 5-kb target DNA

Lambda vectors: 20-kb target DNA

BACs and YACs: 1- to 2-Mb target DNA

Transfection of recombinant vectors should favor incorporation of only one unique vector per cell.

In a genomic DNA library, each clone contains a fragment of the entire genome of an organism.

In a cDNA library, each clone contains an intact gene.

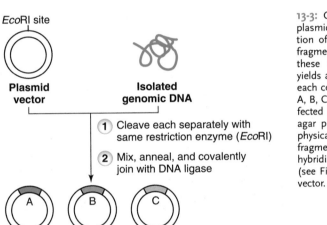
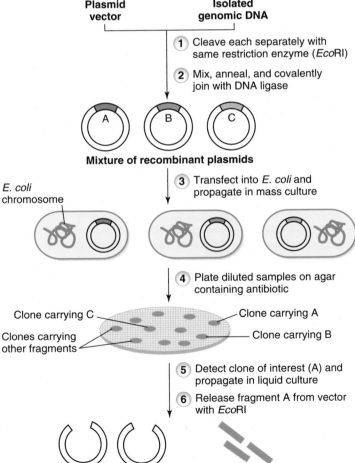

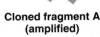

EcoRI site

Plasmid vector   **Isolated genomic DNA**

① Cleave each separately with same restriction enzyme (EcoRI)

② Mix, anneal, and covalently join with DNA ligase

A   B   C

Mixture of recombinant plasmids

③ Transfect into E. coli and propagate in mass culture

E. coli chromosome

④ Plate diluted samples on agar containing antibiotic

Clone carrying C

Clones carrying other fragments

Clone carrying A

Clone carrying B

⑤ Detect clone of interest (A) and propagate in liquid culture

⑥ Release fragment A from vector with EcoRI

Cleaved plasmid   **Cloned fragment A (amplified)**

13-3: Overview of DNA cloning with a plasmid vector. Restriction enzyme digestion of genomic DNA produces multiple fragments with different sequences. Mixing these fragments with cleaved plasmids yields a mixture of recombinant plasmids, each containing a different fragment (e.g., A, B, C). After growth and plating of transfected *E. coli* cells on antibiotic-containing agar plates, the fragments are separated physically into different clones. A specific fragment of interest can be identified by hybridization with a complementary probe (see Fig. 11-4) and then isolated from the vector.

II. Detection of Specific Nucleic Acid Sequences with Probes

A. Overview

1. Probes are labeled single-strand oligonucleotides that are synthesized to hybridize with target DNA.
2. A specific DNA fragment in a DNA library that is immobilized by blotting on a nitrocellulose filter can be identified by incubation with probes.
3. When the type of molecule being detected is DNA, the procedure is called a Southern blot; for RNA detection, the procedure is a Northern blot; and for protein detection, the procedure is a Western blot.
4. Microarrays use immobilized probes in known locations to screen mixtures of fluorescently labeled RNAs produced by individual cell types through hybridization and color analysis.

B. Screening DNA libraries

1. Probes are synthetic single-strand oligonucleotides (20 to 30 bases long) that are most often labeled with a radioactive isotope (e.g., ^{32}P), for easy detection.
2. In a mixture of denatured single-strand DNA fragments (or RNAs), a probe can hybridize (i.e., form base pairs) only with fragments containing a complementary sequence.
3. Screening of DNA libraries for a specific target DNA fragment can be performed with a membrane hybridization assay using a complementary probe (Fig. 13-4).

C. Blotting analysis

1. Blotting analysis can separate and detect specific nucleic acid sequences or proteins in complex mixtures (Fig. 13-5).
   a. Southern blotting detects DNA with labeled DNA probes.
      (1) Useful in identifying a specific restriction fragment out of the millions present in a restriction digest of an individual's genomic DNA
   b. Northern blotting detects RNA with labeled DNA probes.
      (1) Useful in determining whether a specific mRNA is expressed in a particular tissue
   c. Western blotting detects proteins by using labeled antibodies.
      (1) Useful confirmatory test for human immunodeficiency virus (HIV) infection when the enzyme-linked immunosorbent assay (ELISA) screening test result is positive

---

DNA libraries can be searched with probes to obtain sequences containing target DNA.

Probes: synthetic, short, single-strand oligonucleotides often labeled with a radioactive isotope

Immobilization of denatured single-strand DNA fragments permits identification of specific sequences when hybridized with single-stranded probes.

Southern blotting detects DNA; Northern blotting, RNA; Western blotting, proteins; Southwestern blotting, DNA-binding proteins

Southern blotting: useful for finding specific restriction fragments

Northern blotting: useful for determining which mRNAs are expressed

Western blotting: useful for confirming positive HIV test results

Southwestern blotting: useful for detecting transcription factors

---

13-4: Identification of a clone carrying a specific target DNA fragment by membrane hybridization (e.g., fragment A clone in Fig. 13-3). In this technique, a replica of the master plate is transferred to a nitrocellulose filter (i.e., blotting) and immobilized on it. The radioactive probe hybridizes only with a recombinant DNA containing a complementary sequence. An autoradiogram is produced by placing an x-ray film over the filter; radioactivity from the labeled probe exposes the film so that a dark spot appears when the film is developed, revealing the location of the target sequence.

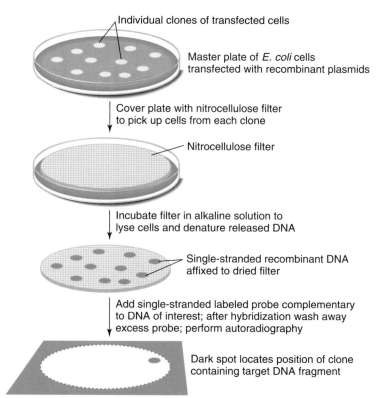

Individual clones of transfected cells

Master plate of *E. coli* cells transfected with recombinant plasmids

Cover plate with nitrocellulose filter to pick up cells from each clone

Nitrocellulose filter

Incubate filter in alkaline solution to lyse cells and denature released DNA

Single-stranded recombinant DNA affixed to dried filter

Add single-stranded labeled probe complementary to DNA of interest; after hybridization wash away excess probe; perform autoradiography

Dark spot locates position of clone containing target DNA fragment

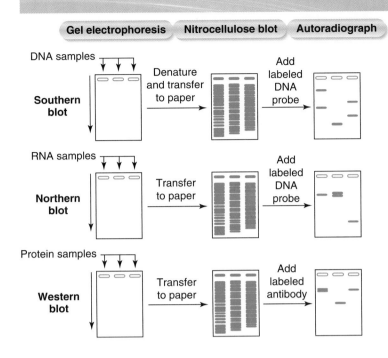

Gel electrophoresis   Nitrocellulose blot   Autoradiograph

Southern blot — DNA samples — Denature and transfer to paper — Add labeled DNA probe

Northern blot — RNA samples — Transfer to paper — Add labeled DNA probe

Western blot — Protein samples — Transfer to paper — Add labeled antibody

13-5: Blotting analysis can detect specific DNA sequences (Southern blot), RNA sequences (Northern blot), and proteins (Western blot). After samples from three different sources are separated by gel electrophoresis, the gel is placed on nitrocellulose paper. The nucleic acids or proteins transfer to the paper (i.e., blotting) and bind to it, maintaining the banding pattern in the gel. The nitrocellulose blot is treated with a radioactive probe that binds to a specific target molecule of interest and then is subjected to autoradiography, eliminating the background interference from all other molecules. In each blot, the three samples exhibit different banding patterns.

    d. Southwestern blotting detects DNA-binding proteins with labeled DNA probes.
       (1) Useful in detecting expression of transcription factors that interact with regulatory sequences controlling specific genes

D. Microarrays
    1. Microarray technology permits the screening for the expression of many genes at one time by arranging fragments of known genes in known positions on a small grid and hybridizing a fluorescently labeled RNA probe obtained from the cell type under study.
    2. Hybridization patterns produce colors that differentiate genes that are expressed in normal cells or mutated cells, or both.
    3. Because the sequence of each fragment is known for each position on the grid, a rapid identification of sequences of interest is possible.

III. Polymerase Chain Reaction
  A. Overview
    1. Polymerase chain reaction (PCR) is an in vitro enzymatic method for amplifying a target DNA sequence located between two short flanking sites whose sequences are known.
    2. Alternating cycles of replication and denaturation initiated with primers of a known sequence that border a target sequence produce large quantities of the target sequence.
    3. PCR can be used to analyze viral infections (RT-PCR) or mutant alleles due to its high degree of sensitivity.
  B. PCR procedure (Fig. 13-6)
    1. The PCR incubation requires several components in addition to DNA containing the target sequence.
      a. Single-strand oligonucleotide primers complementary to the two flanking sites
      b. A heat-stable DNA polymerase that remains functional throughout the heating phases
      c. All four deoxyribonucleoside triphosphates (dNTPs)
    2. Doubling of the target DNA sequence occurs in each heating-cooling cycle.
      a. Within a few cycles, almost all the DNA molecules correspond to the target sequence.
    3. Sensitivity is so great that only one target sequence present in a single cell can be amplified and detected.
      a. PCR is useful in forensics for identifying DNA from tiny blood, saliva, and semen samples.
  C. RT-PCR
    1. RNA virus infection can be detected with RT-PCR at very early stages.
    2. Reverse transcriptase (RT) is used to convert the viral RNA genome to DNA, which is amplified by PCR and identified by Southern blotting.

Microarray technology: immobilization of known DNA fragments on a grid mixed with a fluorescently labeled RNA probe to produce colored dots

Microarray dot colors distinguish between normal and mutant genes, including heterozygotes that produce a unique color blend.

PCR: alternating cycles of replication and denaturation initiated with primers of known sequence

PCR is cloning without a host; primers allow specific selection of target DNA.

Heat-stable DNA polymerase is necessary for PCR because of heating and cooling cycles.

Target DNA sequence is doubled with each heating and cooling cycle in the PCR.

RT-PCR permits early detection of the RNA viruses HIV, enterovirus, and Norwalk virus.

13-6: Polymerase chain reaction (PCR). DNA strands are separated by heating, excess primers are added to begin the replication process, and heat-stable DNA polymerase catalyzes $5' \rightarrow 3'$ synthesis using deoxyribonucleoside triphosphates (dNTPs). Three cycles of amplification are shown. Notice how the strands of the original DNA, labeled A and B, are diluted, whereas the proportion of new molecules containing only the target DNA region rapidly increases. After 30 cycles, there are 10^9 target-region molecules for each beginning DNA molecule.

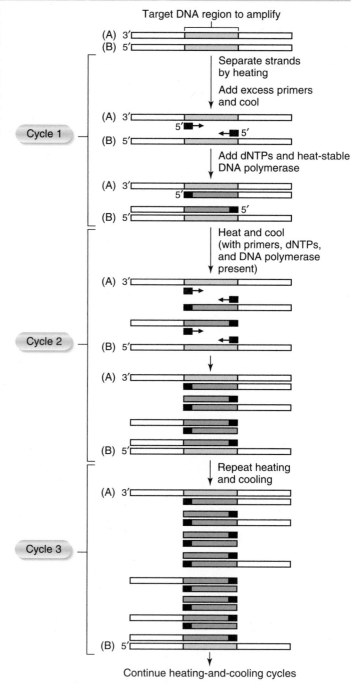

3. HIV, enterovirus, and Norwalk virus infections can be diagnosed with RT-PCR before significant viral replication or detectable antibody production has occurred.

D. PCR analysis of inherited diseases

1. Cystic fibrosis is an autosomal recessive disorder caused by a 3-base deletion coding for phenylalanine in the cystic fibrosis transmembrane regulation (*CFTR*) gene (see Chapter 3).

   a. DNA from normal, carrier, and affected individuals can be differentiated by PCR analysis using probes that flank the mutated region of *CFTR* (Fig. 13-7).

2. Other genetic diseases detectable by PCR analysis include familial hypercholesterolemia, hemophilia, Lesch-Nyhan syndrome, lysosomal and glycogen storage diseases, retinoblastoma, sickle cell anemia, β-thalassemia, and von Willebrand's disease.

Cystic fibrosis: affected, carrier, and normal individuals distinguished by PCR analysis, which is also used for many other genetic diseases

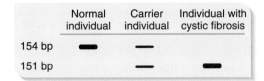

	Normal individual	Carrier individual	Individual with cystic fibrosis
154 bp	▬	—	
151 bp		—	▬

13-7: PCR detection of cystic fibrosis. Use of primers that flank the 3-base deletion in the *CFTR* gene that causes cystic fibrosis (CF) results in different products starting with DNA from normal, carrier, and affected individuals. After samples are amplified, they are analyzed by Southern blotting. The shorter amplification product from the mutant allele (151 base pairs [bp]) moves faster in gel electrophoresis and is easily differentiated from the 154-bp product from the normal allele. The intensity of bands from homozygotes is about twice that of bands from the heterozygous carrier.

IV. Restriction Fragment Length Polymorphisms
A. Overview
1. Target DNA can be analyzed by developing a physical map of restriction sites contained in its sequence.
2. When mutations occur in a restriction site, the lengths of fragments obtained in a restriction digest vary from normal sizes because the corresponding restriction enzyme can no longer cut the DNA.
3. A different length of a known restriction fragment found in a normal population is called a restriction fragment length polymorphism (RFLP).
4. RFLPs help to analyze inheritance patterns for genetic diseases within family pedigrees and to establish genetic identity through DNA fingerprinting.
5. DNA sequencing uses dideoxynucleotides that create a random interruption of synthesis where that nucleotide is specified; produces fragments of unequal length to deduce DNA sequence.

B. Restriction maps
1. After target DNA has been obtained by PCR or cloning, it can be analyzed by developing a physical map of its sequence.
   a. The target DNA is digested in separate samples with different restriction enzymes and then separated on gel electrophoresis.
   b. The size of fragments from each digest allows assembly of a map depicting distances between each restriction site (Fig. 13-8).
2. Knowledge of a restriction map allows removal and analysis of specific segments using the restriction enzymes as landmarks.

> RFLP: a variation within a normal population of restriction fragment length due to mutations in the sequence at the restriction site
>
> RFLPs: basis for detecting certain genetic diseases and for DNA fingerprinting
>
> Restriction sites serve as landmarks in a sample of target DNA.
>
> Restriction mapping identifies the restriction sites in a region of DNA and the distance in kilobases between them.

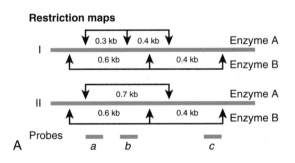

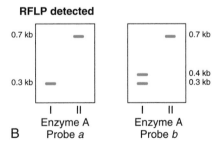

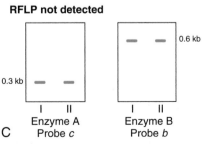

13-8: Restriction fragment length polymorphism detection depends on restriction enzymes and probes. **A,** Restriction maps for enzymes A and B of the same DNA region are from two individuals (I and II). *Small arrows* point to restriction sites. Restriction sites for these individuals differ for enzyme A but not for enzyme B. Oligonucleotide probes *a*, *b*, and *c* are aligned below the DNA sequences with which they hybridize. **B** and **C,** Southern blots were obtained with different enzyme and probe combinations. No polymorphism is detected in this region if the A-digested samples are treated with probe *c*. Likewise, no polymorphism is detected if the samples are digested only with enzyme B and analyzed with any one of the three probes.

Polymorphism: any normal variation in a gene in the same population

There are two types of polymorphisms: single-nucleotide polymorphisms and tandem repeats.

Sickle cell mutation alters a restriction site so that RFLP analysis can differentiate affected, carrier, and normal individuals.

C. RFLPs

1. DNA polymorphisms are small, heritable variations in the DNA sequence found in the general population.
   a. DNA polymorphisms occur every 200 to 500 nucleotides in the human genome, but most cause no phenotypic effect because large portions of the human genome do not code for protein.
   b. If a mutation alters a restriction site, the restriction enzyme can no longer cut the DNA; gel electrophoresis shows one larger fragment replacing two smaller fragments.
2. The appearance of a variable electrophoretic pattern for the restriction fragments produced from the same restriction enzyme is referred to as a polymorphism, specifically an RFLP.
   a. There are two types of RFLPs: single-nucleotide polymorphisms (SNPs) and tandem repeats.
3. SNPs are useful in identifying deleterious mutations.
   a. If an SNP occurs within an exon, it can represent a deleterious point mutation and can be detected on a Southern blot.
   b. RFLP analysis can detect mutant alleles that are consistently associated with a particular polymorphism.
      (1) The mutation that is responsible for sickle cell anemia lies within the restriction site for the *Mst*II restriction enzyme (Fig. 13-9A).
      (2) When patients from an affected family are subjected to RFLP analysis, a pedigree can be constructed to show the inheritance pattern of the mutation (see Fig. 13-9B).

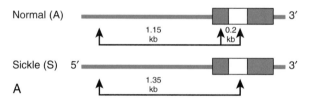

Mst II restriction sites in relation to the β-globin gene

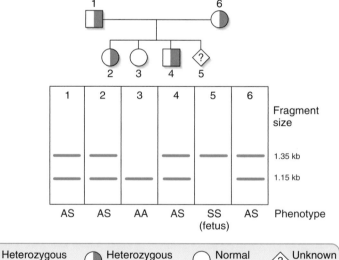

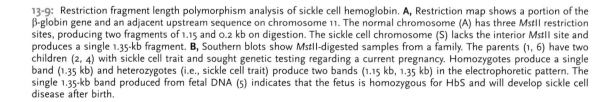

13-9: Restriction fragment length polymorphism analysis of sickle cell hemoglobin. **A,** Restriction map shows a portion of the β-globin gene and an adjacent upstream sequence on chromosome 11. The normal chromosome (A) has three *Mst*II restriction sites, producing two fragments of 1.15 and 0.2 kb on digestion. The sickle cell chromosome (S) lacks the interior *Mst*II site and produces a single 1.35-kb fragment. **B,** Southern blots show *Mst*II-digested samples from a family. The parents (1, 6) have two children (2, 4) with sickle cell trait and sought genetic testing regarding a current pregnancy. Homozygotes produce a single band (1.35 kb) and heterozygotes (i.e., sickle cell trait) produce two bands (1.15 kb, 1.35 kb) in the electrophoretic pattern. The single 1.35-kb band produced from fetal DNA (5) indicates that the fetus is homozygous for HbS and will develop sickle cell disease after birth.

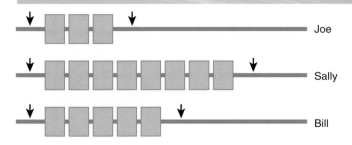

13-10: Polymorphism among three subjects for a specific DNA sequence. Joe, Sally, and Bill can be differentiated by the number of repeats found in their DNA samples. Each box represents a 6-base sequence GATTCC, and the *arrows* represent restriction sites that are located on either side of the repeat sequences. These repeat sequences were generated by unequal recombination and inherited by subsequent generations. Because each of the fragments has a different number of repeats, they migrate differently on electrophoresis (see Fig. 3-11). *(From Pelley JW: Elsevier's Integrated Biochemistry. Philadelphia, Mosby, 2007.)*

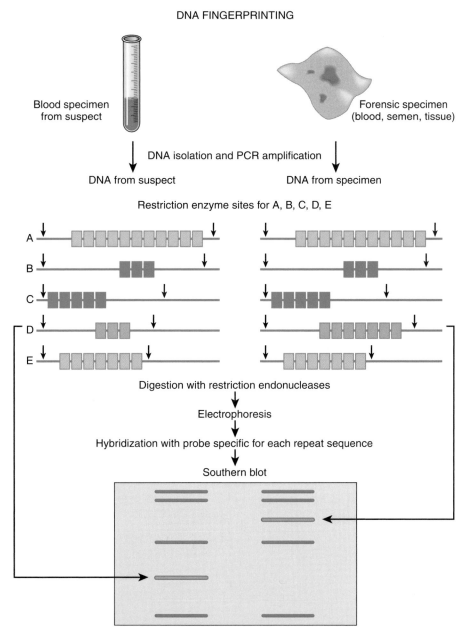

13-11: Comparison of DNA fingerprints from a suspect and a forensic specimen. Restriction digests *(arrows)* of DNA from both samples is amplified to produce samples from selected polymorphic regions; *boxes* indicate repeated sequences. Electrophoresis of samples reveals a match or confirms innocence. *(From Pelley JW: Elsevier's Integrated Biochemistry. Philadelphia, Mosby, 2007.)*

4. Tandem repeats are useful in DNA fingerprinting analysis.
   a. Repetitive DNA in humans is highly polymorphic with respect to *number* of copies of repetitive sequences.
   b. Tandem repeat polymorphisms are caused by dinucleotide or trinucleotide (or longer) sequences that are connected in tandem, but the number of repeats varies (variable number of tandem repeats [VNTRs]) from person to person (Fig. 13-10).
   c. VNTRs produce restriction fragments of different lengths that can be detected on a Southern blot, referred to as a DNA fingerprint (Fig. 13-11).
      (1) This method is 100% effective in eliminating genetic identity if a match between a reference DNA sample and an individual does not occur.
      (2) Proof of identity relies on an estimate of statistical probability of the frequency of a given polymorphism in the population (i.e., the number of false positives).

D. DNA sequencing
1. The Sanger dideoxy method relies on the use of dideoxynucleotides to interrupt DNA synthesis using the unknown DNA as the template.
2. Incubation in separate tubes, each containing one of the dideoxynucleotides, produces fragments of unequal length; they represent a random interruption of synthesis where that nucleotide is specified.
3. Separation of the resulting fragments on gel electrophoresis produces a set of bands that reveal the sequence by their physical position.

VNTRs are repeated sequences in DNA produced by unequal crossing over that can have different lengths within a normal population.

DNA fingerprinting: 100% effective in eliminating genetic identity; conditional probability for positive identity

DNA sequencing: DNA synthesis with labeled dideoxynucleotides producing fragments of various lengths; sequence deduced after gel electrophoresis

COMMON LABORATORY VALUES

TEST	CONVENTIONAL UNITS	SI UNITS
Blood, Plasma, Serum		
Alanine aminotransferase (ALT/GPT at 30° C)	8-20 U/L	8-20 U/L
Amylase, serum	25-125 U/L	25-125 U/L
Aspartate aminotransferase (AST/GOT at 30° C)	8-20 U/L	8-20 U/L
Bilirubin, serum (adult): total; direct	0.1-1.0 mg/dL; 0.0-0.3 mg/dL	2-17 μmol/L; 0-5 μmol/L
Calcium, serum (Ca²⁺)	8.4-10.2 mg/dL	2.1-2.8 mmol/L
Cholesterol, serum	Recommended: <200 mg/dL	<5.2 mmol/L
Cortisol, serum	8:00 AM: 6-23 μg/dL 4:00 PM: 3-15 μg/dL 8:00 PM: ≤50% of 8:00 AM	170-630 nmol/L 80-410 nmol/L Fraction of 8:00 AM: ≤0.50
Creatine kinase, serum	Male: 25-90 U/L Female: 10-70 U/L	25-90 U/L 10-70 U/L
Creatinine, serum	0.6-1.2 mg/dL	53-106 μmol/L
Electrolytes, serum		
Sodium (Na⁺)	136-145 mEq/L	135-145 mmol/L
Chloride (Cl⁻)	95-105 mEq/L	95-105 mmol/L
Potassium (K⁺)	3.5-5.0 mEq/L	3.5-5.0 mmol/L
Bicarbonate (HCO₃⁻)	22-28 mEq/L	22-28 mmol/L
Magnesium (Mg²⁺)	1.5-2.0 mEq/L	1.5-2.0 mmol/L
Estriol, total, serum (in pregnancy)		
24-28 wk; 32-36 wk	30-170 ng/mL; 60-280 ng/mL	104-590 nmol/L; 208-970 nmol/L
28-32 wk; 36-40 wk	40-220 ng/mL; 80-350 ng/mL	140-760 nmol/L; 280-1210 nmol/L
Ferritin, serum	Male: 15-200 ng/mL Female: 12-150 ng/mL	15-200 μg/L 12-150 μg/L
Follicle-stimulating hormone (FSH), serum/plasma	Male: 4-25 mIU/mL Female: Premenopause: 4-30 mIU/mL Midcycle peak: 10-90 mIU/mL Postmenopause: 40-250 mIU/mL	4-25 U/L 4-30 U/L 10-90 U/L 40-250 U/L
Gases, arterial blood (room air)		
pH	7.35-7.45	[H⁺] = 36-44 nmol/L
Pco₂	33-45 mm Hg	4.4-5.9 kPa
Po₂	75-105 mm Hg	10.0-14.0 kPa
Glucose, serum	Fasting: 70-110 mg/dL 2 hr postprandial: <120 mg/dL	3.8-6.1 mmol/L <6.6 mmol/L
Growth hormone-arginine stimulation	Fasting: <5 ng/mL Provocative stimuli: >7 ng/mL	<5 μg/L >7 μg/L
Immunoglobulins, serum		
IgA	76-390 mg/dL	0.76-3.90 g/L
IgE	0-380 IU/mL	0-380 kIU/L

(Continued)

TEST	CONVENTIONAL UNITS	SI UNITS
Blood, Plasma, Serum—cont'd		
IgG	650-1500 mg/dL	6.5-15 g/L
IgM	40-345 mg/dL	0.4-3.45 g/L
Iron	50-170 µg/dL	9-30 µmol/L
Lactate dehydrogenase, serum	45-90 U/L	45-90 U/L
Luteinizing hormone (LH), serum/plasma	Male: 6-23 mIU/mL	6-23 U/L
	Female:	
	Follicular phase: 5-30 mIU/mL	5-30 U/L
	Midcycle: 75-150 mIU/mL	75-150 U/L
	Postmenopause: 30-200 mIU/mL	30-200 U/L
Osmolality, serum	275-295 mOsm/kg	275-295 mOsm/kg
Parathyroid hormone, serum, N-terminal	230-630 pg/mL	230-630 ng/L
Phosphatase (alkaline), serum (*p*-nitrophenyl phosphate at 30° C)	20-70 U/L	20-70 U/L
Phosphorus (inorganic), serum	3.0-4.5 mg/dL	1.0-1.5 mmol/L
Prolactin (hPRL), serum	<20 ng/mL	<20 µg/L
Proteins, serum		
Total (recumbent)	6.0-8.0 g/dL	60-80 g/L
Albumin	3.5-5.5 g/dL	35-55 g/L
Globulin	2.3-3.5 g/dL	23-35 g/L
Thyroid-stimulating hormone (TSH), serum or plasma	0.5-5.0 µU/mL	0.5-5.0 mU/L
Thyroidal iodine (^{123}I) uptake	8-30% of administered dose/24 hr	0.08-0.30/24 hr
Thyroxine (T$_4$), serum	4.5-12 µg/dL	58-154 nmol/L
Triglycerides, serum	35-160 mg/dL	0.4-1.81 mmol/L
Triiodothyronine (T$_3$), serum (radioimmunoassay)	115-190 ng/dL	1.8-2.9 nmol/L
Triiodothyronine (T$_3$) resin uptake	25-38%	0.25-0.38
Urea nitrogen, serum (BUN)	7-18 mg/dL	1.2-3.0 mmol urea/L
Uric acid, serum	3.0-8.2 mg/dL	0.18-0.48 mmol/L
Cerebrospinal Fluid		
Cell count	0-5 cells/mm^3	0-5 × 10^6/L
Chloride	118-132 mEq/L	118-132 mmol/L
Gamma globulin	3-12% of total proteins	0.03-0.12
Glucose	50-75 mg/dL	2.8-4.2 mmol/L
Pressure	70-180 mm H$_2$O	70-180 mm H$_2$O
Proteins, total	<40 mg/dL	<0.40 g/L
Hematology		
Bleeding time (template)	2-7 min	2-7 min
Erythrocyte count	Male: 4.3-5.9 million/mm^3	4.3-5.9 × 10^{12}/L
	Female: 3.5-5.5 million/mm^3	3.5-5.5 × 10^{12}/L
Erythrocyte sedimentation rate (Westergren)	Male: 0-15 mm/hr	0-15 mm/hr
	Female: 0-20 mm/hr	0-20 mm/hr
Hematocrit (Hct)	Male: 40-54%	0.40-0.54
	Female: 37-47%	0.37-0.47
Hemoglobin A$_{1C}$	≤6%	≤0.06%
Hemoglobin (Hb), blood	Male: 13.5-17.5 g/dL	2.09-2.71 mmol/L
	Female: 12.0-16.0 g/dL	1.86-2.48 mmol/L
Hemoglobin, plasma	1-4 mg/dL	0.16-0.62 mmol/L
Leukocyte count and differential		
Leukocyte count	4500-11,000/mm^3	4.5-11.0 × 10^9/L
Segmented neutrophils	54-62%	0.54-0.62
Bands	3-5%	0.03-0.05
Eosinophils	1-3%	0.01-0.03
Basophils	0-0.75%	0-0.0075
Lymphocytes	25-33%	0.25-0.33
Monocytes	3-7%	0.03-0.07
Mean corpuscular hemoglobin (MCH)	25.4-34.6 pg/cell	0.39-0.54 fmol/cell

TEST	CONVENTIONAL UNITS	SI UNITS
Mean corpuscular hemoglobin concentration (MCHC)	31-37% Hb/cell	4.81-5.74 mmol Hb/L
Mean corpuscular volume (MCV)	80-100 μm³	80-100 fL
Partial thromboplastin time, activated (aPTT)	25-40 sec	25-40 sec
Platelet count	150,000-400,000/mm³	150-400 × 10⁹/L
Prothrombin time (PT)	12-14 sec	12-14 sec
Reticulocyte count	0.5-1.5% of red cells	0.005-0.015
Thrombin time	<2 sec deviation from control	<2 sec deviation from control
Volume		
Plasma	Male: 25-43 mL/kg	0.025-0.043 L/kg
	Female: 28-45 mL/kg	0.028-0.045 L/kg
Red cell	Male: 20-36 mL/kg	0.020-0.036 L/kg
	Female: 19-31 mL/kg	0.019-0.031 L/kg
Sweat		
Chloride	0-35 mmol/L	0-35 mmol/L
Urine		
Calcium	100-300 mg/24 hr	2.5-7.5 mmol/24 hr
Creatinine clearance	Male: 97-137 mL/min	
	Female: 88-128 mL/min	
Estriol, total (in pregnancy)		
30 wk	6-18 mg/24 hr	21-62 μmol/24 hr
35 wk	9-28 mg/24 hr	31-97 μmol/24 hr
40 wk	13-42 mg/24 hr	45-146 μmol/24 hr
17-Hydroxycorticosteroids	Male: 3.0-9.0 mg/24 hr	8.2-25.0 μmol/24 hr
	Female: 2.0-8.0 mg/24 hr	5.5-22.0 μmol/24 hr
17-Ketosteroids, total	Male: 8-22 mg/24 hr	28-76 μmol/24 hr
	Female: 6-15 mg/24 hr	21-52 μmol/24 hr
Osmolality	50-1400 mOsm/kg	
Oxalate	8-40 μg/mL	90-445 μmol/L
Proteins, total	<150 mg/24 hr	<0.15 g/24 hr

INDEX

Note: Page numbers followed by *f* indicate figures, *t* indicate tables and *b* indicate boxes.

A

A (adenine)
  in nucleotides, 124, 125*f*
  salvage of, 128
Abetalipoproteinemia, 95, 95*b*
ABO blood group antigens, 79*b*, 80
Accommodation, of β-adrenergic receptors, 30*b*, 31
Acetoacetate, 86, 86*b*
  in liver metabolism in starvation state, 120
Acetoacetyl CoA, formation of, 103
Acetone, 86, 86*b*, 87
Acetyl CoA
  in catabolism, 55, 55*b*, 56*f*
  in citric acid cycle, 56, 56*b*
  in fatty acid and triacylglycerol synthesis, 81, 81*b*
  formation of, 103
  in gluconeogenesis, 70*b*, 71
  in glycolysis, 68, 68*b*
  in liver metabolism in starvation state, 119
Acetyl CoA carboxylase, in fatty acid and triacylglycerol synthesis,
    81, 81*b*, 83
N-Acetylglutamate, in urea cycle, 100, 100*b*, 101
Achlorhydria, 39*b*
Acidic sugars, in glycosaminoglycans, 80, 80*b*
Acidosis, potassium in, 50, 50*b*
Acrodermatitis enteropathica, 52*b*, 53
Actinomycin D, and DNA synthesis, 135, 135*b*
Activated partial thromboplastin time (aPTT), normal values
    for, 161*t*
Activator protein, in eukaryotic control of gene expression,
    142, 142*f*
Active transport
  primary, 25*f*, 25*t*, 27
    Ca⁺-ATPase pumps in, 27, 27*b*
      in tissue hypoxia, 27, 27*b*
    defined, 25*b*, 26*b*
    Na⁺/K⁺/ATPase pump in, 27
      albuterol and, 27, 27*b*
      β-blockers and, 27, 27*b*
      cardiotonic steroids and, 27, 27*b*
      insulin and, 27, 27*b*
      succinylcholine and, 27, 27*b*
  secondary, 25*f*, 25*t*, 27
    cotransporter carrier proteins in, 27
    defined, 25*b*, 27*b*
    of glucose, 27, 27*b*, 28*f*
    Na⁺/K⁺-ATPase pump in, 27, 28*f*
    Na⁺-linked Ca²⁺ antiporter in, 28, 28*b*, 28*f*
    Na⁺-linked symporters in, 27, 27*b*, 28*f*
Activity factor, 36
Acute intermittent porphyria, 109*b*, 109*t*, 110
Adenine (A)
  in nucleotides, 124, 125*f*
  salvage of, 128
Adenine phosphoribosyl transferase (APRT), in adenine salvage, 128
Adenosine deaminase (ADA)
  deficiency of, 127, 127*b*, 128*t*
  in purine degradation, 127
Adenosine monophosphate (AMP)
  in purine degradation, 127
  in purine synthesis, 124

Adenosine monophosphate (AMP) deaminase, in purine
    degradation, 127
Adenosine triphosphate (ATP), in glycolysis, 65, 65*b*, 66*f*
Adenosine triphosphate (ATP) synthase, 59*f*, 60, 60*b*
Adenosine triphosphate–adenosine diphosphate (ATP-ADP) cycle,
    54, 54*b*, 55*f*
Adenosine triphosphate–adenosine diphosphate (ATP-ADP)
    translocase, 59*f*, 60
Adequate intake (AI), 35
Adipose tissue metabolism
  in fasting state, 117, 118*f*
  in starvation state, 119*f*, 120
  in well-fed state, 116, 116*f*
ADMA (asymmetric dimethylarginine), 112, 112*b*
Adrenal cortex, steroid hormones in, 5, 90, 91*f*
Adrenocortical steroids, 5, 90, 91*f*
Adrenogenital syndrome, 91, 91*b*
Adrenoleukodystrophy, 87*b*, 88
Adult hemoglobin (HbA), 16, 20*b*
Agonists, 34
AI (adequate intake), 35
ALA (δ-aminolevulinic acid), in heme synthesis, 109
ALA (δ-aminolevulinic acid) hydratase, in heme synthesis, 109
ALA (δ-aminolevulinic acid) synthase, in heme synthesis, 108*b*, 109
Alanine (Ala), 8*t*
  synthesis of, 99*t*
Alanine aminotransferase (ALT), 98, 98*b*
  in diagnosis, 16*t*
  normal values for, 161*t*
Alanine cycle
  in gluconeogenesis, 72*b*
  in muscle metabolism in fasting state, 118, 118*f*
Albinism, 104*t*, 105, 105*b*
Albumin, 9, 9*b*
  normal values for serum, 161*t*
Albuterol
  as agonist, 34, 34*b*
  and Na⁺/K⁺/ATPase pump, 27, 27*b*
Alcohol dehydrogenase, in alcohol metabolism, 122
Alcohol metabolism, 122
  in higher concentrations, 123
  in low concentrations, 122, 122*f*
  overview of, 122
Alcoholics, thiamine deficiency in, 41, 41*b*
Aldehyde(s), 1
Aldehyde dehydrogenase, in alcohol metabolism, 122
Aldolase A, in glycolysis, 63
Aldolase B, in fructose metabolism, 76*b*, 77
Aldolase reductase, in galactose metabolism, 76*f*
Aldoses, 1, 2*t*
Aldosterone, 6*f*
  synthesis of, 91*f*
Alkaline phosphatase
  normal values for serum, 161*t*
  vitamin D and, 47, 47*b*
Alkalosis
  in calcium regulation, 48*b*, 49
  potassium in, 50, 50*b*
Alkaptonuria, 104*t*, 105, 105*b*
Alkylating agents, and DNA synthesis, 135, 135*b*
Alleles, 145
Allolactose, in prokaryotic control of gene expression, 141

Allopurinol, 127*b*, 128
  as noncompetitive inhibitor, 14
Allosteric regulation
  of enzymes, 15, 15*b*
  of metabolism, 113, 114*t*
Allosterism, of enzymes, 15, 15*b*
All-*trans*-retinoic acid, 45*b*, 47
α-helix, of protein, 10
Alport's syndrome, 22, 22*b*
ALT (alanine aminotransferase), 98, 98*b*
  in diagnosis, 16*t*
  normal values for, 161*t*
Alternative splicing, 143, 143*b*
Alternative tailing, 143
Altitude, and oxygen-binding curve, 18
α-Amanitin, and RNA polymerase, 138*b*
Amino acid(s), 6
  acid-base properties of, 8, 9*b*
  branched-chain
    alanine cycle for disposing of nitrogen from, 118, 118*f*
    metabolism of, 105, 105*b*
  carbon skeletons of, 102*f*, 103
  catabolic pathways of, 102
    carbon skeletons in, 102*f*, 103
    for leucine, isoleucine, and valine (branched chain amino
      acids), 105
    for methionine, 105*f*, 106
    overview of, 102
    for phenylalanine and tyrosine, 103, 103*f*, 104*t*
  essential, 6, 6*b*
  glucogenic, 102, 102*b*
  hydrophilic (polar)
    charged, 7, 8*t*
    uncharged, 7, 8*t*
  hydrophobic (nonpolar), 7, 8*t*
  ketogenic, 102, 103*b*
  modification of residues in proteins with, 9
  nonessential, 98, 98*b*, 99*t*
  structure of, 6
    side chain (R group) in, 7, 7*b*
      modification of, 9
Amino acid derivative(s), 106
  asymmetric dimethylarginine (ADMA) as, 112, 112*b*
  catecholamines as, 106, 107*f*
  creatine as, 112
  γ-aminobutyrate (GABA) as, 112, 112*b*
  heme as, 108, 108*f*, 109*t*, 110*f*
  histamine as, 112, 112*b*
  overview of, 106
  serotonin, melatonin, and niacin as, 111*f*, 112
Amino acid nitrogen, removal and disposal of, 98
  ammonia metabolism in, 101
  overview of, 98
  transamination and oxidative deamination in, 98, 99*f*
  urea cycle in, 99, 100*f*
Amino acid residues, modification in proteins of, 9
Amino sugars, 1
  in glycosaminoglycans, 80, 80*b*
Aminoacyl-tRNA synthetases, 144, 144*b*, 144*f*, 145*f*
Aminoglycosides, protein synthesis inhibition by, 149
δ-Aminolevulinic acid (ALA), in heme synthesis, 109
δ-Aminolevulinic acid (ALA) hydratase, in heme synthesis,
  109
δ-Aminolevulinic acid (ALA) synthase, in heme synthesis,
  108*b*, 109
Aminotransferases, 98
Ammonia (NH₃)
  excess, 101
  metabolism of, 101, 101*b*
  sources of, 101, 101*b*
Ammonium (NH₄+)
  in ammonia metabolism, 101, 101*b*
  in oxidative deamination, 98, 99
  in urea cycle, 100, 101
Amobarbital (Amytal), 62*t*
AMP (adenosine monophosphate)
  in purine degradation, 127
  in purine synthesis, 124
AMP (adenosine monophosphate) deaminase, in purine
  degradation, 127
Amylase
  in diagnosis, 16*t*
  normal values for serum, 161*t*
α-Amylase, starch digestion by, 37, 37*b*
Amylo-α-1,6-glucosidase deficiency, 75*t*

Amylopectin, 3
Amylose, 3
Amytal (amobarbital), 62*t*
Anaerobic glycolysis, 59*b*, 65*b*
Anaplerosis, in citric acid cycle, 58*b*
Andersen's disease, 75*t*
Androgens, 5
  synthesis of, 90*b*, 91
Anemia
  Cooley's, 21, 21*b*
  Fanconi's, 137*t*
  sickle cell, 20, 20*b*
Angelman syndrome, 147
Angiotensin II, in steroid hormone synthesis, 90, 90*b*
Animal fats, 37*b*, 38
Animal proteins, 39
Antagonists, 34
Antibiotics, protein synthesis inhibition by
  eukaryotic, 149, 149*b*
  prokaryotic, 148*f*, 149, 149*b*
Anticancer drugs, inhibition of DNA synthesis by, 126*t*, 127,
  131*f*, 134, 134*b*
Anticipation, 146*b*, 147
Anticodon, 144, 144*f*
Antileukotriene drugs, 6, 6*b*
Antimycin A, 62*t*
Antineoplastic drugs, and cell cycle, 131*f*
Antiporters, 26
Antipsychotics, as antagonists, 34, 34*b*
Antisense strand, of DNA, 138, 138*b*, 139*f*
Anti-topoisomerase antibodies, 132*b*
APC gene, 137*t*
Apolipoprotein(s), functions of, 93
Apolipoprotein A-I (apoA-I), 93, 93*b*
Apolipoprotein B-48 (ApoB-48), 38*b*, 38*f*, 39, 93, 93*b*
Apolipoprotein B-100 (ApoB-100), 93, 93*b*
Apolipoprotein C-II (ApoC-II), 93, 93*b*
Apolipoprotein C-II (ApoC-II) deficiency, 96*t*
Apolipoprotein E (ApoE), 93, 93*b*
Apoprotein B, editing of, 143, 143*b*
APRT (adenine phosphoribosyl transferase), in adenine
  salvage, 128
aPTT (activated partial thromboplastin time), normal values
  for, 161*t*
Arachidonic acid, 3*t*
Arginase, in urea cycle, 101
Arginine (Arg), 7, 7*b*, 8*t*
  charge on, 8
  creatine synthesis from, 112
  synthesis of, 99*t*
  in urea cycle, 100, 100*b*, 101
Arginine stimulation test, for growth hormone, 161*t*
Argininosuccinate, in urea cycle, 100
Argininosuccinate lyase, in urea cycle, 100
Argininosuccinate synthetase, in urea cycle, 100
Arterial blood gases, normal values for, 161*t*
Ascorbate. *See* Ascorbic acid (vitamin C).
Ascorbic acid (vitamin C), 1
  in collagen synthesis, 22*b*
  deficiency of, 41*t*, 44*b*, 45
  excess intake of, 44*b*, 45
  functions of, 44*b*, 45
  sources of, 45
Asn (asparagine), 8*t*
  glycosylation of, 9
  synthesis of, 99*t*
Asparagine (Asn), 8*t*
  glycosylation of, 9
  synthesis of, 99*t*
Aspartate (Asp), 8*t*
  charge on, 9
  synthesis of, 99*t*
  in urea cycle, 100
Aspartate aminotransferase (AST), 98, 98*b*
  in diagnosis, 16*t*
  normal values for, 161*t*
Aspartate transcarbamoylase, in pyrimidine synthesis, 126
Aspirin, as irreversible inhibitor, 14
Asymmetric dimethylarginine (ADMA), 112, 112*b*
Ataxia telangiectasia, 137*t*
Atorvastatin, and cholesterol synthesis, 89
ATP. *See* Adenosine triphosphate (ATP).
Autosomal dominant inheritance, 145
Autosomal recessive inheritance, 145
AZT (zidovudine), and reverse transcriptase, 133, 133*b*

B

Bacterial artificial chromosomes (BACs), 153, 153*b*
Bacterial protein synthesis, antibiotic inhibition of, 148*f*, 149, 149*b*
Bacterial ureases
 in ammonia metabolism, 101, 101*b*
 in urea cycle, 100*b*, 101
Bands, normal values for, 161*t*
Basal metabolic rate (BMR), 35, 35*b*
Base excision repair, 134, 134*b*, 135*f*
Basophils, normal values for, 161*t*
BCAAs (branched-chain amino acids)
 alanine cycle for disposing of nitrogen from, 118, 118*f*
 metabolism of, 105, 105*b*
BCL2 gene, 137*t*
β-adrenergic receptors, accommodation of, 30*b*, 31
β-blockers
 as antagonists, 34, 34*b*
 and Na⁺/K⁺/ATPase pump, 27, 27*b*
β-oxidation, of fatty acids. *See* Fatty acid oxidation.
β-sheet, of protein, 10, 10*b*
BH₂. *See* Dihydrobiopterin (BH₂).
BH₄ (tetrahydrobiopterin)
 in catecholamine synthesis, 107
 in phenylalanine and tyrosine metabolism, 103, 103*b*, 105
Bicarbonate (HCO₃⁻), 9
 in CO₂ transport, 19, 19*b*, 19*f*
 normal values for serum, 161*t*
Bile acids, 5, 89, 89*b*
 synthesis of, 90, 90*f*
Bile salt(s), 89
 in lipid metabolism, 38, 38*b*, 38*f*
 malabsorption due to deficiency of, 39
 synthesis of, 89, 89*b*
Bilirubin, 110*b*, 111
 conjugated (direct), 110*b*, 110*f*, 111
 excess, 111
 normal values for serum, 161*t*
 unconjugated (indirect), 110*b*, 110*f*, 111
Bilirubin diglucuronide, in heme degradation, 111
Biogenic amines, 106, 106*b*, 107*f*
Biotin, 44
 deficiency of, 41*t*, 44, 44*b*
  and gluconeogenesis, 69
 and enzymes, 12, 12*b*
 functions of, 44
 sources of, 44, 44*b*
1,3-Bisphosphoglycerate (1,3-BPG), in glycolysis, 65, 67, 67*b*
2,3-Bisphosphoglycerate (2,3-BPG)
 in glycolysis, 65, 65*b*, 67, 67*b*
 in oxygen binding by hemoglobin, 18, 18*b*, 18*f*
Bleeding time, normal values for, 161*t*
Bleomycin, and cell cycle, 131*f*, 134*b*, 135
Blood, common laboratory values for, 161–164
Blood glucose levels
 in fasting state, 117*b*
 in starving state, 119*b*
 in well-fed state, 115, 115*b*
Blood group antigens, 79*b*, 80
Blood sugar, 1*b*
Blood urea nitrogen (BUN), 100*b*
 normal values for, 161*t*
Bloom's syndrome, 137*t*
Blotting analysis, 154, 155*f*
 Northern, 154, 154*b*, 155*f*
 Southern, 154, 154*b*, 155*f*
 Southwestern, 154*b*, 155
 Western, 154*b*, 155, 155*f*
BMR (basal metabolic rate), 35, 35*b*
Body mass index (BMI), 35*b*, 36, 36*b*
Bohr effect, 18, 18*b*
1,3-BPG (1,3-bisphosphoglycerate), in glycolysis, 65, 67, 67*b*
2,3-BPG (2,3-bisphosphoglycerate)
 in glycolysis, 65, 65*b*, 67, 67*b*
 in oxygen binding by hemoglobin, 18, 18*b*, 18*f*
Brain metabolism
 in fasting state, 118*f*, 119
 in starvation state, 119*f*, 120
 in well-fed state, 116*f*, 117, 117*b*
Brain use, in well-fed, fasting, and starvation state
 of glucose, 115*t*
 of ketones, 115*t*
Branched-chain α-ketoacid dehydrogenase, deficiency, 104*t*, 105*b*, 106

Branched-chain amino acids (BCAAs)
 alanine cycle for disposing of nitrogen from, 118, 118*f*
 metabolism of, 105, 105*b*
Branching enzyme
 disorders of, 75, 75*t*
 in glycogenesis, 72*b*, 73
BRCA1 gene, 137*t*
BRCA2 gene, 137*t*
Brittle bone disease, 22, 22*b*
Buffers, 9*b*
BUN (blood urea nitrogen), 100*b*
 normal values for, 161*t*

C

C (cytosine)
 in base excision repair, 134, 134*b*
 in nucleotides, 124, 125*f*
CAAT box, in RNA transcription, 139, 140*f*
CAC. *See* Citric acid cycle (CAC).
Calcitonin, in calcium regulation, 48*b*, 49
Calcitriol
 in calcium regulation, 49
 formation of, 46*f*, 47
 functions of, 47, 47*b*
Calcium (Ca²⁺), 49, 49*t*
 deficiency of, 49, 49*b*, 49*t*
 excess of, 49, 49*b*, 49*t*
 functions of, 48*b*, 49
 normal values for
  serum, 161*t*
  in urine, 161*t*
 regulation of, 48*b*, 49
 sources of, 49
Calcium (Ca²⁺)-calmodulin complex, 31*f*, 32
Calcium (Ca²⁺) concentration, extracellular *vs.* cytosolic, 32, 32*b*
Calcium homeostasis, vitamin D in, 46*f*
Calcium-adenosine triphosphatase (Ca⁺-ATPase) pumps, 27, 27*b*
 in tissue hypoxia, 27, 27*b*
Calmodulin, 29*b*, 31*f*
cAMP (cyclic adenosine monophosphate) pathway, 30, 30*f*, 31*b*
Cancer, DNA damage and, 135, 135*b*
 due to defects in DNA-repair enzymes, 137, 137*t*
 overview of, 135
 due to protooncogenes, 135, 137*t*
 due to tumor-suppressor genes, 137, 137*t*
CAP (catabolite activator protein), in prokaryotic control of gene expression, 141, 141*f*
Capillary lipoprotein lipase
 in cholesterol metabolism, 94*b*, 95
 in fatty acid and triacylglycerol synthesis, 82
Captopril, as noncompetitive inhibitor, 14
Carbamoyl aspartate, in pyrimidine synthesis, 126
Carbamoyl phosphate, in urea cycle, 100, 100*b*
Carbamoyl phosphate synthetase (CPS), deficiency of, 126*b*
Carbamoyl phosphate synthetase I (CPS I), in urea cycle, 100, 100*b*
Carbamoyl phosphate synthetase I (CPS II), in pyrimidine synthesis, 126
Carbohydrate(s), 1
 dietary, 37
  and fatty acid synthesis, 81*b*
 disaccharides as, 2
 monosaccharide derivatives as, 1
 monosaccharides as, 1, 2*t*
 polysaccharides as, 2, 2*t*, 3*f*
 protein-sparing effect of, 40
Carbohydrate metabolism, 63–80
 fructose metabolism in, 75
  clinical relevance of, 77
  interface with other pathways of, 77
  overview of, 75
  pathway reaction steps in, 76*f*, 77
  regulated steps in, 77
  unique characteristics of, 77
 galactose metabolism in, 75
  clinical relevance of, 77
  interface with other pathways of, 76
  overview of, 75
  pathway reaction steps in, 76, 76*f*
  regulated steps in, 76
  unique characteristics of, 76

Carbohydrate metabolism (*Continued*)
  gluconeogenesis in, 68
    clinical relevance of, 72
    interface with other pathways of, 72
    overview of, 68
    pathway reaction steps in, 68, 70*f*, 71*b*
    regulated steps in, 70, 70*f*
    unique characteristics of, 71
  glycogen metabolism in, 72
    clinical relevance of, 75, 75*t*
    interface with other pathways of, 75
    overview of, 72
    pathway reaction steps in, 72, 72*f*, 73*f*
    regulated steps in, 74, 74*f*
    unique characteristics of, 75
  glycolysis and pyruvate oxidation in, 63
    clinical relevance of, 68, 69*t*
    interface with other pathways of, 67, 67*f*
    overview of, 63
    pathway reaction steps for, 63, 64*f*, 64*t*, 66*f*
    regulated steps in, 65, 66*f*
    unique characteristics of, 67
  glycoproteins in, 78, 79*b*, 80
  pentose phosphate pathway in, 77
    clinical relevance of, 78
    interface with other pathways of, 78
    overview of, 77
    pathway reaction steps in, 77, 78*f*
    regulated steps in, 77
    unique characteristics of, 78
  proteoglycans in, 80
Carbon dioxide partial pressure (PCO$_2$), of arterial blood, 161*t*
Carbon dioxide (CO$_2$) transport, hemoglobin and bicarbonate in, 19, 19*b*, 19*f*
Carbon dioxide/bicarbonate (CO$_2$/HCO$_3^-$) buffer system, 9
Carbon monoxide (CO), 62*t*
  affinity of hemoglobin *vs.* myoglobin to, 18, 18*b*
Carbon skeletons, of amino acids, 102*f*, 103
Carcinoid syndrome, 42, 42*b*
  serotonin in, 112, 112*b*
Cardiotonic steroids, and Na$^+$/K$^+$/ATPase pump, 27, 27*b*
Carnitine acyltransferase
  deficiency of, 87
  in fatty acid and triacylglycerol synthesis, 81
  in triacylglycerol mobilization and fatty acid oxidation, 85
Carnitine acyltransferase I, in triacylglycerol mobilization and fatty acid oxidation, 85, 85*b*, 86
Carnitine acyltransferase II, in triacylglycerol mobilization and fatty acid oxidation, 85
Carnitine deficiency, 4*b*, 87, 87*b*
Carnitine shuttle, in triacylglycerol mobilization and fatty acid oxidation, 85
β-Carotenes, 45, 45*b*
Carrier proteins
  cotransport, 26
  hereditary defects in, 28, 28*b*
  uniport, 26, 26*b*
  Na$^+$-independent glucose transporters as, 26, 26*t*, 27*f*
Catabolic stages, 55, 56*f*
Catabolite activator protein (CAP), in prokaryotic control of gene expression, 141, 141*f*
Catalyzed reactions, 12, 12*f*
Catecholamines, 106, 106*b*, 107*f*
Catechol-*O*-methyltransferase (COMT), in catecholamine synthesis, 107
CDKs (cyclin-dependent protein kinases), 131, 131*b*
cDNA (complementary DNA), 151, 151*b*
cDNA (complementary DNA) libraries, 153*b*, 154
Celiac disease, malabsorption due to, 39
Cell count, in CSF, 161*t*
Cell cycle, 131, 131*b*, 131*f*
Cell surface receptors, general properties of, 29
Cell–cell signaling, 29
  clinical aspects of, 33
  overview of, 29
  sequence of events in, 29, 29*f*
Cellular strategies, for regulating metabolic pathways, 15
Cellulose, 3, 3*b*
Ceramide, 95
Cerebrosides, 5, 5*t*
Cerebrospinal fluid, common laboratory values for, 161–164
Cerebrospinal fluid pressure, normal values for, 161*t*
Ceruloplasmin, 52*b*, 53
CETP (cholesteryl ester transfer proteins), 95, 95*b*
CFTR (cystic fibrosis transmembrane regulator gene), 28, 156, 157*f*

Chain termination signal, in RNA transcription, 138*b*, 139
Checkpoints, in cell cycle, 131*b*, 131*f*, 132
Chemiosmotic coupling, 60, 60*b*
Chenodeoxycholic acid, 90
Chloramphenicol, protein synthesis inhibition by, 148*f*, 149
Chloride (Cl$^-$), 51
  control of, 51
  deficiency of, 49*t*, 51
  excess of, 49*t*, 51
  functions of, 50*b*, 51
  normal values for
    in CSF, 161*t*
    serum, 161*t*
    in sweat, 161*t*
  source of, 51
Chlorpromazine, as antagonist, 34, 34*b*
Cholecalciferol (vitamin D$_3$), 47
Cholera, 33, 33*b*
Cholesterol, 5, 5*b*
  functions of, 88, 88*b*
  in membranes, 24
  normal values for serum, 161*t*
  structure of, 5, 6*f*
  synthesis and regulation of, 88*b*, 88*f*, 89
    allosteric and hormonal, 114*t*
Cholesterol esterase, 38, 38*f*
Cholesterol metabolism, 88
  in adrenogenital syndrome, 91
  bile salts and bile acids in, 89, 90*f*
  cholesterol synthesis and regulation in, 88*b*, 88*f*, 89
  overview of, 88
  steroid hormones in adrenal cortex in, 81, 91*f*
Cholesteryl ester(s), resynthesis of, 38, 38*b*, 38*f*
Cholesteryl ester transfer proteins (CETP), 95, 95*b*
Cholic acid, 6*f*, 90
Choline, 4*t*
Chondroitin sulfate, 80, 80*b*
Chromatin, 129, 129*b*, 130*f*
Chromium, 53, 53*b*
  deficiency of, 52*t*, 53, 53*b*
Chromosome(s), 129, 129*b*
  mutations with gain or loss of entire, 146*b*, 147
Chylomicrons
  functions and metabolism of, 92*t*, 93, 93*f*
  nascent, 38*f*, 39
  structure and composition of, 92*b*, 92*t*, 93
Cirrhosis, hyperammonemia due to, 101, 101*b*
Citrate
  in fatty acid and triacylglycerol synthesis, 81
  in gluconeogenesis, 71, 71*b*
  in glycolysis, 65, 65*b*
Citrate lyase, in fatty acid and triacylglycerol synthesis, 81
Citrate shuttle, in fatty acid and triacylglycerol synthesis, 81
Citrate synthase, 57*f*, 58
  in fatty acid and triacylglycerol synthesis, 81
Citric acid cycle (CAC), 56, 56*f*
  allosteric and hormonal regulation of, 114*t*
  clinical relevance of, 58
  interface with other pathways of, 58, 58*f*
  overview of, 56, 56*b*, 57*f*
  pathway reaction steps in, 57, 57*b*, 57*f*
  regulated steps in, 57, 57*f*
  unique characteristics of, 58
Citrulline, in urea cycle, 100
CK (creatine kinase), in diagnosis, 16*t*
Cl$^-$. *See* Chloride (Cl$^-$).
Clindamycin, protein synthesis inhibition by, 148*f*, 149
Cloning, DNA, 151, 151*b*
  basic steps in, 153, 153*b*, 153*f*
  cloning vectors in, 152, 152*b*
    defined, 151*b*, 152*b*
    other, 153
    plasmid, 152, 152*b*, 152*f*, 153*b*
  DNA libraries in, 153
    cDNA, 153*b*, 154
    genomic, 151*b*, 153, 153*b*
  genomic DNA and complementary DNA in, 151
  overview of, 151
  restriction endonucleases in, 151, 151*b*, 152*f*
  target DNA in, 151, 151*b*, 152*f*
Cloning vectors, 152, 152*b*
  defined, 151*b*, 152*b*
  other, 153
  plasmid, 152, 152*b*, 152*f*, 153*b*
CN$^-$ (cyanide), 62*t*

CO (carbon monoxide), 62*t*
  affinity of hemoglobin *vs.* myoglobin to, 18, 18*b*
CO_2 (carbon dioxide) transport, hemoglobin and bicarbonate
    in, 19, 19*b*, 19*f*
CO_2/HCO_3^- buffer system, 9
Cobalamin (vitamin B_{12}), 43
  in citric acid cycle, 58
  deficiency of, 41*t*, 43*b*, 44
    and homocysteine, 43, 43*b*, 106*b*
    and methylmalonic acid, 43, 43*b*
    and methylmalonyl CoA, 106
    and propionyl CoA, 106
  in fatty acid oxidation, 87, 87*b*
  functions of, 43
    in DNA synthesis, 43, 43*f*
    and homocysteine, 43, 43*b*, 43*f*
    in odd-chain fatty acid metabolism, 43, 43*b*
  metabolism of, 43, 43*b*
  in methionine metabolism, 106
  methylated, 106
  sources of, 42*b*, 43
Codons, 143
Coenzyme(s), 12, 12*b*
Coenzyme Q (CoQ), in oxidative phosphorylation, 59, 59*f*
Colchicine, and cell cycle, 131*f*
Collagen, 21
  assembly of, 21, 22*b*, 22*f*
  cross-links in, 9, 9*b*
  disorders of, 22
  types of, 22
Colorectal cancer, hereditary nonpolyposis, 137*t*
Compartmentation, of metabolic pathways, 55
Complementary DNA (cDNA), 151, 151*b*
Complementary DNA (cDNA) libraries, 153*b*, 154
COMT (catechol-*O*-methyltransferase), in catecholamine
    synthesis, 107
Cooley's anemia, 21, 21*b*
Cooperativity
  of enzymes, 15
  in hemoglobin *vs.* myoglobin, 17, 17*b*
Copper (Cu), 53
  and ceruloplasmin, 52*b*, 53
  deficiency of, 52*t*, 53, 53*b*
  and enzymes, 13
  excess of, 53
  functions of, 52*b*, 53
  sources of, 53
Coproporphyrin I, in heme synthesis, 110
Coproporphyrinogen I, in heme synthesis, 110
Coproporphyrinogen III, in heme synthesis, 110
CoQ (coenzyme Q), in oxidative phosphorylation, 59, 59*f*
Cori cycle, 72, 72*b*
Cori's disease, 75*t*
Cortisol, synthesis of, 91*f*
Cotransport carrier proteins, 26
Coumarin (Warfarin), and vitamin K, 48, 48*b*
Coupled reactions, 54, 54*b*, 55*f*
CpG islands, in eukaryotic control of gene expression, 142
CPS (carbamoyl phosphate synthetase), deficiency of, 126*b*
CPS I (carbamoyl phosphate synthetase I), in urea cycle,
    100, 100*b*
CPS II (carbamoyl phosphate synthetase I), in pyrimidine
    synthesis, 126
Creatine, synthesis of, 112, 112*b*
Creatine kinase (CK), in diagnosis, 16*t*
Creatinine
  normal values for serum, 161*t*
  synthesis of, 112, 112*b*
Creatinine clearance, normal values for, 161*t*
Crigler-Najjar syndrome, 111, 111*b*
Crohn's disease, malabsorption due to, 39
Crossing over, unequal, 143, 143*b*
Cu. *See* Copper (Cu).
Cyanide (CN⁻), 62*t*
Cyclic adenosine monophosphate (cAMP) pathway, 30, 30*f*, 31*b*
Cyclin(s), 131, 131*b*
Cyclin-dependent protein kinases (CDKs), 131, 131*b*
Cycloheximide, protein synthesis inhibition by, 148*f*, 149, 149*b*
Cyclophosphamide, and DNA synthesis, 135, 135*b*
Cystathionine, in methionine metabolism, 106
Cystathionine synthase, deficiency of, 104*t*, 106, 106*b*
Cysteine (Cys), 8*t*
  charge on, 9
  modification of, 9
  synthesis of, 98, 99*t*

Cystic fibrosis, 28, 28*b*
  polymerase chain reaction analysis of, 34, 156*b*, 157*f*
Cystic fibrosis transmembrane regulator gene *(CFTR)*, 28, 156, 157*f*
Cystinuria, 28
Cytochrome oxidase, in oxidative phosphorylation, 59, 59*b*, 59*f*
Cytochrome P450 enzymes, in alcohol metabolism, 123, 123*b*
Cytochrome P450 hydroxylases, in steroid hormone synthesis, 90
Cytokinesis, 130*f*, 131
Cytosine (C)
  in base excision repair, 134, 134*b*
  in nucleotides, 124, 125*f*
Cytosol, in compartmentation in metabolic pathways, 56

D

DAG (diacylglycerol), 4*b*
  in phosphoinositide pathway, 31*f*, 32, 32*b*
Daily energy expenditure, 36, 36*b*
Daughter strand, 132, 132*f*, 133, 133*b*
Debranching enzyme(s)
  disorders of, 75, 75*t*
  in glycogenolysis, 73, 73*b*
Denaturation, of proteins, 11, 11*b*
Deoxy sugars, 1
Deoxycholic acid, 90
Deoxynucleotide, 124*b*
Deoxyribonucleic acid. *See* DNA.
Deoxyribose, 124
2-Deoxyribose, 1, 1*b*
Deoxyuridine monophosphate (dUMP), in pyrimidine
    synthesis, 126
Dephosphorylation, reversible, in regulation of enzymes, 15
Dermatan sulfate, 80, 80*b*
DHAP (dihydroxyacetone phosphate)
  in fructose metabolism, 77
  in glycolysis, 63, 63*b*, 64*f*, 65*b*
Diabetes mellitus (DM), 120
  overview of, 120, 120*t*
  type 1, 120, 120*t*, 121*f*
  type 2, 120*t*, 122
Diacylglycerol (DAG), 4*b*
  in phosphoinositide pathway, 31*f*, 32, 32*b*
Diagnostic enzymology, 16, 16*b*, 16*t*
Dicer protein, 143
Dicer-RNA complex, 143, 143*b*
Diet. *See* Nutrition.
Dietary fuels, 36
  carbohydrates as, 37, 37*b*
  energy generation from, 54–62
    citric acid cycle in, 56, 56*f*
      clinical relevance of, 58
      interface with other pathways of, 58, 58*f*
      overview of, 56, 56*b*, 57*f*
      pathway reaction steps in, 57, 57*b*, 57*f*
      regulated steps in, 57, 57*f*
      unique characteristics of, 58
    electron transport and oxidative phosphorylation in, 59
      ATP synthase and chemiosmotic coupling in, 59*f*, 60
      electron transport chain in, 56*f*, 59, 59*f*
      mitochondrial DNA mutations and, 60, 60*b*
      overview of, 59
      oxidative phosphorylation in, 56*f*, 59*f*, 60
      respiratory control in, 60
    inhibitors of mitochondrial ATP synthesis and, 61, 62*t*
      electron transport blockers as, 61, 62, 62*b*, 62*t*
      due to mitochondrial malfunction, 62, 62*t*
      overview of, 61
      uncouplers as, 56, 61*b*, 62, 62*b*
    intermediary metabolism in, 54
      catabolic stages in, 55, 56*f*
      compartmentation of metabolic pathways in, 55
      five common perspectives for many metabolic
          pathways in, 56
      overview of, 55
    metabolic pathways in, 54
      ATP-ADP cycle in, 54, 54*b*, 55*f*
      change in free energy in, 54, 54*b*
      coupled reactions in, 54, 54*b*, 55*f*
      overview of, 54
      redox coenzymes in, 54, 54*b*, 55*f*
    mitochondrial transport systems in, 60
      ATP-ADP translocase in, 59*f*, 60
      NADH shuttle mechanisms in, 60, 61*f*

Dietary fuels (*Continued*)
    overview of, 60
    specialized inner membrane transporters in, 61
  lipids as, 37, 38*f*, 39*b*
  overview of, 36
  proteins as, 39
Dietary reference intake (DRI), 35, 35*b*
Diffusion
  facilitated, 25*f*, 25*t*, 26
    cotransport carrier proteins in, 26
    defined, 25*b*
    ion channels in, 26
    uniport carrier proteins in, 26, 26*b*
      Na^+-independent glucose transporters as, 26, 26*t*, 27*f*
  simple (passive), 25, 25*f*, 25*t*
    defined, 25*b*
    limits of, 25, 26*b*
Digestive enzymes, 3*b*
Digitalis, mechanism of action of, 28, 28*f*
Dihydrobiopterin (BH_2)
  in catecholamine synthesis, 107
  in phenylalanine and tyrosine metabolism, 103, 105
Dihydrobiopterin (BH_2) reductase
  in catecholamine synthesis, 107
  deficiency of, 104*t*
  in metabolism of phenylalanine and tyrosine, 103, 105
Dihydrofolate (FH_2), 44
Dihydrofolate (FH_2) reductase gene, amplification of, 143, 143*b*
Dihydroorotate, in pyrimidine synthesis, 126*f*
Dihydroorotase, in pyrimidine synthesis, 126
Dihydrotestosterone, 91
Dihydroxyacetone, 2*t*
Dihydroxyacetone phosphate, in liver metabolism in well-fed
    state, 116
Dihydroxyacetone phosphate (DHAP)
  in fructose metabolism, 77
  in glycolysis, 63, 63*b*, 64*f*, 65*b*
1,25-Dihydroxycholecalciferol, 46*f*, 47
Dimethylarginine, asymmetric, 112, 112*b*
Dinitrophenol, 62*b*, 62*t*
Diphtheria toxin, protein synthesis inhibition by, 148*f*, 149, 149*b*
Direct repair, of DNA, 134, 134*b*, 136*f*
Disaccharide(s), 2, 2*b*, 2*t*
  digestion of, 37
Disaccharide lactose, in galactose metabolism, 76
Disaccharide sucrose, in fructose metabolism, 77
Disulfide bond formation, as posttranslational modification, 150*t*
Disulfiram, and alcohol metabolism, 122*b*, 123
DM. *See* Diabetes mellitus (DM).
DNA, 129–137
  antineoplastic drug action on, 131*f*, 134
  and cancer, 135, 135*b*
    defects in DNA-repair enzymes in, 137, 137*t*
    overview of, 135
    protooncogenes in, 135, 137*t*
    tumor-suppressor genes in, 137, 137*t*
  complementary, 151, 151*b*
  genomic, 151, 151*b*
  linker, 129, 130*f*
  mitochondrial, 60, 60*b*
  nucleotides in, 124, 124*b*, 125*f*
  organization of, 129, 129*b*
    double helix in, 129, 129*b*, 130*f*
    linker DNA in, 129, 130*f*
    nucleosomes in, 129, 129*b*, 130*f*
    overview of, 129
    pseudogenes in, 129, 129*b*
    repetition and transposons in, 129, 129*b*
    supercoiling in, 129, 130*f*
  recombinant, 151, 152*f*
  repair of, 133, 133*b*
    base excision, 134, 134*b*, 135*f*
    direct, 134, 134*b*, 136*f*
    double-strand, 134, 134*b*
    mismatch, 133, 133*b*, 134*f*
    nucleotide excision, 134, 134*b*, 136*f*
    overview of, 133
  repetitive, 129, 129*b*
  satellite, 129
  sense strand of, 138, 138*b*, 139*f*
  synthesis of, 131
    cell cycle in, 131, 131*b*, 131*f*
    daughter strand in, 132, 132*f*, 133, 133*b*
    lagging-strand, 132*f*, 133
    leading-strand, 132*f*, 133
    overview of, 131

DNA (*Continued*)
    parental strand in, 132, 132*f*
    replication fork in, 132, 132*f*
    reverse transcriptase in, 133
    telomerase in, 133
  target, 151, 151*b*, 152*f*
  template strand of, 138, 138*b*, 139*f*
DNA cloning, 151, 151*b*
  basic steps in, 153, 153*b*, 153*f*
  cloning vectors in, 152, 152*b*
    defined, 151*b*, 152*b*
    other, 153
    plasmid, 152, 152*b*, 152*f*, 153*b*
  DNA libraries in, 153
    cDNA, 153*b*, 154
    genomic, 151*b*, 153, 153*b*
  genomic DNA and complementary DNA in, 151
  overview of, 151
  restriction endonucleases in, 151, 151*b*, 152*f*
  target DNA in, 151, 151*b*, 152*f*
DNA fingerprinting analysis, 159*f*, 160, 160*b*
DNA gyrase, 132, 132*b*
DNA libraries, 153
  cDNA, 153*b*, 154
  genomic, 151*b*, 153, 153*b*
  screening of, 154, 154*b*, 154*f*
DNA ligase, 132*f*, 151*b*, 152
DNA polymerases (DNAPs), 132, 132*f*, 133*b*
  proofreading by, 133, 133*b*
DNA polymorphisms, 158, 158*b*
  single-nucleotide, 158, 158*f*
  tandem repeats as, 159*f*, 160, 160*b*
DNA sequencing, 160, 160*b*
DNA technology, 151–160
  detection of specific nucleic acid sequences with
    probes as, 154
    blotting analysis in, 154, 155*f*
      Northern, 154, 154*b*, 155*f*
      Southern, 154, 154*b*, 155*f*
      Southwestern, 154*b*, 155
      Western, 154*b*, 155, 155*f*
    microarrays in, 155, 155*b*
    overview of, 154, 154*b*
    screening DNA libraries in, 154, 154*b*, 154*f*
  polymerase chain reaction as, 155
    of inherited diseases, 156, 157*f*
    overview of, 155, 155*b*
    procedure for, 155, 155*b*, 156*f*
    reverse transcriptase, 155*b*, 156
  recombinant DNA and DNA cloning as, 151, 151*b*
    basic steps in, 153, 153*b*, 153*f*
    cloning vectors in, 152, 152*b*
      defined, 151*b*, 152*b*
      other, 153
      plasmid, 152, 152*b*, 152*f*, 153*b*
    DNA libraries in, 153
    genomic DNA and complementary DNA in, 151
    overview of, 151
    restriction endonucleases in, 151, 151*b*, 152*f*
    target DNA in, 151, 151*b*, 152*f*
  restriction fragment length polymorphisms as, 157
    defined, 157, 157*b*
    for DNA sequencing, 160
    overview of, 157, 157*b*
    polymorphisms in, 158, 158*b*
      single-nucleotide, 158, 158*f*
      tandem repeats as, 159*f*, 160, 160*b*
    restriction maps in, 157, 157*b*, 157*f*
DNAPs (DNA polymerases), 132, 132*f*, 133*b*
  proofreading by, 133, 133*b*
DNA-repair enzymes, genetic defects in, and cancer, 137, 137*t*
Dopa, in catecholamine synthesis, 107, 107*b*
Dopa decarboxylase, 107
Dopamine, synthesis and degradation of, 106, 107*b*, 107*f*
Dopamine hydroxylase, in catecholamine synthesis, 107
Double helix, of DNA, 129, 129*b*, 130*f*
Double-strand repair, 134, 134*b*
Down syndrome, 146*b*, 147
Doxorubicin, and DNA synthesis, 135, 135*b*
DRI (dietary reference intake), 35, 35*b*
dTMP (thymidylate), in pyrimidine synthesis, 126
dTMP (thymidylate) synthase
  fluorouracil and, 43*b*, 44
  in pyrimidine synthesis, 126
dUMP (deoxyuridine monophosphate), in pyrimidine synthesis, 126
Dysbetalipoproteinemia, familial, 96*t*

E

Editing, of mRNA, 143, 143*b*
EF(s) (elongation factors), 147*b*, 148*f*, 149
Ehlers-Danlos Syndrome, 22, 22*b*
EI (enzyme-inhibitor) complex, 14
Eicosanoids, 5, 5*b*, 7*f*
Electrolyte(s), 48
   calcium as, 49, 49*t*
   chloride as, 49*t*, 51
   magnesium as, 49, 49*t*
   normal values for serum, 161*t*
   overview of, 48
   phosphorus (phosphate) as, 49*t*, 51
   potassium as, 49*t*, 50
   sodium as, 49*t*, 50
Electron transport, 59, 59*b*
Electron transport blockers, 61, 62, 62*b*, 62*t*
Electron transport chain (ETC), 56*f*, 59, 59*f*
   uncouplers of, 61, 61*b*, 62, 62*b*
Elongation factors (EFs), 147*b*, 148*f*, 149
Energy expenditure, daily, 36, 36*b*
Energy generation, from dietary fuels, 54–62
   citric acid cycle in, 56, 56*f*
      clinical relevance of, 58
      interface with other pathways of, 58, 58*f*
      overview of, 56, 56*b*, 57*f*
      pathway reaction steps in, 57, 57*b*, 57*f*
      regulated steps in, 57, 57*f*
      unique characteristics of, 58
   electron transport and oxidative phosphorylation in, 59
      ATP synthase and chemiosmotic coupling in, 59*f*, 60
      electron transport chain in, 56*f*, 59, 59*f*
      mitochondrial DNA mutations and, 60, 60*b*
      overview of, 59
      oxidative phosphorylation in, 56*f*, 59*f*, 60
      respiratory control in, 60
   inhibitors of mitochondrial ATP synthesis and, 61, 62*t*
      electron transport blockers as, 61, 62, 62*b*, 62*t*
      due to mitochondrial malfunction, 62, 62*t*
      overview of, 61
      uncouplers as, 56, 61*b*, 62, 62*b*
   intermediary metabolism in, 54
      catabolic stages in, 55, 56*f*
      compartmentation of metabolic pathways in, 55
      five common perspectives for many metabolic pathways in, 56
      overview of, 55
   metabolic pathways in, 54
      ATP-ADP cycle in, 54, 54*b*, 55*f*
      change in free energy in, 54, 54*b*
      coupled reactions in, 54, 54*b*, 55*f*
      overview of, 54
      redox coenzymes in, 54, 54*b*, 55*f*
   mitochondrial transport systems in, 60
      ATP-ADP translocase in, 59*f*, 60
      NADH shuttle mechanisms in, 60, 61*f*
      overview of, 60
      specialized inner membrane transporters in, 61
Energy metabolism, hormonal regulation of, 113, 113*b*
   glucagon and epinephrine action in, 113
   insulin action in, 113
   overview of, 113, 114*t*
Enhancers, in eukaryotic control of gene expression, 142, 142*b*, 142*f*
Enterotoxigenic *Escherichia coli*, 33, 33*b*
Enzyme(s), 11
   active site of, 13, 13*b*
   in cellular strategies for regulating metabolic pathways, 15
   co-, 12, 12*b*
   compartmentation of, 15
   cooperativity and allosterism of, 15, 15*b*
   in diagnosis, 16, 16*t*
   general properties of, 12, 12*b*, 12*f*
   induced fit of, 13
   inhibition of, 14, 14*f*
      competitive, 14, 14*b*, 14*f*
      irreversible, 14, 14*b*
      noncompetitive, 14, 14*b*, 14*f*
      overcoming, 14
   isoenzymes and isoforms of, 16
   kinetics of, 13, 13*b*
      K_m in, 12, 12*b*, 13, 13*b*, 13*f*
      Lineweaver-Burk plot of, 13
      maximal velocity in, 12, 12*b*, 13, 13*b*, 13*f*
      Michaelis-Menten model of, 13, 13*b*, 13*f*
      reaction velocity in, 13
      temperature and pH effect on, 14

Enzyme(s) (*Continued*)
   metallo-, 13, 13*b*
   pro-, 15, 16*b*
   prosthetic groups of, 12
   regulated, 15
   transition state of, 13
Enzyme cascades, 15
Enzyme-inhibitor (EI) complex, 14
Enzyme-substrate (ES) complex, 13
Enzymology, diagnostic, 16, 16*b*, 16*t*
Eosinophils, normal values for, 161*t*
Ephedrine, as agonist, 34, 34*b*
Epinephrine, 113
   in fatty acid and triacylglycerol synthesis, 83
   in glycogen metabolism, 74, 74*b*
   metabolic actions of, 113, 114, 114*b*
   receptors for, 113, 114*b*
   in regulation of metabolism, 113
   secretion of, 113, 113*b*
   synthesis and degradation of, 106, 107*f*
   in triacylglycerol mobilization and fatty acid oxidation, 85
ERBB2 gene, 137*t*
Ergocalciferol (vitamin D_2), 47
Erythrocyte count, normal values for, 161*t*
Erythrocyte sedimentation rate, normal values for, 161*t*
Erythromycin, protein synthesis inhibition by, 148*f*, 149
Erythrose, 2*t*
ES (enzyme-substrate) complex, 13
Escherichia coli, enterotoxigenic, 33, 33*b*
Essential fatty acids, 3*b*, 3*t*
   in diet, 38, 38*b*
      deficiency in, 38, 38*b*, 84, 84*b*
      functions of, 38
      sources of, 38, 38*b*
Estradiol, synthesis of, 82*b*, 90*b*, 91
Estradiol-17β, 6*f*
Estriol, total, in pregnancy
   serum, 161*t*
   urine, 161*t*
Estrogens, 5
   synthesis of, 90*b*, 91
ETC (electron transport chain), 56*f*, 59, 59*f*
   uncouplers of, 61, 61*b*, 62, 62*b*
Ethanol metabolism, 122
   in higher concentrations, 123
   in low concentrations, 122, 122*f*
   overview of, 122
Ethanolamine, 4*t*
Ethylene glycol, as competitive inhibitor, 14
Etoposide, and cell cycle, 131*f*, 134*b*, 135
Euchromatin, 142, 142*b*
Eukaryotic control, of gene expression, 142, 142*f*
Eukaryotic protein synthesis, antibiotic inhibition of, 149, 149*b*
Eukaryotic transcription, of mRNA, 139, 140*f*
Exons, 139, 139*b*, 140*f*
Extracellular signals, 29*b*, 29*f*

F

FA(s). *See* Fatty acid(s) (FAs).
Fabry's disease, 97*t*
Facilitated diffusion, 25*f*, 25*t*, 26
   cotransport carrier proteins in, 26
   defined, 25*b*
   ion channels in, 26
   uniport carrier proteins in, 26, 26*b*
      Na^+-independent glucose transporters as, 26, 26*t*, 27*f*
FAD (flavin adenine dinucleotide), 41, 54, 54*b*, 55*f*
$FADH_2$ (flavin adenine dinucleotide, reduced or hydrogenated), 54, 55*f*
Fanconi's anemia, 137*t*
Farnesyl pyrophosphate, in cholesterol synthesis, 89
Fasting state, metabolism in, 117
   adipose tissue, 117, 118*f*
   brain, 118*f*, 119
   liver, 117, 118*f*
   muscle, 118, 118*f*
   overview of, 115*t*, 117
Fat(s)
   dietary, 37, 38*f*
      animal, 37*b*, 38
      plant, 37, 37*b*
   monounsaturated, 37
   polyunsaturated, 38
   saturated, 38

Fat oxidation, in gluconeogenesis, 71*b*
Fatty acid(s) (FAs), 3, 3*b*, 3*t*
  cis, 4
  dietary, 37, 38*f*
  essential, 3*b*, 3*t*
    in diet, 38, 38*b*
      deficiency in, 38, 38*b*, 84, 84*b*
      excess of, 84
      functions of, 38
      sources of, 38, 38*b*
  free, 37
  long-chain, 4, 4*b*
  medium-chain, 4, 4*b*
    absorption of, 38, 38*b*, 38*f*
  monounsaturated, 38
  muscle use of, in well-fed, fasting, and starvation state, 115*t*
  polyunsaturated, 38
  saturated, 38, 84*b*
  short-chain, 4, 4*b*
    absorption of, 38, 38*b*, 38*f*
  trans, 4, 4*b*
  unsaturated, 4, 84, 84*b*
    n-3 (ω-3), 4, 4*b*
    n-6 (ω-6), 4, 4*b*
Fatty acid desaturase, 84, 84*b*
Fatty acid oxidation, 84
  allosteric and hormonal regulation of, 114*t*
  clinical relevance of, 87
  vs. fatty acid synthesis, 86*t*
  interface with other pathways of, 87, 87*f*
  in liver metabolism in fasting state, 117, 117*b*
  overview of, 84, 84*f*
  pathway reaction steps in, 84*f*, 85
  regulated steps in, 85, 86*t*
  unique characteristics of, 86, 86*f*
  in well-fed, fasting, and starvation state, 115*t*
Fatty acid synthase, in fatty acid and triacylglycerol synthesis, 81
Fatty acid synthesis, 81
  allosteric and hormonal regulation of, 114*t*
  clinical relevance of, 84, 84*f*
  vs. fatty acid oxidation, 86*t*
  interface with other pathways of, 84
  in liver metabolism in well-fed state, 116, 116*b*
  overview of, 81
  pathway reaction steps in, 81, 82*f*
  regulated steps in, 82*f*, 83
  unique characteristics of, 83
Fatty acyl CoA, in triacylglycerol mobilization and fatty acid oxidation, 85
Fatty acyl CoA synthetase
  in fatty acid and triacylglycerol synthesis, 82
  in triacylglycerol mobilization and fatty acid oxidation, 85
Fatty liver, in alcoholics, 123
Fe. *See* Iron (Fe).
Feedback inhibition, in regulation of enzymes, 15, 15*b*
Ferritin, 51*b*, 52
  normal values for serum, 161*t*
Ferrochelatase
  in heme synthesis, 110
  in lead poisoning, 110, 110*b*
Fetal hemoglobin (HbF), 18*f*, 19, 20*b*
FH$_2$ (dihydrofolate), 44
FH$_2$ (dihydrofolate) reductase gene, amplification of, 143, 143*b*
FH$_4$ (tetrahydrofolate), 12, 43*b*, 44
Fiber, dietary, 37, 37*b*
Fibronectin, 78*b*, 80
  alternative splicing of, 143
Flavin adenine dinucleotide, reduced or hydrogenated (FADH$_2$), 54, 55*f*
Flavin adenine dinucleotide (FAD), 41, 54, 54*b*, 55*f*
Flavin mononucleotide (FMN), 41
Fluoride, 53, 53*b*
  deficiency of, 52*t*, 53, 53*b*
  excess of, 53, 53*b*
Fluoroacetate, in citric acid cycle, 58, 58*b*
Fluoroquinolone antibiotics, and DNA organization, 129, 129*b*
Fluorosis, 53, 53*b*
5-Fluorouracil (5-FU)
  and cell cycle, 131*f*, 134, 134*b*
  inhibition of nucleotide synthesis by, 126*t*
  as irreversible inhibitor, 14
  and thymidylate synthase, 43*b*, 44
FMN (flavin mononucleotide), 41
Folate. *See* Folic acid.
Folate monoglutamate, 44

Folic acid, 44
  active form of, 43*b*
  deficiency of, 41*t*, 44, 44*b*
    due to cancer, 44, 44*b*
    due to drugs, 44*b*
    and homocysteine, 43, 43*b*
  and enzymes, 12*b*
  functions of, 43*f*, 44
  metabolism of, 44, 44*b*
  in methionine metabolism, 106
  in pregnancy and lactation, 44, 44*b*
  sources of, 44
Follicle-stimulating hormone (FSH), normal values for serum/plasma, 161*t*
Frameshift mutations, 145*b*, 146, 146*b*, 146*f*
Free energy change (△*G*), 54, 54*b*
Fructokinase, in fructose metabolism, 77
Fructose, 2*t*
Fructose 1,6-bisphosphatase, in gluconeogenesis, 69*b*, 70
Fructose 1,6-bisphosphate
  in gluconeogenesis, 69, 70
  in glycolysis, 63
Fructose 1-phosphate, in fructose metabolism, 77
Fructose 2,6-bisphosphate
  in gluconeogenesis, 71, 71*b*
  in glycolysis, 65*b*, 66, 66*f*
Fructose 6-phosphate
  in gluconeogenesis, 70
  in glycolysis, 63, 67
  in pentose phosphate pathway, 77
Fructose intolerance, hereditary, 69*t*, 77
Fructose metabolism, 75
  clinical relevance of, 77
  hereditary defects in, 69*t*
  interface with other pathways of, 77
  overview of, 75
  pathway reaction steps in, 76*f*, 77
  regulated steps in, 77
  unique characteristics of, 77
Fructosuria, essential, 69*t*, 77
FSH (follicle-stimulating hormone), normal values for serum/plasma, 161*t*
5-FU. *See* 5-Fluorouracil (5-FU).
Fuels, dietary. *See* Dietary fuels.
Fumarate
  formation of, 103
  in metabolism of phenylalanine and tyrosine, 105
  in urea cycle, 100
Fumarylacetoacetate, in metabolism of phenylalanine and tyrosine, 105
Fumarylacetoacetate hydrolase
  deficiency of, 104*t*, 105, 105*b*
  in metabolism of phenylalanine and tyrosine, 105
Furanose sugars, 1

G

G (guanine)
  in direct DNA repair, 134, 134*b*, 136*f*
  in nucleotides, 124, 125*f*
  salvage of, 128
G protein(s)
  activated, 30, 30*b*
  trimeric, 30, 30*t*
G protein–coupled receptors (GPCRs), 29, 29*b*, 30*b*, 30*t*
G$_1$ checkpoint, in cell cycle, 131*f*, 132
G$_1$ phase, of cell cycle, 131, 131*b*, 131*f*
G$_2$ checkpoint, in cell cycle, 131*f*, 132
G$_2$ phase, of cell cycle, 131, 131*b*, 131*f*
GABA (γ-aminobutyrate), synthesis of, 112, 112*b*
GAG(s). *See* Glycosaminoglycans (GAGs).
Gain-of-function mutations, 135, 135*b*
Galactitol, 1, 77
Galactokinase deficiency, 69*t*, 77
Galactosamine, 2
Galactose, 2*t*
Galactose 1-phosphate uridyltransferase (GALT), in galactose metabolism, 75*b*, 76, 76*f*
Galactose metabolism, 75
  clinical relevance of, 77
  hereditary defects in, 69*t*
  interface with other pathways of, 76
  overview of, 75

Galactose metabolism (*Continued*)
    pathway reaction steps in, 76, 76*f*
    regulated steps in, 76
    unique characteristics of, 76
Galactosemia, 69*t*, 77
Gallstones, cholesterol synthesis and, 89, 89*b*
GALT (galactose 1-phosphate uridyltransferase), in galactose
        metabolism, 75*b*, 76, 76*f*
Gamma globulin, normal values in CSF for, 161*t*
γ-aminobutyrate (GABA), synthesis of, 112, 112*b*
γ-carboxylation, as posttranslational modification, 150*t*
γ-glutamyltransferase (GGT)
    in alcohol metabolism, 123
    in diagnosis, 16*t*
Gangliosides, 5, 5*b*, 5*t*
GATC sequence, in mismatch repair, 133, 134*f*
Gaucher's disease, 97*t*
GC box, in RNA transcription, 139, 140*f*
Gene(s)
    jumping, 129, 129*b*
    pseudo-, 129, 129*b*
Gene amplification, 143, 143*b*
Gene expression, 138–150
    aminoacyl-tRNA synthesis in, 144, 144*f*, 145*f*
    effects of mutations on, 144, 145*b*, 146*f*
    genetic code in, 143, 144*f*
    protein degradation in, 150
    protein synthesis in, 147
        bacterial antibiotic action in, 149
        eukaryotic antibiotic action in, 149
        overview of, 147
        polyribosomes in, 149
        prokaryotic example of, 147, 148*f*
        ribosomes in, 147
        secreted proteins in, 149*f*, 150, 150*t*
    in regulation of enzymes, 15
    RNA transcription in, 138
        eukaryotic, 139, 140*f*
        overview of, 138
        processing of primary mRNA transcript in, 139, 139*f*
        prokaryotic, 138, 139*f*
        RNA polymerase in, 138
        and types of RNA, 138
    transcriptional control of, 140, 140*b*
        alternative splicing in, 143
        editing of mRNA in, 143
        eukaryotic, 142, 142*f*
        gene amplification in, 143
        overview of, 140
        prokaryotic, 141, 141*f*
        RNA interference and gene silencing in, 143
Gene silencing, 143
Genetic code, 143, 143*b*, 144*f*
Genetic predisposition, to cancer, 137, 137*b*
Genomic DNA, 151, 151*b*
Genomic DNA libraries, 151*b*, 153, 153*b*
GGT (γ-glutamyltransferase)
    in alcohol metabolism, 123
    in diagnosis, 16*t*
Gilbert's disease, 111, 111*b*
Gln. *See* Glutamine (Gln).
Globin(s), 16
Globin chains, hemoglobinopathies due to structural
        alterations in, 20
Globin synthesis, hemoglobinopathies due to altered
        rates of, 20
Globulin, normal values for serum, 161*t*
Glu. *See* Glutamate (Glu).
Glucagon, 113
    in cholesterol synthesis, 88*f*, 89
    in gluconeogenesis, 71, 71*b*
    in glycogen metabolism, 74, 74*b*
    metabolic actions of, 113, 114*b*
    in regulation of metabolism, 113
    secretion of, 113, 113*b*
Glucagon receptors, 113, 114*b*
Glucocorticoids
    synthesis of, 90, 90*b*
    in triacylglycerol mobilization and fatty acid oxidation, 86
Glucogenic amino acids, 72
Glucokinase
    in glycolysis, 63, 63*b*, 64*f*
        vs. hexokinase, 64*t*, 65, 65*b*
    in liver metabolism in well-fed state, 116, 116*b*
Gluconeogenesis, 68
    allosteric and hormonal regulation of, 114*t*

Gluconeogenesis (*Continued*)
    clinical relevance of, 72
    in fasting state, 71*b*, 72
    glycolysis *vs.*, 71*b*
    interface with other pathways of, 72
    in liver metabolism
        in fasting state, 117, 117*b*
        in starvation state, 119, 119*b*
    overview of, 68
    pathway reaction steps in, 68, 70*f*, 71*b*
    regulated steps in, 70, 70*f*
    sites of, 71*b*, 72
    unique characteristics of, 71
Gluconeogenic enzyme deficiencies, 68, 68*b*, 72, 72*b*
Glucosamine, 2
Glucose, 2*t*
    facilitated diffusion of, 26, 26*t*, 27*f*
    in fatty acid and triacylglycerol synthesis, 81
    normal values for
        in CSF, 161*t*
        serum, 161*t*
    phosphorylation of, 2*b*, 63, 63*b*
    secondary active transport of, 27, 27*b*, 28*f*
    in well-fed, fasting, and starvation state
        brain use of, 115*t*
        muscle use of, 115*t*
        red blood cell use of, 115*t*
Glucose 1-phosphate
    in galactose metabolism, 76, 76*f*
    in glycogen metabolism, 75, 75*b*
    in glycogenesis, 72, 73
Glucose 6-phosphatase
    in gluconeogenesis, 70, 69*b*
    in glycogen metabolism, 74*b*, 75
    in liver metabolism in fasting state, 117, 117*b*
Glucose 6-phosphatase deficiency, 75*t*
Glucose 6-phosphate
    in galactose metabolism, 76, 76*b*, 76*f*
    in gluconeogenesis, 70
    in glycogen metabolism, 74, 74*b*
    in glycogenesis, 72
    in glycolysis, 63, 67
    in liver metabolism in well-fed state, 116
    in pentose phosphate pathway, 77
Glucose 6-phosphate dehydrogenase (G6PD), in pentose
        phosphate pathway, 77, 77*b*, 78
Glucose 6-phosphate dehydrogenase (G6PD) deficiency, 11, 11*b*,
        69*t*, 77*b*
Glucose metabolism, hereditary defects in, 69*t*
Glucose polymer, 3
Glucose transporter(s) (GLUTs)
    Na$^+$-independent, 26, 26*t*, 27*f*
    Na$^+$-linked, 27, 27*b*, 28*f*
Glucose transporter 4 (GLUT4) receptors, 113, 113*b*
α-1,4-Glucosidase deficiency, 75*t*
Glucosyl 4,6-transferase
    deficiency of, 75*t*
    in glycogenesis, 72*b*, 73
Glucuronic acid, 1, 1*b*
GLUT(s) (glucose transporters)
    Na$^+$-independent, 26, 26*t*, 27*f*
    Na$^+$-linked, 27, 27*b*, 28*f*
GLUT4 (glucose transporter 4) receptors, 113, 113*b*
Glutamate (Glu), 8*t*
    ammonia derived from, 101
    charge on, 9
    GABA synthesis from, 112, 112*b*
    oxidative deamination of, 98, 98*b*, 99*f*
    synthesis of, 99*t*
    in urea cycle, 99*b*, 100
Glutamate dehydrogenase, 99
Glutamate oxaloacetate transaminase (GOT), normal
        values for, 161*t*
Glutamate pyruvate transaminase (GPT), normal
        values for, 161*t*
Glutamine (Gln), 8*t*
    ammonia derived from, 101, 101*b*
    in ammonia metabolism, 101, 101*b*
    in purine synthesis, 124
    synthesis of, 99*t*
Glutathione (GSH), 53, 53*b*
    in pentose phosphate pathway, 77*b*, 78
Glutathione (GSH) peroxidase, 53
Gly (glycine), 8*t*
    creatine synthesis from, 112
    synthesis of, 99*t*

Glyceraldehyde, 2*t*
  in fructose metabolism, 77
Glyceraldehyde 3-phosphate
  in fructose metabolism, 77
  in glycolysis, 63, 65
  in pentose phosphate pathway, 77
Glycerol, 1
  in gluconeogenesis, 72
  linked to second phosphatidic acid, 4*t*
  in liver metabolism in fasting state, 117
  in well-fed, fasting, and starvation state, 115*t*
Glycerol 3-phosphate, 1*b*
  in adipose tissue metabolism in well-fed state, 116
  in fatty acid and triacylglycerol synthesis, 82, 83
  in glycolysis, 63, 67, 67*b*
  in triacylglycerol mobilization and fatty acid oxidation, 85
Glycerol 3-phosphate dehydrogenase, in glycolysis, 63
Glycerol kinase, 1
  in fatty acid and triacylglycerol synthesis, 82*b*
  in gluconeogenesis, 72, 72*b*
  in triacylglycerol mobilization and fatty acid oxidation, 85
Glycerol phosphate, 1
Glycerol phosphate shuttle, 60, 60*b*, 61*f*
Glycine (Gly), 8*t*
  creatine synthesis from, 112
  synthesis of, 99*t*
Glycochenodeoxycholic acid, 90*f*
Glycocholic acid, 90*f*
Glycogen, 3, 3*b*, 72, 72*b*
  structure of, 3*f*
Glycogen metabolism, 72
  clinical relevance of, 75, 75*t*
  interface with other pathways of, 75
  overview of, 72
  pathway reaction steps in, 72, 72*f*, 73*f*
  regulated steps in, 73*b*, 74, 74*f*
  unique characteristics of, 75
Glycogen phosphorylase, 3, 3*b*
  in glycogenolysis, 73, 73*b*
Glycogen synthase
  in glycogen metabolism, 74*b*, 75
  in glycogenesis, 72*b*, 73
  in liver metabolism in well-fed state, 116
Glycogenesis, 72, 72*f*
  allosteric and hormonal regulation of, 114*t*
  in well-fed, fasting, and starvation state, 115*t*
Glycogenolysis, 73, 73*f*
  allosteric and hormonal regulation of, 114*t*
  in liver, 73, 73*b*
  in liver metabolism in fasting state, 117
  in muscle, 73*b*, 74
  in well-fed, fasting, and starvation state, 115*t*
Glycogenoses, 75, 75*b*, 75*t*
Glycolysis, 63
  aerobic *vs.* anaerobic, 59*b*, 65*b*, 67, 67*b*
  allosteric and hormonal regulation of, 114*t*
  clinical relevance of, 68, 69*t*
  vs. gluconeogenesis, 71*b*
  interface with other pathways of, 67, 67*f*
  overview of, 63
  pathway reaction steps for, 63, 64*f*, 64*t*, 66*f*
  regulated steps in, 65, 66*f*
  unique characteristics of, 67
Glycoproteins, 80
  clinically important, 79*b*, 80
  defined, 67, 78, 78*b*
  synthesis of, 80
Glycosaminoglycans (GAGs), 1, 3, 3*b*
  clinically important, 80
  defined, 80, 80*b*
  repeated disaccharide units in, 80, 80*b*
Glycosylated hemoglobin (HbA$_{1c}$), 2*b*, 19, 19*b*
  normal values for, 161*t*
Glycosylation, 2, 2*b*
  of amino acids, 9
  nonenzymatic, 121, 121*b*
  as posttranslational modification, 150*t*
GMP (guanosine monophosphate), in purine synthesis, 124
Goodpasture's syndrome, 22, 22*b*
GOT (glutamate oxaloacetate transaminase), normal values for, 161*t*
Gout, 127*b*, 128, 128*t*
GPCRs (G protein–coupled receptors), 29, 29*b*, 30*b*, 30*t*
G6PD (glucose 6-phosphate dehydrogenase), in pentose phosphate pathway, 77, 77*b*, 78

G6PD (glucose 6-phosphate dehydrogenase) deficiency, 11, 11*b*, 69*t*, 77*b*
GPT (glutamate pyruvate transaminase), normal values for, 161*t*
Graves' disease, 33, 33*b*
Growth hormone
  arginine stimulation test for, 161*t*
  in fatty acid and triacylglycerol synthesis, 83
  in triacylglycerol mobilization and fatty acid oxidation, 86
GSH (glutathione), 53, 53*b*
  in pentose phosphate pathway, 77*b*, 78
GSH (glutathione) peroxidase, 53
Guanine (G)
  in direct DNA repair, 134, 134*b*, 136*f*
  in nucleotides, 124, 125*f*
  salvage of, 128
Guanosine monophosphate (GMP), in purine synthesis, 124

H

Hairpin loop, in RNA transcription, 138*b*, 139
Haloperidol, as antagonist, 34, 34*b*
Hartnup's disease, 28, 42, 42*b*
Hb. *See* Hemoglobin (Hb).
HCO$_3^-$ (bicarbonate), 9
  in CO$_2$ transport, 19, 19*b*, 19*f*
  normal values for serum, 161*t*
HDL (high-density lipoprotein)
  in cholesterol synthesis, 89
  functions and metabolism of, 92*t*, 94*f*, 95, 95*b*
  structure and composition of, 92*t*
Helicase, 132, 132*b*, 132*f*
Hematocrit (Hct), normal values for, 161*t*
Hematology, normal values for, 161*t*
Heme, 17, 17*f*, 108
  degradation of, 110, 110*f*
    and hyperbilirubinemia, 111
  in hemoglobin *vs.* myoglobin, 17, 17*b*
  overview of, 108
  synthesis of, 108, 108*b*, 108*f*
    allosteric and hormonal regulation of, 114*t*
    genetic disorders involving, 109*t*
Heme iron, 51*b*, 52
Hemochromatosis, 51*b*, 52
Hemoglobin (Hb), 16
  adult (HbA), 16, 20*b*
  carbon monoxide affinity to, 18
  in CO$_2$ transport, 19, 19*f*
  cooperativity in, 17, 17*b*
  fetal (HbF), 18*f*, 19, 20*b*
  functional differences between myoglobin and, 17, 17*t*, 18*f*
  glycosylated (Hb$_{1c}$), 2*b*, 19, 19*b*
    normal values for, 161*t*
  normal values for
    in blood, 161*t*
    mean corpuscular, 161*t*
    plasma, 161*t*
  other normal, 19, 20*f*
  oxygen binding by
    curve for, 17, 18*f*
    factors affecting, 18, 18*f*
  R form of, 17, 17*b*
  sickle cell (HbS), 20, 20*b*
    single-nucleotide polymorphisms of, 158*b*, 158*f*, 160
  structure of, 16, 17*f*
  T form of, 17, 17*b*
Hemoglobin A (HbA), 16, 20*b*
Hemoglobin A$_{1c}$ (HbA$_{1c}$), 2*b*, 19, 19*b*
  normal values for, 161*t*
Hemoglobin A$_2$ (HbA$_2$), 20*b*
Hemoglobin (Hb) Bart's disease, 21
Hemoglobin C (HbC), 20
Hemoglobin F (HbF), 18*f*, 19, 20*b*
Hemoglobin H (HbH) disease, 21, 21*b*
Hemoglobin S (HbS), 20, 20*b*
  single-nucleotide polymorphisms of, 158*b*, 158*f*, 160
Hemoglobinopathies
  due to altered rates of globin synthesis, 20
  due to structural alterations in globin chains, 20
Hemosiderin, 51*b*, 52
Hemosiderosis, 51*b*, 52
Henderson-Hasselbalch equation, 8, 8*b*
Heparan sulfate, 80, 80*b*
Heparin, 80, 80*b*
Hepatitis, viral, mixed hyperbilirubinemia due to, 111, 111*b*

Heptose, 2*t*
Hereditary nonpolyposis colorectal cancer (HNPCC), 137*t*
Hers' disease, 75*t*
Heterochromatin, 142, 142*b*
Heterogeneous nuclear RNA (hnRNA), 139
    processing of, 139, 140*f*
Heterotropic effect, 15*b*
Hexokinase, in glycolysis, 63, 63*b*, 64*f*
    vs. glucokinase, 64*t*, 65, 65*b*
Hexose, 2*t*
Hexose monophosphate (HMP) pathway. *See* Pentose phosphate
    pathway.
Hexose transport proteins, 26*t*
HGPRT (hypoxanthine-guanine phosphoribosyl transferase)
    deficiency of, 128, 128*t*
    in purine salvage, 128
5-HIAA (5-hydroxyindoleacetic acid), 112
High-density lipoprotein (HDL)
    in cholesterol synthesis, 89
    functions and metabolism of, 92*t*, 94*f*, 95, 95*b*
    structure and composition of, 92*t*
Histamine, synthesis of, 112, 112*b*
Histidine (His), 7, 7*b*, 8*t*
    charge on, 9
    histamine synthesis from, 112
Histones, in DNA, 129, 129*b*, 130*f*
HIV (human immunodeficiency virus), reverse transcriptase
    in, 133
HMG CoA (3-hydroxy-3-methylglutaryl coenzyme A)
    in cholesterol synthesis, 88*f*, 89
    in ketone body synthesis, 86
HMG CoA (3-hydroxy-3-methylglutaryl coenzyme A) reductase,
    in cholesterol synthesis, 89, 89*b*
HMP (hexose monophosphate) pathway. *See* Pentose phosphate
    pathway.
HNPCC (hereditary nonpolyposis colorectal cancer), 137*t*
hnRNA (heterogeneous nuclear RNA), 139
    processing of, 139, 140*f*
Homocysteine
    cobalamin and, 43, 43*b*, 43*f*, 106*b*
    in methionine metabolism, 106, 106*b*
Homocystinuria, 104*t*, 106, 106*b*
Homogentisate, in metabolism of phenylalanine and tyrosine, 105
Homogentisate oxidase
    deficiency of, 104*t*, 105
    in metabolism of phenylalanine and tyrosine, 105
Homotropic effect, 15*b*
Homovanillic acid (HVA), in catecholamine synthesis, 107,
    107*b*, 108
Hormonal regulation, of metabolism, 113, 113*b*
    glucagon and epinephrine action in, 113
    insulin action in, 113
    overview of, 113, 114*t*
Hormone-sensitive lipase
    in adipose tissue metabolism in starvation state, 120, 120*b*
    in fatty acid and triacylglycerol synthesis, 83, 83*b*
    in triacylglycerol mobilization and fatty acid oxidation,
        85, 85*b*
Human immunodeficiency virus (HIV), reverse transcriptase
    in, 133
Human prolactin (hPRL), normal values for serum, 161*t*
Hunter's disease, 79
Hurler's disease, 79
HVA (homovanillic acid), in catecholamine synthesis, 107,
    107*b*, 108
Hyaluronic acid, 3, 3*b*, 80, 80*b*
Hydrops fetalis, 21
3-Hydroxy-3-methylglutaryl coenzyme A (HMG CoA)
    in cholesterol synthesis, 88*f*, 89
    in ketone body synthesis, 86
3-Hydroxy-3-methylglutaryl coenzyme A (HMG CoA) reductase,
    in cholesterol synthesis, 89, 89*b*
Hydroxyapatite, 47*b*
β-Hydroxybutyrate, 86, 86*b*
    in liver metabolism in starvation state, 120
25-Hydroxycholecalciferol, 47
17-Hydroxycorticosteroids, normal values in urine for, 161*t*
5-Hydroxyindoleacetic acid (5-HIAA), 112
11α-Hydroxylase deficiency, 91*f*, 92, 92*b*
11β-Hydroxylase deficiency, 91*f*, 92, 92*b*
21α-Hydroxylase deficiency, 91*f*, 92, 92*b*
Hydroxylation
    of amino acids, 9
    as posttranslational modification, 150*t*
Hydroxylysine, 9
    in collagen assembly, 22, 22*b*, 22*f*

Hydroxymethylbilane, in heme synthesis, 110
Hydroxyproline, 9
    in collagen assembly, 22, 22*f*
5-Hydroxytryptamine
    in carcinoid syndrome, 112, 112*b*
    deficiency of, 112, 112*b*
    functions of, 112, 112*b*
    synthesis of, 111*b*, 111*f*, 112, 112*b*
5-Hydroxytryptophan, 112
Hydroxyurea, inhibition of nucleotide synthesis by, 126*t*
Hyperammonemia, 101, 101*b*
    acquired, 101
    hereditary, 101
    signs and symptoms of, 102
    treatment for
        nonpharmacologic, 101*b*, 102
        pharmacologic, 102
Hyperbilirubinemia, 111
    conjugated, 111*b*, 112
    mixed, 111, 111*b*
Hypercalcemia, 49, 49*b*, 49*t*
Hyperchloremia, 49*t*, 51
Hypercholesterolemia
    familial, 28, 96*t*
    treatment of, 89, 89*b*
Hypercupremia, 53
Hyperglycemia, in diabetes mellitus type 1, 121, 121*b*
Hyperkalemia, 49*t*, 50, 50*b*
Hyperlipoproteinemias, 95, 96*t*
Hypermagnesemia, 49*b*, 49*t*, 50
Hypernatremia, 49*b*, 49*t*, 50
Hyperphosphatemia, 49*t*, 50*b*, 51
Hypertriglyceridemia
    in diabetes mellitus type 1, 121, 121*b*
    familial, 96*t*
Hyperuricemia, 127*b*, 128, 128*t*
Hypoalbuminemia, in calcium regulation, 49
Hypocalcemia, 49, 49*b*, 49*t*
Hypochloremia, 49*t*, 51
Hypocupremia, 52*t*, 53, 53*b*
Hypoglycemia, 75, 75*b*
    fasting, due to alcohol, 123
Hypokalemia, 49*t*, 50, 50*b*
Hypomagnesemia, 49*b*, 49*t*, 50
Hyponatremia, 49*b*, 49*t*, 50
Hypophosphatemia, 49*t*, 50*b*, 51
Hypoxanthine
    in purine degradation, 127
    salvage of, 128
Hypoxanthine-guanine phosphoribosyl transferase (HGPRT)
    deficiency of, 128, 128*t*
    in purine salvage, 128

I
^{123}I (iodine, radioactive), normal values for thyroidal uptake
    of, 161*t*
I (inclusion) cell disease, 150, 150*b*
IDL (intermediate-density lipoprotein), 94, 94*f*
IF(s) (initiation factors), 147, 148*f*
IF (intrinsic factor), in vitamin B$_{12}$ metabolism, 43, 43*b*
Ile (isoleucine), 7, 7*b*, 8*t*
    metabolism of, 105
Immunodeficiency, severe combined, 128*t*
Immunoglobulin(s) (Ig), normal values for serum, 161*t*
Immunoglobulin A (IgA), normal values for serum, 161*t*
Immunoglobulin E (IgE), normal values for serum, 161*t*
Immunoglobulin G (IgG), normal values for serum, 161*t*
Immunoglobulin M (IgM), normal values for serum, 161*t*
IMP (inosine monophosphate)
    in purine degradation, 127
    in purine synthesis, 124
Inclusion (I) cell disease, 79, 150, 150*b*
Inheritance
    autosomal dominant, 145
    autosomal recessive, 145
    Mendelian, 144*b*, 145, 145*b*
    polygenic, 144*b*
    X-linked recessive, 145
Initiation complex
    in protein synthesis, 147, 147*b*, 148*f*
    in RNA transcription, 139, 139*b*
Initiation factors (IFs), 147, 148*f*
Inner membrane transporters, specialized, 61, 61*b*

Inner mitochondrial membrane, 57f, 60b
Inosine, in purine degradation, 127
Inosine monophosphate (IMP)
    in purine degradation, 127
    in purine synthesis, 124
Inositol, 4t
Inositol triphosphate (IP₃), 4b, 31f, 32b
Insulin, 113
    in cholesterol synthesis, 89, 89b
    in fatty acid and triacylglycerol synthesis, 83
    in gluconeogenesis, 71, 71b
    in glycogen metabolism, 74, 74b
    metabolic actions of, 113, 113b
    and $Na^+/K^+/ATPase$ pump, 27, 27b
    in regulation of metabolism, 113
    regulation of secretion of, 113, 113b
    synthesis of, 113, 113b
    in triacylglycerol mobilization and fatty acid
        oxidation, 85
Insulin receptor, 113, 113b
    signal transduction from, 32, 32f, 33b
Insulin receptor substrate 1 (IRS-1), 32, 32f
Insulin therapy, for diabetes mellitus type 1, 121
Insulin-to-glucagon ratio
    in fasting state, 117b
    in starving state, 119b
    in well-fed state, 115, 115b
Intermediary metabolism, 54
    catabolic stages in, 55, 56f
    compartmentation of metabolic pathways in, 55
    five common perspectives for many metabolic pathways
        in, 56
    overview of, 55
Intermediate-density lipoprotein (IDL), 94, 94f
Intermembrane space, 57f
Interphase, 131, 131b, 131f
Intracellular receptors, for lipophilic hormones, 29b, 33, 33f
Intracellular signal(s), 29b
Intracellular signal transduction, 29
    clinical aspects of, 33
    overview of, 29
    sequence of events in, 29, 29f
Intrinsic factor (IF), in vitamin B₁₂ metabolism, 43, 43b
Introns, 139, 139b, 140f
Iodine, 53, 53b
    deficiency of, 52t, 53, 53b
    radioactive (¹²³I), normal values for thyroidal uptake
        of, 161t
Ion channels, in facilitated diffusion, 26
IP₃ (inositol triphosphate), 4b, 31f, 32b
IPP (isopentyl pyrophosphate), in cholesterol synthesis, 89
Iron (Fe), 51, 52t
    deficiency of, 52, 52b, 52t
    and enzymes, 13
    excess of, 52, 52b
    in ferritin, 51b, 52
    functions of, 51b, 52
    heme, 51b, 52
    in hemosiderin, 51b, 52
    nonheme, 52
    normal values for serum, 161t
    sources of, 51
    and transferrin, 51b, 52
Iron overload diseases, 52, 52b
Iron poisoning, 52
IRS-1 (insulin receptor substrate 1), 32, 32f
Isocitrate dehydrogenase, 57f, 58
Isoelectric point (pI), 9
Isoenzymes, 16
Isoforms, of enzymes, 16
Isoleucine (Ile), 7, 7b, 8t
    metabolism of, 105
Isopentyl pyrophosphate (IPP), in cholesterol synthesis, 89
Isoprene, in cholesterol synthesis, 89, 89b
Isotretinoin, 45b, 47
Isozymes, 16

J

Jamaican vomiting sickness, 88
Jaundice, 111, 111b
    obstructive, 112
Jumping genes, 129, 129b

K

K⁺. *See* Potassium (K⁺).
Kearns-Sayre syndrome, 60
Keratan sulfate, 80
Ketoacidosis
    in diabetes mellitus type 1, 121
    in liver metabolism in starvation state, 120
Ketogenesis, in liver metabolism in fasting state, 117
α-Ketoglutarate, formation of, 103
α-Ketoglutarate dehydrogenase, 57f, 58
    in glycolysis, 65
Ketone(s), 1
    in well-fed, fasting, and starvation state
        brain use of, 115t
        muscle use of, 115t
Ketone bodies, 86, 86b
    due to alcohol, 123
    degradation of, 87
    excess production of, 87
    in liver metabolism in starvation state, 120, 120b
    measurement of, 87
    synthesis of, 86, 86f
        site of, 86, 86b
    in well-fed, fasting, and starvation state, 115t
Ketoses, 1, 2t
17-Ketosteroids, normal values for total urine, 161t
K_m, in enzyme kinetics, 12, 12b, 13, 13b, 13f
Krabbe's disease, 97t
Kwashiorkor, 40, 40b

L

Laboratory values, common, 161–164
lac operon, in prokaryotic control of gene expression,
    141, 141f
Lactase, in galactose metabolism, 76, 76b
Lactate
    in gluconeogenesis, 72
    in glycolysis, 65, 65b
Lactate dehydrogenase (LDH)
    in diagnosis, 16t
    in glycolysis, 65
    normal values for serum, 161t
Lactation, folic acid in, 44
Lactic acidosis, 65b, 68
    due to alcohol, 123
    in diabetes mellitus type 1, 121
Lactose, 2, 2b
Lactose intolerance, 37, 37b
Lactulose, for hyperammonemia, 102
Lagging-strand synthesis, 132f, 133
Lambda-phage vectors, 153, 153b
Laminin, 78b, 80
LCAT (lecithin-cholesterol acyltransferase), 95
LDH (lactate dehydrogenase)
    in diagnosis, 16t
    in glycolysis, 65
    normal values for serum, 161t
LDL (low-density lipoprotein)
    in cholesterol synthesis, 89
    functions and metabolism of, 92t, 94, 94f, 95
    structure and composition of, 92t
Lead poisoning
    ALA dehydratase in, 109, 109b, 109t
    ferrochelatase in, 110, 110b
Leading-strand synthesis, 132f, 133
Leaflets, of membranes, 24, 24b
Leber's hereditary optic neuropathy, 60
Lecithin-cholesterol acyltransferase (LCAT), 95
Lesch-Nyhan syndrome, 128, 128b, 128t
Leucine (Leu), 7, 7b, 8t
    metabolism of, 105
Leucine zippers, 10b, 11
Leukocyte count, normal values for, 161t
Leukocyte differential count, normal values for, 161t
Leukodystrophy, metachromatic, 97t
Leukotrienes (LTs), 6, 6b, 7f
LH (luteinizing hormone), normal values for serum/plasma, 161t
Lineweaver-Burk plot, of enzyme kinetics, 13
Linker DNA, 129, 130f
Linoleic acid, 3t
Linolenic acid, 3t
Lipase, in diagnosis, 16t

Lipid(s), 3
  dietary, 37, 38f
    digestion of, 38, 38f
    eicosanoids as, 5, 7f
    fatty acids as, 3, 3t
    malabsorption of, 39b
    in membranes, 24, 24b
    phospholipids as, 4, 4t
    sphingolipids as, 4, 5t
    steroids as, 5, 6f
    triacylglycerols as, 4
Lipid bilayers, of membranes, 24, 24b
Lipid metabolism, 81–97
  cholesterol and steroid metabolism in, 88
    in adrenogenital syndrome, 91
    bile salts and bile acids in, 89, 90f
    cholesterol synthesis and regulation in, 88f, 89
    overview of, 88
    steroid hormones in adrenal cortex in, 81, 91f
  fatty acid and triacylglycerol synthesis in, 81
    clinical relevance of, 84, 84f
    interface with other pathways of, 84
    overview of, 81
    pathway reaction steps in, 81, 82f, 83f
    regulated steps in, 82f, 83
    unique characteristics of, 83
  plasma lipoproteins in, 92
    functions and metabolism of, 93, 93f, 94f
    hereditary disorders related to defective metabolism of, 95, 96t
    overview of, 92
    structure and composition of, 92, 92t
  sphingolipids in, 95
    ceramide as, 95
    degradation of, 96, 96f
    disorders of, 96, 97t
    overview of, 95
  triacylglycerol mobilization and fatty acid oxidation in, 84
    clinical relevance of, 87
    interface with other pathways of, 87, 87f
    overview of, 84, 84f
    pathway reaction steps in, 84f, 85
    regulated steps in, 85, 86t
    unique characteristics of, 86, 86f
Lipoic acid, in glycolysis, 65
Lipolysis, 84
  allosteric and hormonal regulation of, 114t
  clinical relevance of, 87
  interface with other pathways of, 87, 87f
  overview of, 84, 84f
  pathway reaction steps in, 84f, 85
  regulated steps in, 85, 86t
  unique characteristics of, 86, 86f
  in well-fed, fasting, and starvation state, 115t
Lipophilic hormones, intracellular receptors for, 29b, 33, 33f
Lipoprotein(s), 92
  functions and metabolism of, 93, 93f, 94f
    hereditary disorders related to defective, 95, 96t
  high-density
    in cholesterol synthesis, 89
    functions and metabolism of, 92t, 94f, 95, 95b
    structure and composition of, 92t
  intermediate-density, 94, 94f
  low-density
    in cholesterol synthesis, 89
    functions and metabolism of, 92t, 94, 94f, 95
    structure and composition of, 92t
  overview of, 92
  structure and composition of, 92, 92t
  very-low-density
    functions and metabolism of, 92t, 94, 94f
    structure and composition of, 92t
Lipoprotein lipase, in adipose tissue metabolism in well-fed state, 116, 116b
Lipoprotein lipase deficiency, familial, 96t
Lithocholic acid, 90
Liver glycogen phosphorylase deficiency, 75t
Liver metabolism
  in fasting state, 117, 118f
  in starvation state, 119, 119f
  in well-fed state, 115, 116f
Losartan, as antagonist, 34, 34b
Loss-of-function mutations, 137, 137b
Low-density lipoprotein (LDL)
  in cholesterol synthesis, 89

Low-density lipoprotein (LDL) (Continued)
  functions and metabolism of, 92t, 94, 94f, 95
  structure and composition of, 92t
LTs (leukotrienes), 6, 6b, 7f
Lung surfactant, 4, 4b
Lupus erythematosus, systemic, 140, 140b
Luteinizing hormone (LH), normal values for serum/plasma, 161t
Lymphocytes, normal values for, 161t
Lysine (Lys), 7, 8t
  charge on, 8
  hydroxylation of, 9
Lysosomal diseases, 79b
Lysosomal enzymes, 79b, 80
Lysosomal storage diseases, 79, 80, 80b, 97t, 150
Lysyl oxidase, in collagen assembly, 22

M
M checkpoint, in cell cycle, 131f, 132, 132b
Magnesium (Mg²⁺), 49, 49t
  deficiency of, 49b, 49t, 50
  and enzymes, 13
  excess of, 49b, 49t, 50
  functions of, 49, 49b
  normal values for serum, 161t
  sources of, 49
Malabsorption, 39b
Malate
  in fatty acid and triacylglycerol synthesis, 81
  and gluconeogenesis, 69
Malate-aspartate shuttle, 60b, 61, 61f
Maleylacetoacetate, in metabolism of phenylalanine and tyrosine, 105
Malic enzyme, in fatty acid and triacylglycerol synthesis, 81
Malnutrition, protein-energy, 40
Malonyl CoA
  in fatty acid and triacylglycerol synthesis, 81, 81b, 83
  in triacylglycerol mobilization and fatty acid oxidation, 86
Maltose, 2, 2b
Mannose 6-phosphate residues, as posttranslational modification, 150
MAO (monoamine oxidase), in catecholamine synthesis, 107
MAP (mitogen-activated protein) kinase, 32f, 33
Maple syrup urine disease, 104t, 105b, 106
Marasmus, 40, 40b
Maximal velocity (V_{max}), of enzymes, 12, 12b, 13, 13b, 13f
McArdle's disease, 75b, 75t
Mean corpuscular hemoglobin (MCH), normal values for, 161t
Mean corpuscular hemoglobin concentration (MCHC), normal values for, 161t
Mean corpuscular volume (MCV), normal values for, 161t
Medium-chain acyl CoA dehydrogenase (MCAD) deficiency, 87, 87b
Melanin, derivation of, 105, 105b
MELAS syndrome, 60
Melatonin, synthesis of, 111b, 111f, 112, 112b
Membrane(s)
  basic properties of, 24
  components of, 24, 24b
    leaflets as, 24, 24b
    lipids as, 24
    proteins as, 24
      integral (intrinsic), 24, 24b
      lipid-anchored, 24
      peripheral (extrinsic), 24, 24b
      transmembrane, 24
  fluid properties of, 24, 25b
  movement of molecules and ions across, 25, 25f, 25t
    via active transport
      primary, 25b, 25f, 25t, 26b, 27
      secondary, 25b, 25f, 25t, 27
    via facilitated diffusion, 25f, 25t, 26
      cotransport carrier proteins in, 26
      defined, 25b
      ion channels in, 26
      uniport carrier proteins in, 26, 26b, 26t, 27f
    hereditary defects in, 28, 28b
    via simple (passive) diffusion, 25, 25b, 25f, 25t
Membrane hybridization, for screening of DNA libraries, 154, 154b, 154f
Mendelian inheritance, 144b, 145, 145b
MERRF syndrome, 60

Messenger RNA (mRNA), 138, 138*b*
 editing of, 143, 143*b*
 eukaryotic transcription of, 139, 140*f*
 processing of primary transcript of, 139, 140*f*
Met (methionine), 7, 8*t*
 cobalamin and, 43, 43*b*, 43*f*
 metabolism of, 105*f*, 106
Metabolic fuel(s), 1–9
 amino acids as, 6
  acid-base properties of, 8, 9*b*
  hydrophilic (polar)
   charged, 7, 8*t*
   uncharged, 7, 8*t*
  hydrophobic (nonpolar), 7, 8*t*
  modification of residues in proteins with, 9
  structure of, 6
 carbohydrates as, 1
  disaccharides as, 2
  monosaccharide derivatives as, 1
  monosaccharides as, 1, 2*t*
  polysaccharides as, 2, 2*t*, 3*f*
 lipids as, 3
  eicosanoids as, 5, 7*f*
  fatty acids as, 3, 3*t*
  phospholipids as, 4, 4*t*
  sphingolipids as, 4, 5*t*
  steroids as, 5, 6*f*
  triacylglycerols as, 4
Metabolic pathways
 compartmentation of, 55
 energetics of, 54
  ATP-ADP cycle in, 54, 54*b*, 55*f*
  change in free energy in, 54, 54*b*
  coupled reactions in, 54, 54*b*, 55*f*
  overview of, 54
  redox coenzymes in, 54, 54*b*, 55*f*
 five common perspectives for many, 56
 processes that affect flow through, 55*f*
Metabolic rate, basal, 35, 35*b*
Metabolism
 adipose tissue
  in fasting state, 117, 118*f*
  in starvation state, 119*f*, 120
  in well-fed state, 116, 116*f*
 of alcohol, 122
  in higher concentrations, 123
  in low concentrations, 122, 122*f*
  overview of, 122
 allosteric regulation of, 113, 114*t*
 brain
  in fasting state, 118*f*, 119
  in starvation state, 119*f*, 120
  in well-fed state, 116*f*, 117, 117*b*
 in diabetes mellitus, 120
  overview of, 120, 120*t*
  type 1, 120, 120*t*, 121*f*
  type 2, 120*t*, 122
 in fasting state, 117
  adipose tissue, 117, 118*f*
  brain, 118*f*, 119
  liver, 117, 118*f*
  muscle, 118, 118*f*
  overview of, 115*t*, 117
 hormonal regulation of, 113, 113*b*
  glucagon and epinephrine action in, 113
  insulin action in, 113
  overview of, 113, 114*t*
 integration of, 113–123
 intermediary, 54
  catabolic stages in, 55, 56*f*
  compartmentation of metabolic pathways in, 55
  five common perspectives for many metabolic pathways
   in, 56
  overview of, 55
 liver
  in fasting state, 117, 118*f*
  in starvation state, 119, 119*f*
  in well-fed state, 115, 116*f*
 muscle
  in fasting state, 118, 118*f*
  in starvation state, 119*f*, 120
  in well-fed state, 116*f*, 117, 117*b*
 in starvation state, 119
  adipose tissue, 119*f*, 120
  brain, 119*f*, 120
  liver, 119, 119*f*

Metabolism (*Continued*)
  muscle, 119*f*, 120
  overview of, 115*t*, 119
 in well-fed state, 115
  adipose tissue, 116, 116*f*
  brain, 116*f*, 117, 117*b*
  liver, 115, 116*f*
  muscle, 116*f*, 117, 117*b*
  overview of, 115, 115*t*
Metachromatic leukodystrophy, 97*t*
Metal ion cofactors, 13, 13*b*
Metalloenzymes, 13, 13*b*
Metanephrine, in catecholamine synthesis, 107, 107*b*, 108
Methanol, as competitive inhibitor, 14
Methemoglobin, 17, 18*b*
Methemoglobin reductase system, in glycolysis, 65, 65*b*, 67, 67*b*
Methemoglobinemia
 acquired, 20
 hereditary, 20
Methionine (Met), 7, 8*t*
 cobalamin and, 43, 43*b*, 43*f*
 metabolism of, 105*f*, 106
Methotrexate (MTX)
 and cell cycle, 131*f*, 134, 134*b*
 as competitive inhibitor, 14
 inhibition of nucleotide synthesis by, 126*t*
Methylated guanine, in direct DNA repair, 134, 134*b*, 136*f*
Methylene tetrahydrofolate, in pyrimidine synthesis, 126
*O*⁶-Methylguanine-DNA methyltransferase (MGMT), in direct
  DNA repair, 134, 136*f*
7-Methylguanosine cap, 140, 140*b*, 140*f*
Methylmalonic acid, cobalamin and, 43, 43*b*
Methylmalonic acidemia, 104*t*
Methylmalonyl CoA
 cobalamin and, 43
 in fatty acid oxidation, 87, 87*b*
 in methionine metabolism, 106
 vitamin B₁₂ and, 106
Methylmalonyl CoA mutase
 deficiency of, 104*t*, 106
 in methionine metabolism, 106
Methyltetrahydrofolate, 44
Mevalonate, in cholesterol synthesis, 89
Mg²⁺. *See* Magnesium (Mg²⁺).
MGMT (*O*⁶-Methylguanine-DNA methyltransferase), in direct
  DNA repair, 134, 136*f*
Micelles, 38, 38*b*, 38*f*
Michaelis-Menten model, of enzyme kinetics, 13, 13*b*, 13*f*
Microarrays, 155, 155*b*
Microdeletion, 147, 147*b*
Mineral(s), 48
 calcium as, 49, 49*t*
 chloride as, 49*t*, 51
 magnesium as, 49, 49*t*
 overview of, 48
 phosphorus (phosphate) as, 49*t*, 51
 potassium as, 49*t*, 50
 RDA for, 48*b*
 sodium as, 49*t*, 50
Mineralocorticoids, synthesis of, 90, 90*b*
Mismatch repair, 133, 133*b*, 134*f*
Missense mutations, 145*b*, 146, 146*f*
Mitochondrial ATP synthesis, inhibitors of, 61, 61*b*, 62*b*, 62*t*
Mitochondrial DNA (mtDNA), mutations in, 60, 60*b*
Mitochondrial matrix, 57*f*
Mitochondrial membrane
 inner, 57*f*, 60*b*
 outer, 57*f*, 60*b*
Mitochondrion(ia), 57*f*
 in compartmentation in metabolic pathways, 55
Mitogen-activated protein (MAP) kinase, 32*f*, 33
Mitosis, 130*f*, 131
Mitral valve prolapse (MVP), dermatan sulfate in, 80, 80*b*
MLC (myosin light chain) kinase, 31*f*, 32
2-Monoacylglycerol, 38, 38*f*
Monoamine(s), ammonia derived from, 101
Monoamine oxidase (MAO), in catecholamine synthesis, 107
Monocytes, normal values for, 161*t*
Monosaccharide(s), 1, 2*t*
Monosaccharide derivatives, 1
Monounsaturated fats, 37
Monounsaturated fatty acids (MUFA), 38
Montelukast, 6
Motifs, in structure of protein, 11
mRNA. *See* Messenger RNA (mRNA).
mtDNA (mitochondrial DNA), mutations in, 60, 60*b*

MTX (methotrexate)
  and cell cycle, 131*f*, 134, 134*b*
  as competitive inhibitor, 14
  inhibition of nucleotide synthesis by, 126*t*
MUFA (monounsaturated fatty acids), 38
Muscle catabolism, in well-fed, fasting, and starvation state, 115*t*
Muscle glycogen phosphorylase deficiency, 75*t*
Muscle glycogenoses, 75, 75*b*
Muscle metabolism
  in fasting state, 118, 118*f*
  in starvation state, 119*f*, 120
  in well-fed state, 116*f*, 117, 117*b*
Muscle use, in well-fed, fasting, and starvation state
  of fatty acids, 115*t*
  of glucose, 115*t*
  of ketones, 115*t*
Muscle wasting, in diabetes mellitus type 1, 121
Mutation(s), 133, 143, 143*b*
  defined, 144*b*, 145
  effects of, 144
  frameshift, 145*b*, 146, 146*b*, 146*f*
  with gain or loss of entire chromosomes, 146*b*, 147
  gain-of-function, 135, 135*b*
  loss-of-function, 137, 137*b*
  microdeletion, 147, 147*b*
  missense, 145*b*, 146, 146*f*
  nonsense, 145*b*, 146, 146*f*
  point, 145, 145*b*, 146*f*
  silent, 145*b*, 146, 146*f*
  translocation, 146*b*, 147
  trinucleotide repeat, 146, 146*b*
MVP (mitral valve prolapse), dermatan sulfate in, 80, 80*b*
MYC gene, 137*t*
Myoglobin, 16
  carbon monoxide affinity to, 18
  functional differences between hemoglobin and, 17, 17*t*, 18*f*
  lack of cooperativity in, 17, 17*b*
  oxygen-binding curve for, 17, 18*f*
  structure of, 16
Myosin light chain (MLC) kinase, 31*f*, 32
Myxedema, 80*b*

N

Na$^+$. *See* Sodium (Na$^+$).
NAD$^+$ (nicotinamide adenine dinucleotide), 12, 42
  in metabolic pathways, 54, 54*b*, 55*f*
  in glycolysis, 65, 65*b*
NADH. *See* Nicotinamide adenine dinucleotide, reduced or hydrogenated (NADH).
NADP$^+$ (nicotinamide adenine dinucleotide phosphate), 42, 54*b*, 55*f*
NADPH. *See* Nicotinamide adenine dinucleotide, reduced or hydrogenated, phosphorylated derivative (NADPH).
Native conformation, of proteins, 11
Neomycin, for hyperammonemia, 102
Neuroblastomas, 107*b*, 108
NF1 gene, 137*t*
NF2 gene, 137*t*
NH$_3$ (ammonia)
  excess, 101
  metabolism of, 101, 101*b*
  sources of, 101, 101*b*
NH$_4$+ (ammonium)
  in ammonia metabolism, 101, 101*b*
  in oxidative deamination, 98, 99
  in urea cycle, 100, 101
Niacin (vitamin B$_3$), 42
  active forms of, 42
  in citric acid cycle, 58
  deficiency of, 42
    causes of, 42, 42*b*
    signs and symptoms of, 41*b*, 41*t*, 42*b*
  and enzymes, 12*b*
  excessive intake of, 42
  functions of, 41*b*
  sources of, 42
  synthesis of, 111*b*, 111*f*, 112
Nicotinamide adenine dinucleotide
  reduced or hydrogenated, phosphorylated derivative (NADPH), 54, 54*b*, 55*f*
    in fatty acid and triacylglycerol synthesis, 81, 81*b*

Nicotinamide adenine dinucleotide (*Continued*)
  in liver metabolism in well-fed state, 116
  in pentose phosphate pathway, 77, 77*b*, 78
  reduced or hydrogenated (NADH), 54, 55*f*
    cytosolic, 60, 60*b*
    in glycolysis, 65
    reduced oxidation of, 62, 62*b*
    reduced or hydrogenated (NADH) shuttle mechanisms, 60, 61*f*
Nicotinamide adenine dinucleotide (NAD$^+$), 12, 42
  in metabolic pathways, 54, 54*b*, 55*f*
    in glycolysis, 65, 65*b*
Nicotinamide adenine dinucleotide phosphate (NADP$^+$), 42, 54*b*, 55*f*
Nicotinic acid, 42. *See also* Niacin (vitamin B$_3$).
Niemann-Pick disease, 97*t*
Nitrogen balance, 40, 40*b*
Nitrogen metabolism, 98–112
  amino acid derivatives in, 106
    asymmetric dimethylarginine (ADMA) in, 112, 112*b*
    catecholamines as, 106, 107*f*
    creatine synthesis from arginine, glycine, and SAM in, 112
    γ-aminobutyrate (GABA) synthesis from glutamate in, 112
    heme synthesis and metabolism in, 108, 108*f*, 109*t*, 110*f*
    histamine synthesis from histidine in, 112, 112*b*
    overview of, 106
    serotonin, melatonin, and niacin synthesis from tryptophan in, 111*f*, 112
  biosynthesis of nonessential amino acids in, 98, 98*b*, 99*t*
  catabolic pathways of amino acids in, 102
    carbon skeletons in, 102*f*, 103
    for leucine, isoleucine, and valine (branched chain amino acids), 105
    for methionine, 105*f*, 106
    overview of, 102
    for phenylalanine and tyrosine, 103, 103*f*, 104*t*
  removal and disposal of amino acid nitrogen in, 98
    ammonia metabolism in, 101
    overview of, 98
    transamination and oxidative deamination in, 98, 99*f*
    urea cycle in, 99, 100*f*
Nitrogen oxide (NO) synthesis, ADMA and, 112, 112*b*
p-Nitrophenyl phosphate, normal values for, 161*t*
Nitrosoureas, and DNA synthesis, 135, 135*b*
N-methyltransferase, in catecholamine synthesis, 107
Nondisjunction, 147
Nonenzymatic glycosylation, 121, 121*b*
Nonheme iron, 52
Nonsense mutations, 145*b*, 146, 146*f*
Norepinephrine
  synthesis and degradation of, 106, 107*f*
  in triacylglycerol mobilization and fatty acid oxidation, 85
Northern blotting, 154, 154*b*, 155*f*
Nucleic acid sequences, detection with probes of specific, 154
  blotting analysis in, 154, 155*f*
    Northern, 154, 154*b*, 155*f*
    Southern, 154, 154*b*, 155*f*
    Southwestern, 154*b*, 155
    Western, 154*b*, 155, 155*f*
  microarrays in, 155, 155*b*
  overview of, 154, 154*b*
  screening DNA libraries in, 154, 154*b*, 154*f*
Nucleoside analogues, 62
Nucleosomes, 129, 129*b*, 130*f*
5'-Nucleotidase, in purine degradation, 127
Nucleotide(s), 124–128
  degradation of, 127, 127*f*
  genetic disorders involving, 128*t*
  overview of, 124
  purines in
    degradation and salvage of, 125, 127, 127*f*
    structure of, 125*f*
    synthesis of, 124, 124*b*, 125*f*
  pyrimidines in
    degradation of, 128, 128*b*
    structure of, 125*f*
    synthesis of, 125, 125*b*, 126*f*
  structure of, 124, 125*f*
  synthesis of
    anticancer drugs inhibiting, 126*t*, 127
    for purines, 124, 124*b*, 125*f*
    for pyrimidines, 125, 125*b*, 126*f*
Nucleotide base, 124, 125*f*
Nucleotide excision repair, 134, 134*b*, 136*f*
Nucleotide pentose, 124, 125*f*

Nutrition, 35–54
    dietary fuels for, 36
        carbohydrates as, 37, 37*b*
        lipids as, 37, 38*f*, 39*b*
        overview of, 36
        proteins as, 39
    minerals and electrolytes in, 48
        calcium as, 49, 49*t*
        chloride as, 49*t*, 51
        magnesium as, 49, 49*t*
        overview of, 48
        phosphorus (phosphate) as, 49*t*, 51
        potassium as, 49*t*, 50
        RDA for, 48*b*
        sodium as, 49*t*, 50
    terminology for, 35
        basal metabolic rate (BMR) in, 35, 35*b*
        body mass index (BMI) in, 35*b*, 36, 36*b*
        dietary reference intake (DRI) in, 35, 35*b*
        overview of, 35
        recommended daily allowance (RDA) in, 35, 35*b*
        respiratory exchange rate (RER) in, 35*b*, 36, 36*b*
    trace elements in, 51
        chromium as, 52*t*, 53
        copper as, 52*t*, 53
        fluoride as, 52*t*, 53
        iodine as, 52*t*, 53
        iron as, 51, 52*t*
        overview of, 51
        RDA for, 48*b*
        selenium as, 52*t*, 53
        zinc as, 52*t*, 53
    vitamins in
        classification and function of, 40*f*
        fat-soluble, 45
            absorption and transport of, 45*b*
            function of, 40*f*
            overview of, 45
            vitamin A (retinol) as, 45, 45*f*, 46*t*
            vitamin D as, 46*f*, 46*t*, 47
            vitamin E as, 46*t*, 48
            vitamin K as, 46*t*, 48
        water-soluble, 40
            ascorbic acid (vitamin C) as, 41*t*, 45
            biotin as, 41*t*, 44
            classification and function of, 40*f*
            cobalamin (vitamin B_{12}) as, 41*t*, 43, 43*f*
            deficiency of, 41*t*
            folic acid as, 41*t*, 44
            niacin (vitamin B_3, nicotinic acid) as, 41*t*, 42
            overview of, 40
            pantothenic acid (vitamin B_5) as, 41*t*, 42, 42*b*
            pyridoxine (vitamin B_6) as, 41*t*, 42
            riboflavin (vitamin B_2) as, 41, 41*t*
            thiamine (vitamin B_1) as, 41, 41*t*
            toxicity of, 41*b*

O

O_2. *See* Oxygen (O_2).
OAA (oxaloacetate)
    in fatty acid and triacylglycerol synthesis, 81
    formation of, 103
    in gluconeogenesis, 69, 69*b*
Okazaki fragments, 132*f*, 133, 133*b*
Oleic acid, 3*t*
Oligomycin, 62*t*
Oligosaccharides, 2*t*
Oncoproteins, 135*b*, 137
Organophosphates, as irreversible inhibitors, 14
Ornithine, in urea cycle, 100, 101
Orotic acid, in pyrimidine synthesis, 126
Orotic acidemia, 126*b*
Osmolality, normal values for
    serum, 161*t*
    of urine, 161*t*
Osteoarthritis, 80*b*
Osteogenesis imperfecta, 22, 22*b*
Outer mitochondrial membrane, 57*f*, 60*b*
Oxalate, normal values for urine, 161*t*
Oxaloacetate (OAA)
    in fatty acid and triacylglycerol synthesis, 81
    formation of, 103
    in gluconeogenesis, 69, 69*b*

Oxidation
    of cysteine, 9
    of fatty acids. (*See* Fatty acid oxidation).
Oxidative deamination, 98, 98*b*, 99*f*
Oxidative phosphorylation, 56*f*, 59*f*, 60
Oxygen (O_2), in oxidative phosphorylation, 59, 59*b*
Oxygen (O_2) binding, by hemoglobin and myoglobin, 16
    factors affecting, 18, 18*f*
Oxygen (O_2)-binding curve, for hemoglobin and myoglobin,
        17, 18*f*
    shifts in, 18
        left, 19, 19*b*
        right, 18, 19*b*
Oxygen partial pressure (PO_2), of arterial blood, 161*t*

P

Paclitaxel, and cell cycle, 135, 135*b*
Palmitate, in fatty acid and triacylglycerol synthesis, 81, 83*b*,
        84, 84*b*
Palmitic acid, 3*t*
    in fatty acid and triacylglycerol synthesis, 81
Palmitoleic acid, 3*t*
Pancreatic cholesterol esterase, 38, 38*f*
Pancreatic insufficiency, malabsorption due to, 39
Pancreatic lipase, 38, 38*b*, 38*f*
Pancreatic proteases, 39
Pantothenic acid (vitamin B_5), 41*t*, 42, 42*b*
    in citric acid cycle, 58
Parathyroid hormone (PTH)
    in calcium regulation, 48*b*, 49
    normal values for serum N-terminal, 161*t*
    and vitamin D, 47, 47*b*
Parental strand, 132, 132*f*
Partial thromboplastin time (PTT), normal values for
        activated, 161*t*
Passive transport, 26*b*
PCO_2 (carbon dioxide partial pressure), of arterial blood, 161*t*
PCR. *See* Polymerase chain reaction (PCR).
Pellagra, 42
    causes of, 42, 42*b*
    signs and symptoms of, 41*b*, 41*t*, 42*b*
Pentose, 2*t*
Pentose phosphate pathway, 77
    allosteric and hormonal regulation of, 114*t*
    clinical relevance of, 78
    function of, 77*b*, 78
    hereditary defects in, 69*t*
    interface with other pathways of, 78
    nonoxidative branch of, 77, 77*b*, 78*f*
    overview of, 77
    oxidative branch of, 77, 77*b*, 78*f*
    pathway reaction steps in, 77, 78*f*
    regulated steps in, 77
    unique characteristics of, 78
PEP (phosphoenolpyruvate)
    in gluconeogenesis, 69
    in glycolysis, 65, 65*b*
PEP (phosphoenolpyruvate) carboxykinase, in gluconeogenesis, 69
Pepsinogen, 39
Peptidases, 39
Peptide bond, 10
Peptidyl transferase, 147, 147*b*, 148
Perilipin, in triacylglycerol mobilization and fatty acid oxidation, 85
Pertussis toxin, 33, 33*b*
PFK-1 (phosphofructokinase 1), in glycolysis, 63, 65, 66*f*
PFK-2 (phosphofructokinase 2), in glycolysis, 66, 66*b*, 66*f*
PGs (prostaglandins), 5, 5*b*, 6*b*, 7*f*
pH
    of arterial blood, 161*t*
    control of, 9*b*
    and enzyme kinetics, 14
    and oxygen binding by hemoglobin, 18
    physiologic, 9*b*
Phenylacetate, for hyperammonemia, 102
Phenylalanine (Phe), 7, 8*t*
    metabolism of, 103, 103*f*
Phenylalanine hydroxylase, deficiency of, 103, 103*b*, 104*t*
Phenylketonuria (PKU), 7, 7*b*, 103
    cause of, 103*b*, 104*t*
    classic, 104*t*
    clinical associations with, 104*t*
    malignant, 103*b*, 104*t*

Pheochromocytomas, 107*b*, 108
Phosphate, 51
  control of, 51
  deficiency of, 49*t*, 50*b*, 51
  excess of, 49*t*, 50*b*, 51
  functions of, 50*b*, 51
  sources of, 51
Phosphatidyl inositol 4,5-bisphosphonate (PIP$_2$), 31*f*, 32
Phosphatidylcholine, in membranes, 24
Phosphatidylethanolamine, in membranes, 24
Phosphatidylserine, in membranes, 24
Phosphoenolpyruvate (PEP)
  in gluconeogenesis, 69
  in glycolysis, 65, 65*b*
Phosphoenolpyruvate (PEP) carboxykinase, in gluconeogenesis, 69
Phosphofructokinase 1 (PFK-1), in glycolysis, 63, 65, 66*f*
Phosphofructokinase 2 (PFK-2), in glycolysis, 66, 66*b*, 66*f*
Phosphoglucomutase
  in galactose metabolism, 76
  in glycogenesis, 72
  in glycogenolysis, 73
6-Phosphogluconolactone, in pentose phosphate pathway, 77
2-Phosphoglycerate, in glycolysis, 65
3-Phosphoglycerate, in glycolysis, 65
Phosphoglycerate kinase, in glycolysis, 65
Phosphoglycerate mutase, in glycolysis, 65
Phosphoinositide pathway, 29*b*, 31, 31*f*
Phospholipase(s), 4, 4*b*
Phospholipase C, 31*f*, 32
Phospholipids, 4, 4*b*, 4*t*
  in membranes, 24
5-Phosphoribosyl 1-pyrophosphate (PRPP), in purine synthesis, 124
5-Phosphoribosyl 1-pyrophosphate (PRPP) synthetase, in purine synthesis, 124
5-Phosphoribosylamine, in purine synthesis, 124, 124*b*
Phosphorus. *See also* Phosphate.
  normal values for serum, 161*t*
Phosphorylation
  of amino acids, 9
  of glucose, 2*b*, 63, 63*b*
  as posttranslational modification, 150*t*
  reversible, in regulation of enzymes, 15
Physostigmine, as noncompetitive inhibitor, 14
pI (isoelectric point), 9
PIP$_2$ (phosphatidyl inositol 4,5-bisphosphonate), 31*f*, 32
PKB (protein kinase B) pathway, 32*f*
PKU. *See* Phenylketonuria (PKU).
Plant fats, 37, 37*b*
Plant proteins, 39
Plasma, common laboratory values for, 161–164
Plasma volume, normal values for, 161*t*
Plasmid(s), 152, 152*b*
Plasmid vectors, 152, 152*b*, 152*f*, 153*b*
Platelet count, normal values for, 161*t*
PLP (pyridoxal phosphate), 98
p-nitrophenyl phosphate, normal values for, 161*t*
Point mutations, 145, 145*b*, 146*f*
Poly(A) tail, 140, 140*b*, 140*f*
Polygenic inheritance, 144*b*
Polymerase chain reaction (PCR), 155
  of inherited diseases, 156, 157*f*
  overview of, 155, 155*b*
  procedure for, 155, 155*b*, 156*f*
  reverse transcription, 155*b*, 156
Polymorphisms, 158, 158*b*
  single-nucleotide, 158, 158*f*
  tandem repeats as, 159*f*, 160, 160*b*
Polyols, 1
Polypeptide synthesis, prokaryotic example of, 147, 148*f*
Polyribosomes, 148*b*, 149
Polysaccharides, 2, 2*t*, 3*f*
Polysomes, 148*b*, 149
Polyunsaturated fats, 38
Polyunsaturated fatty acids (PUFAs), 38
Pompe's disease, 75*b*, 75*t*
Porphobilinogen, in heme synthesis, 109, 110
Porphyria
  acute intermittent, 109*b*, 109*t*, 110
  congenital erythropoietic, 109*t*, 110, 110*b*
  cutanea tarda, 109*t*, 110, 110*b*
Porphyrin(s), synthesis of, 108
  genetic disorders involving, 109*t*
Porphyrinogens, 108
Postprandial thermogenesis, 35

Posttranslational modifications, in proteins, 150, 150*b*, 150*t*
Potassium (K$^+$), 50, 50*b*
  control of, 50
  deficiency of, 49*t*, 50, 50*b*
  excess of, 49*t*, 50, 50*b*
  functions of, 50, 50*b*
  normal values for serum, 161*t*
  sources of, 50
Prader-Willi syndrome, 147
Pravastatin, and cholesterol synthesis, 89
Pregnancy
  folic acid in, 44, 44*b*
  total estriol in
    serum, 161*t*
    urine, 161*t*
Pregnenolone, 90, 90*b*
Primase, 132*f*
Prions, 11, 11*b*
PRL (prolactin), normal values for serum, 161*t*
Pro. *See* Proline (Pro).
Probes
  defined, 154, 154*b*
  detection of specific nucleic acid sequences with, 154
    blotting analysis for, 154, 155*f*
      Northern, 154, 154*b*, 155*f*
      Southern, 154, 154*b*, 155*f*
      Southwestern, 154*b*, 155
      Western, 154*b*, 155, 155*f*
    microarrays for, 155, 155*b*
    overview of, 154, 154*b*
    screening DNA libraries for, 154, 154*b*, 154*f*
Proenzymes, 15, 16*b*
Progesterone, 5, 6*f*
  synthesis of, 90, 91*f*
Proinsulin, 113, 113*b*
Prokaryotic control, of gene expression, 141, 141*f*
  negative, 141, 141*b*
  positive, 141, 141*b*
Prokaryotic example, of polypeptide synthesis, 147, 148*f*
Prokaryotic protein synthesis, antibiotic inhibition of, 148*f*, 149, 149*b*
Prokaryotic transcription, 138, 139*f*
Prolactin (PRL), normal values for serum, 161*t*
Proline (Pro), 8*t*
  and α-helix, 10, 10*b*
  in collagen assembly, 22, 22*f*
  hydroxylation of, 9
  synthesis of, 99*t*
Promoter sequences, in eukaryotic control of gene expression, 142, 142*b*, 142*f*
Proofreading, by DNA polymerases, 133, 133*b*
Propionic acidemia, 104*t*, 106
Propionyl CoA
  cobalamin and, 43, 43*b*
  in fatty acid oxidation, 87, 87*b*
  in methionine metabolism, 106
  vitamin B$_{12}$ and, 106
Propionyl CoA carboxylase
  deficiency of, 104*t*, 106
  in methionine metabolism, 106
Prostaglandins (PGs), 5, 5*b*, 6*b*, 7*f*
Prosthetic group(s)
  attachment of, as posttranslational modification, 150*t*
  in electron transport chain, 59, 59*b*
  of enzymes, 12, 12*b*
Proteasome, 150, 150*b*
Protein(s), 10–23
  denaturation of, 11, 11*b*
  dietary, 39
    ammonia derived from, 101
    animal, 39
    biologic value of, 39, 39*b*
    degradation and resynthesis of, 39
    digestion of, 39
    plant, 39
    RDA for, 39
  hierarchical structure of, 10
    primary, 10, 10*b*
    quaternary, 11, 11*b*
    secondary, 10
      super-, 11
    tertiary, 11, 11*b*
      fibrous, 11*b*
      globular, 11*b*
  major functions of, 10

Protein(s) (*Continued*)
  membrane, 24
    integral (intrinsic), 24, 24*b*
    lipid-anchored, 24
    peripheral (extrinsic), 24, 24*b*
    trans-, 24
  modification of amino acid residues in, 9
  normal values for
    in CSF, 161*t*
    in serum, 161*t*
    in urine, 161*t*
  posttranslational modifications in, 150, 150*b*, 150*t*
Protein chain, elongation of, 148, 148*f*
Protein degradation, 150
Protein kinase B (PKB) pathway, 32*f*
Protein phosphatase, in triacylglycerol mobilization and fatty
    acid oxidation, 85
Protein secretion, 149*f*, 150, 150*t*
Protein synthesis, 147
  bacterial antibiotic action in, 149
  eukaryotic antibiotic action in, 149
  initiation of, 147, 148*f*
  overview of, 147
  polyribosomes in, 149
  prokaryotic example of, 147, 148*f*
  ribosomes in, 147
  on rough endoplasmic reticulum, 149*f*, 150
  secreted proteins in, 149*f*, 150, 150*t*
  termination of, 148*f*, 149
Protein-energy malnutrition, 40
Protein-sparing effect, of carbohydrates, 40
Proteoglycans, 80
  defined, 78*b*, 80
Proteolytic cleavage, as posttranslational modification, 150*t*
Prothrombin time (PT), 48
  normal values for, 161*t*
Proton gradient, 59*f*, 60, 60*b*
Protooncogenes, 135, 137*t*
Protoporphyrin IX, in heme synthesis, 110
Protoporphyrinogen IX, in heme synthesis, 110
PRPP (5-phosphoribosyl 1-pyrophosphate), in purine synthesis, 124
PRPP (5-phosphoribosyl 1-pyrophosphate) synthetase, in purine
    synthesis, 124
Pseudogenes, 129, 129*b*
PT (prothrombin time), 48
  normal values for, 161*t*
PTH (parathyroid hormone)
  in calcium regulation, 48*b*, 49
  normal values for serum N-terminal, 161*t*
  and vitamin D, 47, 47*b*
PTT (partial thromboplastin time), normal values for activated,
    161*t*
PUFAs (polyunsaturated fatty acids), 38
Purine(s)
  ammonia derived from, 101
  degradation and salvage of, 125, 127, 127*f*
  structure of, 125*f*
  synthesis of, 124, 124*b*, 125*f*
    allosteric and hormonal regulation of, 114*t*
Purine nucleoside phosphorylase, in purine degradation, 127
Pyranose sugars, 1
Pyridoxal phosphate, 12
Pyridoxal phosphate (PLP), 98
Pyridoxine (vitamin B$_6$), 41*t*
  active form of, 42
  deficiency of, 41*t*, 42, 42*b*
  and enzymes, 12*b*
  functions of, 42, 42*b*
  sources of, 42
Pyrimidine(s)
  ammonia derived from, 101
  degradation of, 128, 128*b*
  structure of, 125*f*
  synthesis of, 125, 125*b*, 126*f*
    allosteric and hormonal regulation of, 114*t*
Pyrimidine dimers, in nucleotide excision repair, 134, 134*b*
Pyrimidine phosphoribosyl transferase, in pyrimidine synthesis, 127
Pyruvate
  in catabolism, 56*f*
  in fatty acid and triacylglycerol synthesis, 81
  formation of, 103
  in gluconeogenesis, 69
  in glycolysis, 65, 65*b*, 67*f*, 68
    in fasting state, 67*f*, 68, 68*b*
    in fed state, 67*f*, 68, 68*b*

Pyruvate carboxylase
  in citric acid cycle, 58, 58*b*
  in gluconeogenesis, 69
Pyruvate dehydrogenase
  deficiency of, 68, 68*b*, 69*t*
  in glycolysis, 65, 66, 66*b*
  in liver metabolism in well-fed state, 116
Pyruvate kinase
  deficiency of, 68, 69*t*
  in glycolysis, 65, 66, 66*b*
  in liver metabolism, in fasting state, 117
Pyruvate kinase bypass, in gluconeogenesis, 69, 69*b*
Pyruvate metabolism, hereditary defects in, 69*t*
Pyruvate oxidation, 63
  allosteric and hormonal regulation of, 114*t*
  clinical relevance of, 68, 69*t*
  interface with other pathways of, 67, 67*f*
  overview of, 63
  pathway reaction steps for, 63, 64*f*, 64*t*, 66*f*
  regulated steps in, 65, 66*f*
  unique characteristics of, 67

R

R factor, in vitamin B$_{12}$ metabolism, 43
RAS gene, 137*t*
RAS protein, mutant, 34, 34*b*
RAS protooncogene, mutation of, 34, 34*b*
RAS-dependent pathway, 32*f*, 33, 33*b*
RAS-independent pathway, 33
RB1 gene, 137*t*
RBCs (red blood cells), use of glucose in well-fed, fasting, and
    starvation state by, 115*t*
RDA (recommended daily allowance), 35, 35*b*
Reaction velocity (v), of enzymes, 13
Receptor(s), 29
  activation of, 29, 29*f*
  cell surface, general properties of, 29
  G protein–coupled, 29, 29*b*, 30*b*, 30*t*
  intracellular, for lipophilic hormones, 29*b*, 33, 33*f*
  phosphoinositide coupled, 29*b*, 31, 31*f*
Receptor tyrosine kinases (RTKs), 29*b*, 32, 32*b*, 32*f*
Recessive inheritance
  autosomal, 145
  X-linked, 145
Recombinant DNA, 151, 152*f*
Recombinant plasmid vectors, 152, 152*b*, 152*f*, 153*b*
Recommended daily allowance (RDA), 35, 35*b*
Red blood cells (RBCs), use of glucose in well-fed, fasting, and
    starvation state by, 115*t*
Red cell volume, normal values for, 161*t*
Redox coenzymes, 54, 54*b*, 55*f*
Reducing sugars, 1, 2*b*
Refsum's disease, 88
Release factors (RFs), 148*b*, 148*f*, 149
Repetitive DNA, 129, 129*b*
Replication fork, 132, 132*f*
Replication origins, 132, 132*b*
RER (rough endoplasmic reticulum), synthesis of proteins on,
    149*f*, 150
Respiratory control, of electron transport chain, 60, 60*b*
Respiratory exchange rate (RER), 35*b*, 36, 36*b*
Respiratory quotient, 35*b*, 36, 36*b*
Restriction endonucleases, 151, 151*b*, 152*f*
Restriction fragment length polymorphisms (RFLPs), 157
  defined, 157, 157*b*
  for DNA sequencing, 160
  overview of, 157, 157*b*
  polymorphisms in, 158, 158*b*
    single-nucleotide, 158, 158*f*
    tandem repeats as, 159*f*, 160, 160*b*
  restriction maps in, 157, 157*b*, 157*f*
Restriction maps, 157, 157*b*, 157*f*
Restriction sites, 151, 151*b*, 152*f*, 157*b*
Reticulocyte count, normal values for, 161*t*
Retinoic acid, 46
Retinoic acid receptors, 33, 33*b*, 33*f*
Retinol. *See* Vitamin A (retinol).
Retinol esters, 45
Retinol-binding protein, 45*b*
Retrotransposons, 129, 129*b*
Reverse transcriptase, 133, 133*b*
Reverse transcription polymerase chain reaction (RT-PCR),
    155*b*, 156

Reye's syndrome, hyperammonemia due to, 101, 101*b*, 102
RF(s) (release factors), 148*b*, 148*f*, 149
RFLPs. *See* Restriction fragment length polymorphisms (RFLPs).
Riboflavin (vitamin B₂), 41
    active forms of, 41, 41*b*
    in citric acid cycle, 58
    deficiency of, 41*b*, 41*t*, 42
    sources of, 41
Ribonucleic acid. *See* RNA.
Ribonucleotide reductase, 124
Ribose, 2*t*, 124
Ribose 5-phosphate, in pentose phosphate pathway, 77, 78
Ribosomal RNA (rRNA), 138, 138*b*
Ribosomes, 147
    large (60S) subunits of, 147, 147*b*
    small (40S) subunits of, 147, 147*b*
Ribulose, 2*t*
Ribulose 5-phosphate, in pentose phosphate pathway, 77
Ricin, protein synthesis inhibition by, 149
Rifampin, and RNA polymerase, 138, 138*b*
RNA
    defined, 138, 138*b*
    heterogeneous nuclear, 139
        processing of, 139, 140*f*
    messenger, 138, 138*b*
        editing of, 143, 143*b*
        eukaryotic transcription of, 139, 140*f*
        processing of primary transcript of, 139, 140*f*
    nucleotides in, 124, 124*b*, 125*f*
    ribosomal, 138, 138*b*
    transcription of, 138
        eukaryotic, 139, 140*f*
        overview of, 138
        processing of primary mRNA transcript in, 139, 139*f*
        prokaryotic, 138, 139*f*
        RNA polymerase in, 138
        and types of RNA, 138
    transfer, 138, 138*b*
    types of, 138
RNA interference, 143
RNA polymerase, 138, 138*b*
    α-amanitin and, 138*b*
    in eukaryotic transcription, 139
    in prokaryotic transcription, 138, 139*f*
    rifampin and, 138, 138*b*
    types of, 138, 138*b*
Rotenone, 62*t*
Rough endoplasmic reticulum (RER), synthesis of proteins on, 149*f*, 150
rRNA (ribosomal RNA), 138, 138*b*
RTKs (receptor tyrosine kinases), 29*b*, 32, 32*b*, 32*f*
RT-PCR (reverse transcription polymerase chain reaction), 155*b*, 156

S

S phase, of cell cycle, 131, 131*b*, 131*f*
S-adenosylhomocysteine, derivation of, 106
S-adenosylmethionine (SAM)
    creatine synthesis from, 112
    derivation of, 106
Sanger dideoxy method, 160
Satellite DNA, 129
Saturated fats, 38
Saturated fatty acids (SFAs), 38, 82*b*
SCID (severe combined immunodeficiency), 128*t*
Scurvy, 1, 1*b*, 23, 23*b*
Sedoheptulose, 2*t*
Segmented neutrophils, normal values for, 161*t*
Selenium (Se), 53, 53*b*
    deficiency of, 52*t*, 53
    and enzymes, 13
Sense strand, of DNA, 138, 138*b*, 139*f*
Serine (Ser), 4*t*, 8*t*
    glycosylation of, 9
    phosphorylation of, 9
    synthesis of, 99*t*
Serotonin
    in carcinoid syndrome, 112, 112*b*
    deficiency of, 112, 112*b*
    functions of, 112, 112*b*
    synthesis of, 111*b*, 111*f*, 112, 112*b*
Serum, common laboratory values for, 161–164
Serum enzyme markers, in diagnosis, 16, 16*b*, 16*t*

Severe combined immunodeficiency (SCID), 128*t*
SFAs (saturated fatty acids), 38, 82*b*
SGLUT₁, 26*t*
    in secondary active transport, 27, 27*b*, 28*f*
Shiga toxin, protein synthesis inhibition by, 148*f*, 149
Sickle cell anemia, 20, 20*b*
Sickle cell hemoglobin (HbS), 20, 20*b*
    single nucleotide polymorphisms of, 158*b*, 158*f*, 160
Sickle cell trait, 20, 20*b*
Signal amplification, 29, 29*f*
Signal molecules, 29, 29*b*
Signal recognition particle (SRP), in protein secretion, 149*b*, 149*f*, 150
Signal recognition particle (SRP) receptor, in protein secretion, 149*f*, 150
Signal sequence, in protein secretion, 149*b*, 149*f*, 150
Signal transduction cascades, 29, 29*b*, 29*f*
Silencers, in eukaryotic control of gene expression, 142, 142*b*
Silent mutations, 145*b*, 146, 146*f*
Simvastatin, and cholesterol synthesis, 89
Single-gene defects, 144*b*, 145, 145*b*
Single-nucleotide polymorphisms (SNPs), 158, 158*f*
Slow-reacting substance of anaphylaxis (SRS-A), 6
Small bowel disease, malabsorption due to, 39
Small nuclear ribonucleoproteins (snRNPs), 140, 140*b*
Sodium (Na⁺), 50
    control of, 50
    deficiency of, 49*b*, 49*t*, 50
    excess of, 49*b*, 49*t*, 50
    functions of, 49*b*, 50
    normal values for serum, 161*t*
    source of, 50
Sodium benzoate, for hyperammonemia, 102
Sodium (Na⁺)-independent glucose transporters, 26, 26*t*, 27*f*
Sodium (Na⁺)-linked calcium (Ca²⁺) antiporter, in secondary active transport, 28, 28*b*, 28*f*
Sodium (Na⁺)-linked symporters, in secondary active transport of glucose, 27, 27*b*, 28*f*
Sodium/potassium (Na⁺/K⁺) symporter, 26*t*
Sodium/potassium/adenosine triphosphatase (Na⁺/K⁺/ATPase) pump
    in primary active transport, 27
        albuterol and, 27, 27*b*
        β-blockers and, 27, 27*b*
        cardiotonic steroids and, 27, 27*b*
        insulin and, 27, 27*b*
        succinylcholine and, 27, 27*b*
    in secondary active transport, 27, 28*f*
Sorbitol, 1, 1*b*
Southern blotting, 154, 154*b*, 155*f*
Southwestern blotting, 154*b*, 155
Spectinomycin, protein synthesis inhibition by, 149
Spherocytosis
    congenital, 111, 111*b*
    hereditary, 111
Sphingolipid(s), 4, 5*t*, 95
    ceramide as, 95
    degradation of, 96, 96*f*
    disorders of, 5*b*, 96, 97*t*
    overview of, 95
Sphingolipidoses, 5, 5*b*, 95, 95*b*, 96
    types of, 95*b*, 97*t*
Sphingomyelin, 5, 5*b*, 5*t*
    in membranes, 24
Splicing
    alternative, 143, 143*b*
    in RNA transcription, 140, 140*b*, 140*f*
Sprue, celiac, malabsorption due to, 39
Squalene, in cholesterol synthesis, 89
SRP (signal recognition particle), in protein secretion, 149*b*, 149*f*, 150
SRP (signal recognition particle) receptor, in protein secretion, 149*f*, 150
SRS-A (slow-reacting substance of anaphylaxis), 6
Starch, 3
    digestion of, 37, 37*b*
Start codon, 147
Starvation state, metabolism in, 119
    adipose tissue, 119*f*, 120
    brain, 119*f*, 120
    liver, 119, 119*f*
    muscle, 119*f*, 120
    overview of, 115*t*, 119
Statin drugs, and cholesterol synthesis, 89, 89*b*
Stearate, in fatty acid and triacylglycerol synthesis, 84

Stearic acid, 3*t*
Steatorrhea, 39
Steroid hormone(s), 5, 6*f*
　in adrenal cortex, 90, 91*f*
Steroid hormone receptors, 33, 33*b*, 33*f*
Steroid hormone-receptor complex, in eukaryotic control of
　gene expression, 142, 142*b*
Steroid metabolism, 88
　in adrenogenital syndrome, 91
　bile salts and bile acids in, 89, 90*f*
　cholesterol synthesis and regulation in, 88*f*, 89
　overview of, 88
Stop codons, 149
Streptomycin, protein synthesis inhibition by, 148*f*, 149
Succinyl CoA
　in fatty acid oxidation, 87, 87*b*
　formation of, 103
　in methionine metabolism, 106
Succinylcholine, and Na^+/K^+/ATPase pump, 27, 27*b*
Sucrase, in fructose metabolism, 77
Sucrose, 2, 2*b*
Sugar(s), 1, 2*t*
　amino, 1
　blood, 1*b*
　deoxy, 1
　furanose, 1
　hereditary defects in catabolism of, 69*t*
　pyranose, 1
　reducing, 1, 2*b*
Sugar acids, 1
Sugar alcohols, 1
Sugar esters, 2
Supercoiling, of DNA, 129, 130*f*
Surfactant, 4, 4*b*
Sweat, common laboratory values for, 161–164
Symporters, 26, 26*t*
Systemic lupus erythematosus, 140, 140*b*

T

T (thymine)
　in nucleotides, 124, 125*f*
　salvage of, 127
T_3 (triiodothyronine)
　derivation of, 105
　normal values for serum, 161*t*
T_3 (triiodothyronine) resin uptake, normal values for, 161*t*
T_4 (thyroxine)
　derivation of, 105
　normal values for serum, 161*t*
T_4 (thyroxine) receptors, 33, 33*b*, 33*f*
Tailing, alternative, 143
Tamoxifen, 142*b*, 143
Tandem repeats (TRs), 159*f*, 160, 160*b*
　variable number of, 159*f*, 160, 160*b*
Target DNA, 151, 151*b*, 152*f*
TATA box, in RNA transcription, 139, 140*f*
Taurochenodeoxycholic acid, 90*f*
Taurocholic acid, 90*f*
Tay-Sachs disease, 97*t*, 146, 146*b*
Telomerase, 133, 133*b*
Temperature
　and enzyme kinetics, 14
　and oxygen-binding curve, 19
Template strand, of DNA, 138, 138*b*, 139*f*
Terbutaline, as agonist, 34, 34*b*
Testosterone, 6*f*
　synthesis of, 82*b*, 91
Tetracycline, protein synthesis inhibition by, 148*f*, 149
Tetrahydrobiopterin (BH_4)
　in catecholamine synthesis, 107
　in phenylalanine and tyrosine metabolism, 103, 103*b*, 105
Tetrahydrofolate (THF, FH_4), 12, 43*b*, 44
Tetrose, 2*t*
TG. *See* Triacylglycerol (TG).
Thalassemia, 20
　α-, 21
　　major, 21
　　mild, 21, 21*b*
　　silent carrier state of, 21
　β-, 21
　　intermedia, 21
　　major, 21, 21*b*
　　minor, 21, 21*b*

Thalassemia trait
　α-, 21, 21*b*
　β-, 21, 21*b*
Thermogenesis, postprandial, 35
Thermogenin, 62*b*, 62*t*
THF (tetrahydrofolate), 12, 43*b*, 44
Thiamine (vitamin B_1), 41
　in citric acid cycle, 58
　deficiency of, 41, 41*b*, 41*t*
　　in alcoholics, 41, 41*b*
　and enzymes, 12*b*
　functions of, 41, 41*b*
　sources of, 41
Thiamine pyrophosphate, 12, 41
Thioredoxin, 124
Thioredoxin reductase, 124
Threonine (Thr), 7, 8*t*
　glycosylation of, 9
　phosphorylation of, 9
Thrombin time, normal values for, 161*t*
Thromboplastin time, normal values for activated partial, 161*t*
Thromboxane A_2 (TXA_2), 5, 5*b*, 7*f*
Thymidylate (dTMP), in pyrimidine synthesis, 126
Thymidylate (dTMP) synthase
　fluorouracil and, 43*b*, 44
　in pyrimidine synthesis, 126
Thymine (T)
　in nucleotides, 124, 125*f*
　salvage of, 127
Thyroid hormone(s)
　derivation of, 105, 105*b*
　in triacylglycerol mobilization and fatty acid oxidation, 86
Thyroid hormone receptors, 33, 33*b*, 33*f*
Thyroidal iodine uptake, normal values for, 161*t*
Thyroid-stimulating hormone (TSH), normal values for serum
　or plasma, 161*t*
Thyroxine (T_4)
　derivation of, 105
　normal values for serum, 161*t*
Thyroxine (T_4) receptors, 33, 33*b*, 33*f*
Tissue hypoxia, Ca^+-ATPase pumps in, 27, 27*b*
α-Tocopherol, 48
Topoisomerase I, 132, 132*b*
Topoisomerase II, 132, 132*b*
TP53 gene, 132, 132*b*, 137*t*
TR(s) (tandem repeats), 159*f*, 160, 160*b*
　variable number of, 159*f*, 160, 160*b*
Trace element(s), 51
　chromium as, 52*t*, 53
　copper as, 52*t*, 53
　fluoride as, 52*t*, 53
　iodine as, 52*t*, 53
　iron as, 51, 52*t*
　overview of, 51
　RDA for, 48*b*
　selenium as, 52*t*, 53
　zinc as, 52*t*, 53
Transaldolase reactions, in pentose phosphate pathway, 77
Transaminases, 98
Transamination, 98, 98*b*, 99*f*
　in citric acid cycle, 58*b*
Transcription, 138
　eukaryotic, 139, 140*f*
　overview of, 138
　processing of primary mRNA transcript in, 139, 139*f*
　prokaryotic, 138, 139*f*
　RNA polymerase in, 138
　and types of RNA, 138
Transcription bubble, 138, 138*b*, 139*f*
Transcription initiation, regulation of, 142, 142*f*
Transcriptional control, of gene expression, 140, 140*b*
　alternative splicing in, 143
　editing of mRNA in, 143
　eukaryotic, 142, 142*f*
　gene amplification in, 143
　overview of, 140
　prokaryotic, 141, 141*f*
　RNA interference and gene silencing in, 143
Transducin, 30*t*
Transfer RNA (tRNA), 138, 138*b*
Transferrin, 51*b*, 52
Transketolase reactions, in pentose phosphate pathway, 77, 77*b*
Translation, 147
　bacterial antibiotic action in, 149
　eukaryotic antibiotic action in, 149

Translation (*Continued*)
  initiation of, 147, 148*f*
  overview of, 147
  polyribosomes in, 149
  prokaryotic example of, 147, 148*f*
  ribosomes in, 147
  secreted proteins in, 149*f*, 150, 150*t*
  termination of, 148*f*, 149
Translocation, 148, 150
Translocation mutations, 146*b*, 147
Transport proteins
  co-, 26
  hereditary defects in, 28, 28*b*
  uniport, 26, 26*b*
    Na$^+$-independent glucose transporters as, 26, 26*t*, 27*f*
Transposons, 129, 129*b*
Tretinoin, 47
Triacylglycerol(s) (TGs), 4, 4*b*
  dietary, 37
    composition of, 37
    digestion of, 38, 38*f*
    resynthesis of, 38, 38*b*, 38*f*
Triacylglycerol (TG) mobilization, 84
  clinical relevance of, 87
  interface with other pathways of, 87, 87*f*
  overview of, 84, 84*f*
  pathway reaction steps in, 84*f*, 85
  regulated steps in, 85, 86*t*
  unique characteristics of, 86, 86*f*
Triacylglycerol (TG) synthesis, 81
  clinical relevance of, 84, 84*f*
  interface with other pathways of, 84
  overview of, 81
  pathway reaction steps in, 81, 83*f*
  regulated steps in, 82*f*, 83
  unique characteristics of, 83
  in well-fed, fasting, and starvation state, 115*t*
Triglycerides, normal values for serum, 161*t*
Triiodothyronine (T$_3$)
  derivation of, 105
  normal values for serum, 161*t*
Triiodothyronine (T$_3$) resin uptake, normal values for, 161*t*
Trinucleotide repeat mutations, 146, 146*b*
Triose, 2*t*
Triose phosphate isomerase, in glycolysis, 63
Trisomy 21, 146*b*, 147
tRNA (transfer RNA), 138, 138*b*
Tropocollagen, 21, 22*b*, 22*f*
Trypsin, 39
Tryptophan (Trp), 7, 8*t*
  serotonin, melatonin, and niacin synthesis from, 111*b*, 111*f*, 112
Tryptophan hydroxylase, 112
TSH (thyroid-stimulating hormone), normal values for serum or plasma, 161*t*
Tumor-suppressor genes, 137, 137*t*
TXA$_2$ (thromboxane A$_2$), 5, 5*b*, 7*f*
Tyrosinase, 105
  deficiency of, 104*t*, 105
Tyrosine (Tyr), 7, 8*t*
  in catecholamine synthesis, 107, 107*f*
  metabolism of, 103, 103*b*, 103*f*
  phosphorylation of, 9
  synthesis of, 98, 99*t*
Tyrosine hydroxylase, 105
  in catecholamine synthesis, 107
Tyrosinosis, 104*t*, 105, 105*b*

U

U (uracil)
  in base excision repair, 134, 134*b*, 135*f*
  in nucleotides, 124, 125*f*
  salvage of, 127
Ubiquitin, 150, 150*b*
Ubiquitin-proteasome system, 150, 150*b*
UDP-galactose (uridine diphosphate-galactose), in galactose metabolism, 76
UDP-glucose (uridine diphosphate-glucose), in glycogenesis, 72*f*, 73
UGT (uridine diphosphate glucuronyltransferase), in heme degradation, 110*b*, 110*f*, 111
UL (upper intake levels), 35
UMP (uridine monophosphate), in pyrimidine synthesis, 126
Uncouplers, of electron transport chain, 56, 61*b*, 62, 62*b*

Unequal crossing over, 143, 143*b*
Uniport carrier proteins, 26, 26*b*
  Na$^+$-independent glucose transporters as, 26, 26*t*, 27*f*
Uniporters, 26, 26*b*
  Na$^+$-independent glucose transporters as, 26, 26*t*, 27*f*
Unsaturated fatty acids, 4, 84, 84*b*
  n-3 (ω-3), 4, 4*b*
  n-6 (ω-6), 4, 4*b*
Untranslated regions (UTRs), in RNA transcription, 140*f*
Upper intake levels (UL), 35
Uracil (U)
  in base excision repair, 134, 134*b*, 135*f*
  in nucleotides, 124, 125*f*
  salvage of, 127
Urea, measurement of, 100*b*
Urea cycle, 99, 100*f*
  allosteric and hormonal regulation of, 114*t*
Urea excretion, in well-fed, fasting, and starvation state, 115*t*
Urea synthesis, in well-fed, fasting, and starvation state, 115*t*
Uric acid
  degradation of purine nucleotides to, 127, 127*b*, 127*f*
  normal values for serum, 161*t*
Uridine diphosphate glucuronyltransferase (UGT), in heme degradation, 110*b*, 110*f*, 111
Uridine diphosphate-galactose (UDP-galactose), in galactose metabolism, 76
Uridine diphosphate-glucose (UDP-glucose), in glycogenesis, 72*f*, 73
Uridine monophosphate (UMP), in pyrimidine synthesis, 126
Uridine triphosphate (UTP), in glycogenesis, 73
Urine, common laboratory values for, 161–164
Urobilin, in heme degradation, 110*f*, 111
Urobilinogen, in heme degradation, 110*f*, 111, 111*b*
Uroporphyrin I, in heme synthesis, 110
Uroporphyrinogen decarboxylase
  deficiency of, 110, 110*b*
  in heme synthesis, 110
Uroporphyrinogen I, in heme synthesis, 110
Uroporphyrinogen I synthase, in heme synthesis, 110
Uroporphyrinogen III, in heme synthesis, 110
Uroporphyrinogen III cosynthase
  deficiency of, 110, 110*b*
  in heme synthesis, 110
UTP (uridine triphosphate), in glycogenesis, 73
UTRs (untranslated regions), in RNA transcription, 140*f*

V

v (reaction velocity), of enzymes, 13
Valine (Val), 7, 7*b*, 8*t*
  metabolism of, 105
Vanillylmandelic acid (VMA), in catecholamine synthesis, 107, 107*b*, 108
Variable number of tandem repeats (VNTRs), 159*f*, 160, 160*b*
Vectors, cloning, 152, 152*b*
  defined, 151*b*, 152*b*
  other, 153
  plasmid, 152, 152*b*, 152*f*, 153*b*
Very-low-density lipoprotein (VLDL)
  functions and metabolism of, 92*t*, 94, 94*f*
  structure and composition of, 92*t*
Vinca alkaloids, and cell cycle, 131*f*, 135, 135*b*
Vincristine, and cell cycle, 135, 135*b*
Vitamin(s)
  in citric acid cycle, 58, 58*b*
  classification and function of, 40*f*
  fat-soluble, 45
    absorption and transport of, 45*b*
    function of, 40*f*
    overview of, 45
    vitamin A (retinol) as, 45, 45*f*, 46*t*
    vitamin D as, 46*f*, 46*t*, 47
    vitamin E as, 46*t*, 48
    vitamin K as, 46*t*, 48
  water-soluble, 40
    ascorbic acid (vitamin C) as, 41*t*, 45
    biotin as, 41*t*, 44
    classification and function of, 40*f*
    cobalamin (vitamin B$_{12}$) as, 41*t*, 43, 43*f*
    deficiency of, 41*t*
    folic acid as, 41*t*, 44
    niacin (vitamin B$_3$, nicotinic acid) as, 41*t*, 42
    overview of, 40
    pantothenic acid (vitamin B$_5$) as, 41*t*, 42, 42*b*

Vitamin(s) (*Continued*)
  pyridoxine (vitamin B_6) as, 41t, 42
  riboflavin (vitamin B_2) as, 41, 41t
  thiamine (vitamin B_1) as, 41, 41t
  toxicity of, 41b
Vitamin A (retinol), 45, 46t
  absorption and transport of, 45f
  active forms of, 45, 45b
  deficiency of, 45b, 46t, 47
  excess of, 46t, 47
  functions of, 45b, 46
  sources of, 45
Vitamin B_1. *See* Thiamine (Vitamin B_1).
Vitamin B_2. *See* Riboflavin (vitamin B_2).
Vitamin B_3. *See* Niacin (vitamin B_3).
Vitamin B_5 (pantothenic acid), 41t, 42, 42b
  in citric acid cycle, 58
Vitamin B_6. *See* Pyridoxine (vitamin B_6).
Vitamin B_{12}. *See* Cobalamin (vitamin B_{12}).
Vitamin C. *See* Ascorbic acid (vitamin C).
Vitamin D, 47
  active form of, 46f, 47, 47b
  in calcium regulation, 49
  deficiency of, 46t, 47, 47b
  excess of, 46t, 48
  functions of, 47, 47b
  sources of, 47
Vitamin D_2 (ergocalciferol), 47
Vitamin D_3 (cholecalciferol), 47
Vitamin E, 48
  deficiency of, 46t, 47b, 48
  excess of, 46t, 48
  function of, 47b, 48
Vitamin K, 48
  deficiency of, 46t, 48
  excess of, 46t
  functions of, 47b, 48
  sources of, 47b, 48
VLDL (very-low-density lipoprotein)
  functions and metabolism of, 92t, 94, 94f
  structure and composition of, 92t
VMA (vanillylmandelic acid), in catecholamine synthesis, 107, 107b, 108
V_{max} (maximal velocity), of enzymes, 12, 12b, 13, 13b, 13f
VNTRs (variable number of tandem repeats), 159f, 160, 160b
Von Gierke's disease, 75b, 75t

W

Warfarin (coumarin), and vitamin K, 48, 48b
Well-fed state, metabolism in, 115
  adipose tissue, 116, 116f
  brain, 116f, 117, 117b
  liver, 115, 116f
  muscle, 116f, 117, 117b
  overview of, 115, 115t
Wernicke-Korsakoff syndrome (WKS), 41
Western blotting, 154b, 155, 155f
Wilson's disease, 53
WT1 gene, 137t

X

Xanthine, in purine degradation, 127
Xanthine oxidase, in purine degradation, 127, 128
Xeroderma pigmentosum, 137t
X-linked recessive inheritance, 145

Y

Yeast artificial chromosomes (YACs), 153, 153b

Z

Zafirlukast, 6
Zellweger syndrome, 88
Zidovudine (AZT), and reverse transcriptase, 133, 133b
Zileuton, 6, 6b
Zinc (Zn), 52t, 53
  deficiency of, 52b, 52t, 53
  and enzymes, 13
  functions of, 52b, 53
  sources of, 53
Zinc fingers, 10b, 11
Zona fasciculata, glucocorticoid synthesis in, 90, 90b
Zona glomerulosa, mineralocorticoid synthesis in, 90, 90b
Zona reticularis, androgen synthesis in, 90b, 91
Zymogens, 15